# Engineering Properties of Thermoplastics

# ENGINEERING PROPERTIES OF THERMOPLASTICS

*A Collective Work Produced by*
IMPERIAL CHEMICAL INDUSTRIES LIMITED, PLASTICS DIVISION

*Edited by*

R. M. OGORKIEWICZ

M.Sc.(Eng.), A.C.G.I., D.I.C., C.Eng., M.I.Mech.E.

WILEY—INTERSCIENCE
A Division of John Wiley & Sons Ltd.
London New York Sydney Toronto

Library of Congress catalog card No. 72-83219

SBN 471 65301 2

*Reprinted October 1970*

Set on Monophoto Filmsetter and printed by
J. W. Arrowsmith Ltd., Bristol, England

# FOREWORD

More and more, plastics materials are being used in technically exacting applications and it is axiomatic that the designer must be provided with data about these materials in a form which he finds useful. A difficulty immediately arises: unlike many traditional materials, plastics have properties which are strongly dependent on time and temperature. Only when the engineer has learnt to appreciate these distinctive characteristics of plastics will he be able to understand the data and use the specific properties of the materials to assist in material selection and ensure sound design.

Considerations such as these are behind the concept of this book which sets out to give engineering designers a general background to the behaviour of thermoplastics, and also characteristic data for some main types. As in every rapidly developing sphere of human endeavour, nothing is static. Tomorrow there may be some completely new and different materials but it is more likely that new variants with, in some respects, improved properties will appear. They will fit into the general scheme which is outlined.

The scope is limited to a range of thermoplastic products with which the authors are familiar. It is hoped that it will fulfil its objective of improving communication between raw material manufacturers and design engineers.

During the preparation of the book, the authors have had the benefit of the advice, guidance, and encouragement of the Editor, Mr. R. M. Ogorkiewicz of the Imperial College of Science and Technology, to whom I offer my sincere thanks for his most valuable contribution.

E. G. WILLIAMS
*Chairman, ICI Plastics Division*

# PREFACE

The purpose of this book is to provide information about the engineering properties of thermoplastics and to do so in a form which will facilitate the design of plastics components.

The book is made up of four distinct but complementary parts. The first provides an introduction to the characteristics and uses of thermoplastics. Its three chapters outline, in turn, the nature of polymeric materials, the composition of plastics and the main methods of processing plastics materials into finished products.

In the second part thermoplastics are considered generally as a class of engineering materials. Their deformational behaviour, strength characteristics and other physical properties are examined, and ways are introduced in which thermoplastics can be characterized for engineering purposes; it is in these ways that the information about the particular materials with which the book is concerned is subsequently presented.

The third part consists of eight chapters each devoted to a particular type of thermoplastic. For ease of reference, all the chapters follow, as far as possible, the same pattern. Thus, each starts with a general description of the material. This is followed by the main section of the text which discusses in detail the properties of the thermoplastic, mainly by reference to the data which are presented in graphical form at the end of the chapter. The text is completed by a short section on applications. Thus, collectively, the eight chapters contain a very comprehensive body of data on several important thermoplastic materials.

The fourth part of the book consists of a single chapter. This deals with some aspects of the economics of using plastics, and with some of the more practical aspects of plastics product design, particularly the design of the important class of plastics components produced by injection moulding.

Finally, there is a selected bibliography, divided into sections to facilitate the choice of further reading. There is no index, as the arrangement of the book makes one superfluous.

R.M.O.
*London, 1970*

# CONTENTS

# Part I

## INTRODUCTION TO PLASTICS

# 1

# THE NATURE OF PLASTICS

## INTRODUCTION

Most engineering materials have been used and developed over a great many years. Consequently, many of their characteristics are widely known and there is much experience of their application. This is not so, however, with plastics. Even the oldest plastics are still new enough for there to be, as yet, no large store of accumulated experience and, therefore, no true appreciation of them as engineering materials comparable to that which exists for, say, ferrous metals. This first chapter attempts to remedy this lack of historical appreciation, inevitable with new materials; to look at plastics as a whole; to see why and how they are related and can be treated as a class; and to describe how they behave as a class. It thus provides a frame of reference into which the detailed discussions that make up the bulk of the book can be fitted.

However, before plastics can be treated as such, it is necessary to consider the materials upon which they are based, namely, polymers, and to say something about the relationship between polymers and plastics.

## MACROMOLECULES AND POLYMERS

Substances such as gases, water, acids and many organic compounds, consist of small molecules. However, during the last one hundred years, and more particularly in the last fifty years, it has come to be realized that there are a large number of very important materials which are composed of extremely large molecules, also described as 'giant molecules' or 'macromolecules'. Some of these materials occur in nature, some are made by modifying naturally occurring products, while others are completely man-made (synthetic).

Many materials of this kind are made from substances consisting of relatively small molecules, which combine together to give a pattern of repeated groups of atoms in the structure of the macromolecules, in much the same way as links make up a chain. Materials containing macromolecules made up of large numbers of repeated units are called 'polymers', and the substances from which they are made are called 'monomers'. If only one kind of monomer is used, the resulting polymer is known as a 'homopolymer'; if two kinds are used, the product is a 'copolymer'; and if three kinds are used, the product is a 'terpolymer'.

Polymers are not of technological interest to the plastics industry unless their molecular weight is high, say above 5000 (for comparison, sulphuric acid has a molecular weight of 98); such polymers are often called 'high polymers'. In Britain the word polymer can refer to either the pure chemical substance or the material produced as a direct result of the polymerization process; this material, although it may contain small amounts of impurities, is essentially one chemical entity, e.g. polythene or polyvinyl chloride (PVC). Although some polymers can be used as they are, most of them require mixing with additives of one sort or another before they can be processed into useful end products: typical additives are materials such as plasticizers, pigments, fillers, lubricants, extenders, antioxidants, and heat and light stabilizers. Usually, therefore, it is this physical mixture of polymer and additives that forms the plastic. It is important, therefore, to distinguish between a polymer and a plastic, because, although these terms are often used as if they were synonymous, in practice they often refer to different materials.

It is relevant at this point to mention the word 'resin', a term originally applied to certain natural products, e.g. rosin. When the polymeric nature of these natural resins was recognized, synthetic polymers began to be called 'synthetic resins', and later, particularly in the United States, simply 'resins'. Today, the terms 'resin' and 'base resin' are often used to denote a commercial polymer, i.e. a material containing no deliberately added ingredients.

### *The Range of High Polymers*

The significance of high polymers may be judged by looking at some of the more important of them. They occur in the inanimate world as bitumen; in the vegetable world as wood, rubber, gutta-percha, jute, hemp and cotton; and in the animal world as shellac, or proteins such as casein, silk, wool, hair, horn and flesh. Among materials based on synthetic polymers are synthetic rubbers, man-made fibres, synthetic glues, synthetic paints and varnishes, cinematograph and packaging films, and most of the plastics of commercial importance.

Unfortunately, polymers are not neatly divisible into subsections, such as rubbers, plastics, fibres, coatings, films, etc. For example, nylon is perhaps best known as a textile fibre, but it is used as a plastics moulding material, a synthetic gut, and a film; phenolic and amino resins are widely used as adhesives, as binders in surface coatings, and as the bases for certain types of thermosetting plastics.

It follows that there is no simple, unambiguous definition of the term 'plastics', and that this is simply a convenient term to apply to certain polymeric materials in certain circumstances.

## THE MOLECULAR STRUCTURE OF POLYMERS

### *The Prototype Polymer—Polythene*

In order to understand the relationship of one polymer to another and to establish the features responsible for the characteristics which plastics as a group exhibit, it is convenient to consider polythene as having the simplest chemical structure from which the structure of all other polymers can be derived by simple modification. It must be emphasized that these relationships are completely theoretical and have no practical basis whatsoever.

The typical simple polythene molecule can be represented as a long chain of carbon atoms linked to each other, each carbon atom being linked to two hydrogen atoms (Figure 1.1). For the sake of simplicity, the carbon chain has been shown as straight, but it must be remembered that in fact it is not straight and has a zig-zag configuration in one plane, the hydrogen atoms being arranged above and below it.

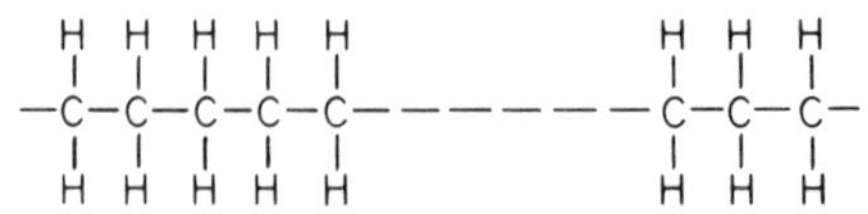

```
  H H H H H             H H H
  | | | | |             | | |
 -C-C-C-C-C- - - - - - -C-C-C-
  | | | | |             | | |
  H H H H H             H H H
```

**Figure 1.1.** Polythene molecule

*Molecular weight and molecular weight distribution.* It will be observed that in Figure 1.1 the length of the chain is unspecified—that is, the polythene molecule has not been assigned any definite molecular weight. It is one of the distinguishing features of high polymers that they do not have a unique molecular weight or chain length and that in any one sample there might be no two molecules of exactly the same weight or length. In general, it is found that from a given natural source or a given set of reaction conditions there will be a typical distribution of molecular weights throughout the sample. Figure 1.2 illustrates some of the possibilities: it will be seen that the distribution of molecular weights can vary and so can the average molecular weight. Both these factors are important technologically, and are strictly controlled during manufacture.

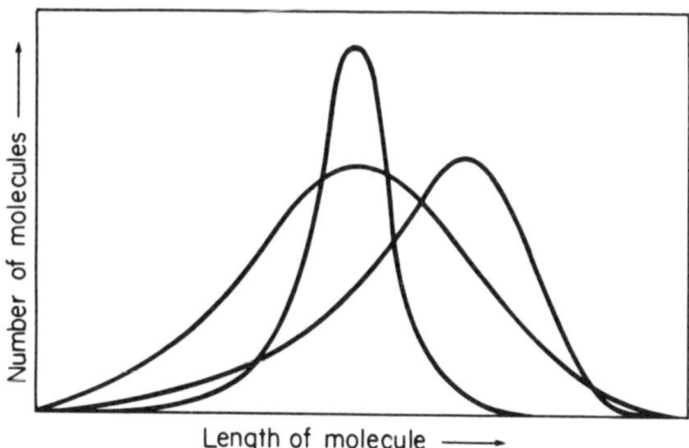


**Figure 1.2.** Molecular weight distribution

*Molecular configuration.* The schematic diagrams (a) and (b) of Figure 1.3 illustrate possible ways in which the long molecules in a mass of polymer can arrange themselves. The approach is not entirely in keeping with recent research findings but it is illustrative of the effects found.

In (a) the molecules are arranged in a completely random fashion, somewhat like a bowl of spaghetti or a ball of cotton wool, and the material is said to be amorphous. In (b), parts of the molecular chains participate in ordered arrangements or crystals while other parts of the same

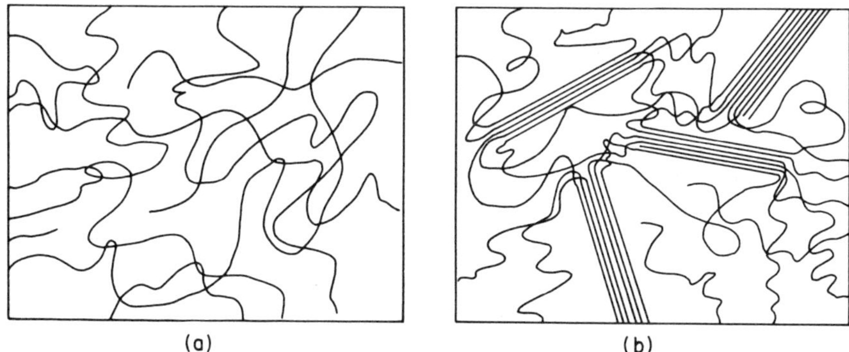


**Figure 1.3.** Possible arrangements of polymer molecules

molecules exist in the disordered state. The polymer is partially crystalline and is prevented from becoming wholly crystalline, in contrast to common salt, washing soda, etc., by the long molecules becoming entangled with one another. Structures such as (b) are possible only with long chain molecules, but the fact that a molecule is long does not in itself mean that structures such as (b) are inevitable; crystallization depends on other factors such as the size and spatial arrangement of side groups attached to the main chain, as well as the temperature.

One of the factors restricting crystallinity in high polymers is 'branching'. Figure 1.1 illustrated a polythene molecule in which each carbon atom was attached to two other carbon atoms and to two hydrogen atoms. This picture represents only the very simplest case and in practice a carbon atom may be attached to three, instead of two, other carbon atoms, thus giving rise to side chains or branches such as those shown in Figure 1.4. Irregularly placed, short branches are not able to fit into a crystalline structure, and therefore a material which is highly branched will be less crystalline than an unbranched material.

```
                            H
                            |
                          H-C-H
                            |
  H   H   H   H   H      H-C-H    H   H   H   H   H   H   H   H   H   H   H   H
  |   |   |   |   |         |     |   |   |   |   |   |   |   |   |   |   |   |
 -C - C - C - C - C ------- C --- C - C - C - C - C - C - C - C - C - C - C - C -
  |   |   |   |   |         |     |   |   |   |   |   |   |   |   |   |   |   |
  H H-C-H H   H   H         H     H   H   H   H   H   H   H  H-C-H H   H   H
      |                                                        |
    H-C-H                                                    H-C-H
      |                                                        |
    H-C-H                                                      H
      |
    H-C-H
      |
      H
```

**Figure 1.4.** Branched polythene molecule

*Chemical Relationship of other Polymers to Polythene*

If the structural formulae of other polymers are compared with that of polythene, it is seen at once that there are certain similarities and differences which can be related to similarities and differences in properties, processing and uses. It is therefore useful, in a general review of the nature of plastics, to stress these similarities and differences in structure by grouping polymers according to the hypothetical reactions required to derive them from polythene. The first hypothetical chemical reaction that can be considered is to replace some or all of the hydrogen atoms in polythene by other atoms or groups. This procedure is illustrated in structures (b)—(e) in Figure 1.5 in which the following replacements have been made:

(b) One hydrogen atom on every alternate carbon atom has been replaced by the methyl group, $CH_3$, giving polypropylene.
(c) One hydrogen atom on every alternate carbon atom has been replaced by a chlorine atom, giving polyvinyl chloride (PVC).
(d) All the hydrogen atoms have been replaced by fluorine atoms, giving polytetrafluoroethylene (PTFE).
(e) The two hydrogen atoms on every alternate carbon atom have been replaced, one by a methyl group and the other by the group —$COOCH_3$ giving poly(methyl methacrylate) (PMMA).

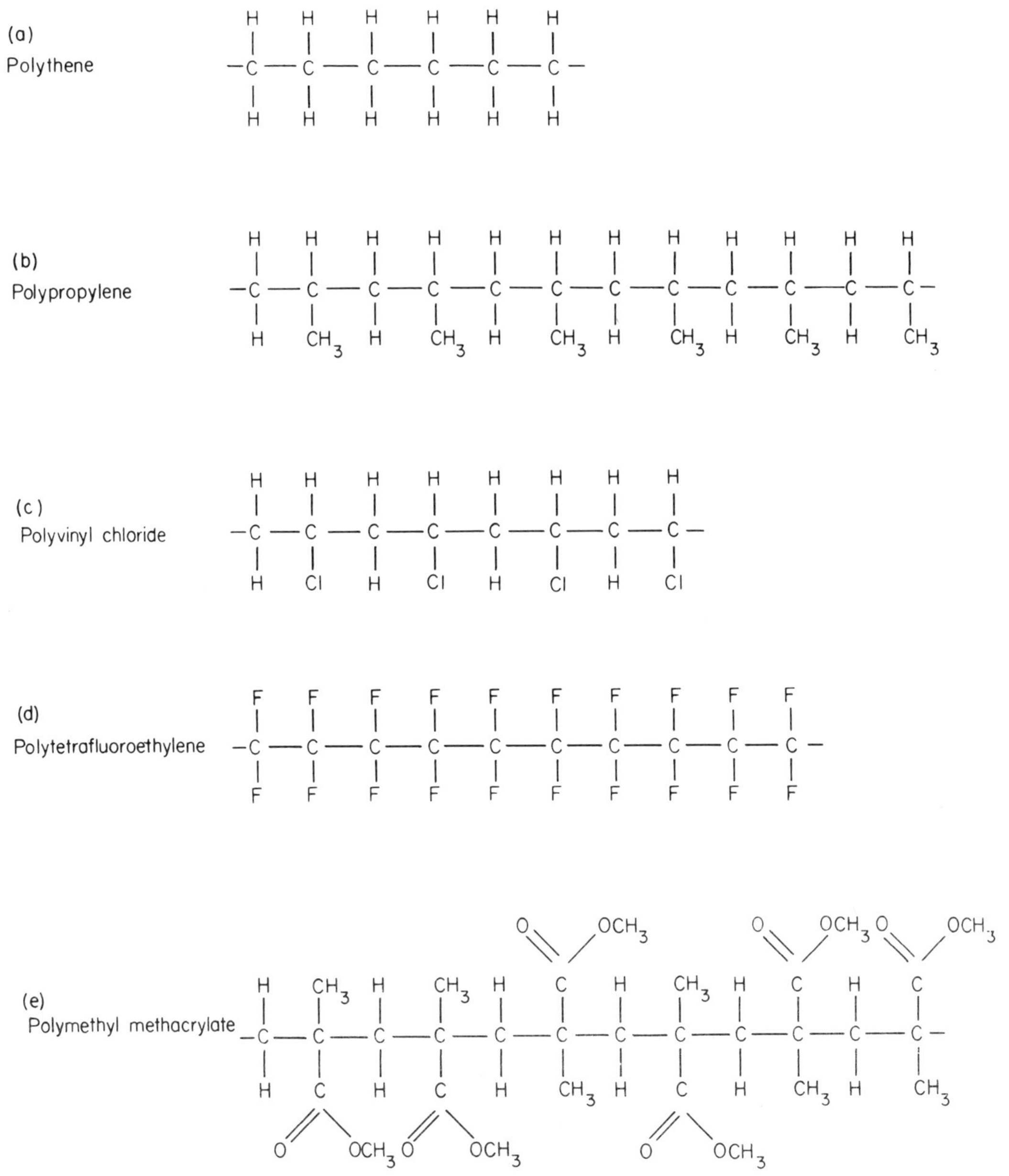

(a) Polythene

(b) Polypropylene

(c) Polyvinyl chloride

(d) Polytetrafluoroethylene

(e) Polymethyl methacrylate

**Figure 1.5.** Some thermoplastic molecules

Because these different materials all have molecules which are long and threadlike they will have certain properties in common with polythene, but because the substituent atom or group ($CH_3$—, Cl—, F—, $CH_3COO$—) is different from the hydrogen atom which it has replaced, they will have other special properties not possessed by polythene. The relative sizes of the different substituent atoms or groups compared with the size of the hydrogen atom and the way in which they are arranged in space around the carbon chain will also affect the properties and, in particular, the possibilities of forming regular crystalline structures.

*Stereoregular Polymers*

From consideration of a polythene molecule in which one hydrogen atom on every alternate carbon atom in the main chain has been replaced by an atom or group, denoted by X, three arrangements for the resulting molecule are possible: these are shown three-dimensionally in Figure 1.6 and, for simplicity, in two dimensions in Figure 1.7.

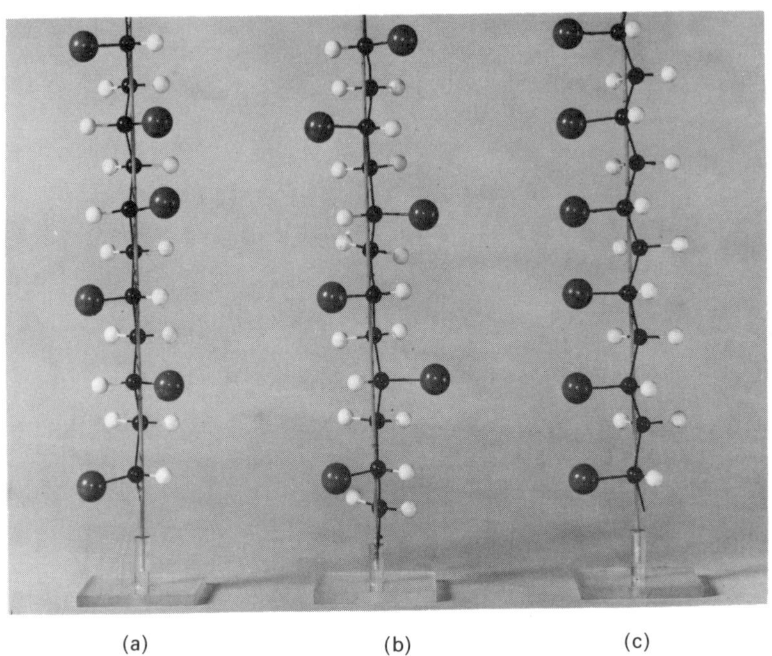

(a) (b) (c)

**Figure 1.6.** Possible molecular structures

```
         H    H    H    H    X    H    H
         |    |    |    |    |    |    |
(a)    — C —  C —  C —  C —  C —  C —  C —
         |    |    |    |    |    |    |
         X    H    X    H    H    H    X
                      atactic

         H    H    X    H    H    H    X
         |    |    |    |    |    |    |
(b)    — C —  C —  C —  C —  C —  C —  C —
         |    |    |    |    |    |    |
         X    H    H    H    X    H    H
                    syndiotactic

         H    H    H    H    H    H    H
         |    |    |    |    |    |    |
(c)    — C —  C —  C —  C —  C —  C —  C —
         |    |    |    |    |    |    |
         X    H    X    H    X    H    X
                      isotactic
```

**Figure 1.7.** Possible molecular structures

In structure (a) the arrangement is completely irregular and unless the group X is very small no close packing of the molecular chains can occur and the material will be amorphous: this arrangement is known as 'atactic'. Structure (b) shows a regular arrangement in which the groups X are oriented alternately on opposite 'sides' of the carbon chain, an arrangement known as 'syndiotactic'; because of its order a syndiotactic material can crystallize. A third structure is shown in (c) where all the groups X are arranged on one 'side' of the carbon chain: this arrangement is called 'isotactic' and again crystallization is possible. Structures such as those shown in (b) and (c) are known as stereoregular.

It is sometimes possible to control the orientation of side groups during the manufacture of polymers so that the material obtained consists of molecules which have predominantly an isotactic or a syndiotactic structure. For example, polypropylene in the atactic form ($X = CH_3$ in Figure 1.7(a)) is an amorphous material of no commercial value, but isotactic polypropylene ($X = CH_3$ in Figure 1.7(c)), which can be readily produced, is a very important polymer for plastics, fibres and films. At the moment, because of the polymerization methods used, most polymers of the type shown in Figures 1.6 and 1.7 are obtained in the atactic form, but there is no reason why regular structures cannot be obtained, although there may be no commercial incentive to do so. For instance, PVC is used almost entirely in the atactic form, which is almost completely amorphous ($X = Cl$ in Figure 1.7(a)), but syndiotactic PVC ($X = Cl$ in Figure 1.7(b)) has been prepared and is a crystalline material, albeit of little use because it is virtually impossible to process.

A second hypothetical chemical reaction to change polythene into other polymers is to replace some of the carbon in the main chain by other atoms such as nitrogen, giving nylon (Figure 1.8(a)) or oxygen, giving polyacetals (Figure 1.8(b)).

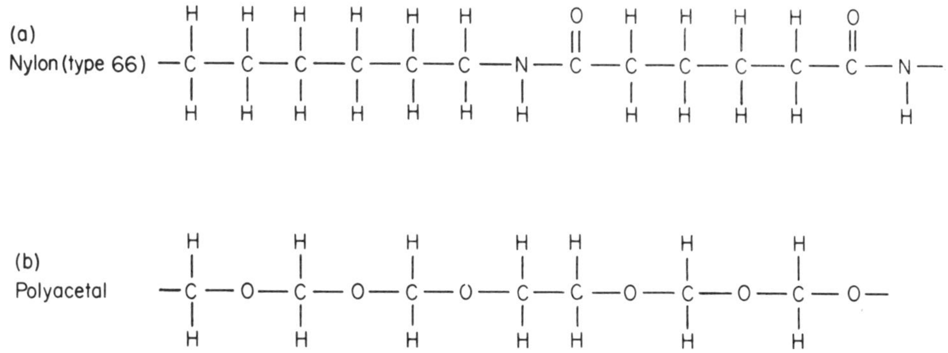

```
                   H   H   H   H   H   H       O   H   H   H   H   O
(a)                |   |   |   |   |   |       ||  |   |   |   |   ||
Nylon (type 66)  —C — C — C — C — C — C — N — C — C — C — C — C — C — N—
                   |   |   |   |   |   |   |       |   |   |   |       |
                   H   H   H   H   H   H   H       H   H   H   H       H

                   H       H       H       H   H       H       H
(b)                |       |       |       |   |       |       |
Polyacetal       —C — O — C — O — C — O — C — C — O — C — O — C — O—
                   |       |       |       |   |       |       |
                   H       H       H       H   H       H       H
```

**Figure 1.8.** Nylon and polyacetal molecules

It can be seen that the nylon molecule consists essentially of polythene-like segments which are interrupted and joined by the $-\overset{\overset{\displaystyle O}{\|}}{C}-\overset{\overset{\displaystyle H}{|}}{N}-$ group. One CONH group in one molecule will exert a very strong attraction for a similar group in a neighbouring molecule and this attraction will have a marked effect on the properties of nylon.

In the polyacetals the replacement of alternate carbon atoms in the main chain by oxygen atoms increases the interchain forces, giving a material with high rigidity and crystalline melting point.

*Cross-linking*

The polymers so far considered in this chapter consist of long, threadlike molecules, either lying closely packed together or entangled and intertwined with one another, which exert physical forces of attraction and repulsion on each other. When substances with such structures are heated, the various forces which hold the molecules together are effectively weakened, the material begins to soften and becomes less stiff, and eventually a stage is reached at which it becomes a viscous, elastic liquid. If the material is now cooled it will gradually become stiffer as the temperature decreases until it returns to its former state. This cycle of softening and hardening can, in theory, be repeated indefinitely and materials which behave in this way are known as 'thermoplastics'. It is with thermoplastics that most of this book is concerned.

In contrast to these are those materials which are derived from polymer chains by a chemical reaction, namely, that of chemically linking neighbouring chains together by means of primary chemical (covalent) bonds. These bonds make it impossible for molecules to slide over one another, or become separated to any appreciable extent. Such materials, which have the ability to cross-link, are described as 'thermosetting', i.e. they can be softened once for a short time and then they become hard, rigid, infusible and insoluble; in other words, they become 'thermoset'. In effect, a component with this kind of structure consists of one single, giant molecule. Cross-linking is often effected by heating and has therefore become known as 'thermosetting' or 'thermohardening'. A similar reaction in rubbers, in which an added material such as sulphur is used to form the links between one molecular chain and another, has been given the name 'vulcanization'.

## THE PROPERTIES OF POLYMERS

*The Properties of Polymers Related to their Structure*

Some of the properties of polymers are almost unique to materials consisting of long chain molecules which are held together by comparatively weak interchain forces; they thus deserve special consideration at this early stage. The most important concept in this context is that of the solid amorphous phase, the glass-like state, which can best be treated in terms of an example.

The rigidity or stiffness of poly(methyl methacrylate) changes markedly over a narrow temperature range at around 105°C. Below this temperature the material is rigid; above it, rubbery. This temperature is not a melting point, because poly(methyl methacrylate) is not crystalline, and it has been given a variety of names such as softening point, glass–rubber transition, glass transition, and relaxation transition. In this transition region, the position of which depends on many factors, there occurs a phenomenon which is peculiar to high polymers and which is an important factor governing their mechanical properties.

At low temperatures the molecules behave like a large number of rigid rods joined together, the overall shape of the molecule being determined by the van der Waals forces and molecular entanglement. As the temperature is increased, the thermal energy in the molecules increases and the individual atoms oscillate with greater amplitude. In the transition region these vibrations become large enough to overcome the restricting intermolecular forces and the shape of the molecules can then change in response to an applied stress. At temperatures above the transition region, the thermal energy is so large that the forces restricting changes in shape become very small and the stiffness falls to a very low level.

It is unusual to find only one transition—most polymers have two or more. For example, polyvinyl chloride begins to soften at a small transition around −40°C but does not become rubbery until +80°C. Polythene has one transition around −110°C, and another around 0°C but it does not become liquid until above its main transition, the crystalline melting point, at 110 to 130°C. It is believed that at minor transitions, segments of the molecules, and sometimes side chains, become free to move, but that the whole molecule becomes substantially free only at the main transition. Even then entanglements render the material rubbery.

It is evident that the magnitude of the various transitions and the temperatures at which they occur influence considerably the mechanical properties of polymers, and that at room temperature these properties can best be understood by considering the transitions. For example:

1. Rubber is well above its transition. Therefore it is very flexible and has a high elongation to break.
2. Poly(methyl methacrylate) is well below its transitions. Therefore it is relatively stiff and has a low elongation to break; in fact, it is brittle.
3. Polyvinyl chloride is below its main transition, but above its secondary transition. Therefore it is relatively stiff and has a moderate elongation to break. It is not normally brittle but it may be so under some conditions.
4. Polythene is above its transition but below its crystalline melting point. Therefore it is soft (although not as soft as rubber), has a high elongation to break and is brittle only under extremely severe conditions.

*Some Properties of Polythene and PVC*

It must be stressed that in the present state of knowledge it is not possible to make a complete quantitative prediction of properties from a knowledge of chemical constitution and molecular configuration, but nevertheless broad, qualitative correlations can be attempted.

For example, because polythene is a long-chain hydrocarbon it is thermoplastic, chemically rather inert and electrically, a first-class insulator. Because the molecules of polythene can be made with different degrees of branching the degree of crystallinity (and hence the density) of polythene can be altered within certain limits: the less branched materials are stiffer and have a higher melting point than the more highly branched materials, because greater lengths of the molecules can be readily accommodated in a crystalline structure and so exert greater forces of attraction on one another, and the crystals are more perfect. Also, in general, as the molecular weight of polythene increases, so do the tensile strength, elongation at break and resistance to impact.

As seen from Figure 1.5(c) the structure of PVC differs from that of polythene in having one chlorine atom substituted for one hydrogen atom in every four in polythene. Compared with the hydrogen atom, the chlorine atom is very large, and also effectively electronegative; the molecule or repeat unit thus behaves as a dipole. This large atom, if distributed in a random fashion (atactic PVC), prevents crystallization and so leads to a substantially amorphous material: because of the lack of crystallinity PVC has a softening region but no melting point. The dipole moment of the carbon–chlorine bond makes PVC an inferior dielectric to polythene (i.e. it has a higher permittivity and, at certain frequencies, a much higher power factor) and, although PVC is resistant to acids, alkalis and water, it is attacked or even dissolved by some polar organic liquids. The lack of electrical symmetry in the PVC molecule leads to stronger intermolecular forces of attraction which, in turn, contribute to greater stiffness. Again, whereas polythene can be ignited in air and will continue to burn, PVC, which contains hydrogen and chlorine, decomposes, liberating hydrochloric acid which tends to put out the flame and render the material self-extinguishing.

*Controlling the Properties of Polymers*

In making polymers, certain variables can, to some extent, be controlled. They are:

chemical constitution,
average molecular weight,
molecular weight distribution,
chain branching,
spatial arrangement of constituent groups in the molecule.

Thus, by suitable choice of these variables it is possible, within limits, to 'design' or 'tailor-make' a polymer with certain desired properties. As more and more information becomes available and techniques of polymerization become more refined, it may be possible to make quantitative, as well as qualitative, forecasts of the properties of given possible molecular structures. It must, however, be remembered that although properties of all kinds can be controlled to some extent, requirements are sometimes mutually exclusive—for example, chemical inertness implies that there will be cementing and adhesion problems.

## THE PROPERTIES OF PLASTICS AS A CLASS

So far in this chapter discussion has been confined to polymers, but as already pointed out, these are seldom used on their own, and it is with plastics that the engineer is concerned. It is therefore necessary from now on to consider the properties of plastics. Because of the great variety of polymers and additives available, one plastic will differ significantly in properties from any other, but nonetheless there are certain similarities among all plastics which enable them to be grouped together and treated as one class. These similarities can conveniently be discussed under the headings of:

General properties,
Thermal properties,
Mechanical properties,
Optical properties,
Electrical properties,
Chemical and weathering resistance.

*General Properties*

All plastics have low densities—generally in the range 0·83 to 2·5 g/cm³, although these figures can be extended upwards or downwards—for example, foamed materials can have densities as low as 0·01 g/cm³ and filled plastics can have densities up to 3·5 g/cm³. All have a fairly low surface hardness: the hardest thermoplastic commercially available, poly(methyl methacrylate), is no harder than aluminium, although some thermosetting plastics are harder.

*Thermal Properties*

Plastics have high coefficients of thermal expansion—many times those of metals. An important point to note is that their dimensional stability is often dependent on their previous history and when heated they may not expand uniformly in all directions. All plastics are good heat insulators.

*Mechanical Properties*

Compared with metals, plastics have high strength to weight ratios and low stiffnesses. They exhibit a very wide variation in impact behaviour—some are brittle whereas others are very tough. The mechanical properties of plastics are very dependent on the rate and duration of loading and the temperature. One important consequence of this is that plastics are subject to 'creep', which must be distinguished from creep in metals (see Chapter 4). Mechanical properties can also be affected by the previous history of the material.

*Optical Properties*

Many polymers are inherently transparent, some are translucent and a few are opaque. Similarly, the optical properties of plastics derived from these polymers cover the same range.

*Electrical Properties*

All plastics are normally good electrical insulators and some of them are excellent.

*Chemical and Weathering Resistance*

In general, plastics are complementary to metals in their chemical resistance in that they are resistant to acids, alkalis and water, but are affected by some organic liquids. Their water absorption can be low but any absorbed water may have a marked effect on other properties. Most plastics have low permeabilities to water vapour, gases and odours. The resistance of plastics to sunlight and outdoor weathering varies considerably, from outstanding to poor, although many plastics can be formulated so as to have good resistance.

# 2

# FROM POLYMER TO PLASTIC

A fuller appreciation of the differences between polymers and plastics can best be obtained by considering one particular group of plastics, that based on polyvinyl chloride.

Commercial vinyl chloride polymers usually consist of pure polymer and traces of other materials such as monomer, catalyst, and solvent, which it is uneconomic to remove. The effect of these impurities is generally to impair the properties of the pure polymer.

In contrast to these undesirable additives, other materials are deliberately added to vinyl chloride polymers to perform one or more specific functions in the resulting plastic or 'compound'. The precise nature of these other materials will depend on the processing method to be used to convert the plastic into a finished article and on the properties required in the finished article. The main classes of such additives used in PVC formulations are discussed below.

### *Heat Stabilizers*

Every PVC plastic contains a heat stabilizing system, the main function of which is to improve the thermal stability, both during processing and during the service life of the article. In addition, some systems confer other desirable properties, such as good weathering performance.

### *Lubricants*

When PVC is softened during processing it becomes sticky. To ease its flow through the equipment and thus minimise local decomposition and tearing and marking of the surface of the article, a lubricant may be added. In addition to exerting this 'external' effect, most lubricants show 'internal' lubricating behaviour by reducing frictional heat.

### *Plasticizers*

A compound of polymer, stabilizer and lubricant will produce a hard, rigid article. Such compounds are often referred to as rigid PVC compounds, but more correctly they are described as unplasticized PVC compounds (UPVC). The addition of plasticizers (usually liquids of high boiling point) results in compositions which, in general, produce articles of greater softness and flexibility, lower strength and electrical insulation properties, better elongation and low temperature performance, and with a lower maximum service temperature. The effect on chemical and weathering resistance, colour, fire-resistance and toxicological behaviour will depend very much also on the exact nature and amount of all the other ingredients present. Because the presence of a plasticizer lowers the temperature at which a PVC composition softens and can be processed, there is less risk with plasticized PVC of decomposition during processing.

*Fillers*

Fillers are often used to reduce the volume cost of a PVC compound. Generally, the mechanical properties and low temperature performance are impaired, the hardness is increased, and a matt finish is produced on extrusions. Sometimes, however, fillers are used for technical reasons, e.g. to improve insulation resistance.

*Pigments*

Pigments are easily incorporated in PVC. The colour possibilities are limited only by the ingredients other than the polymer that are present, e.g. lustrous metallic effects cannot be achieved with filled compounds.

From what has been said, it follows that statements about the properties of PVC compounds must either be accompanied by a full specification of the composition or be of so general a nature as to be of limited value. The position is further complicated by the fact that because of the variety of additives available, and the differing quantities in which they can be used, many hundreds of grades of PVC compounds are available commercially.

PVC has been chosen to exemplify the practical reasons for distinguishing clearly between a polymer and a plastic, but what has been said applies equally to most other materials classed as plastics. If it appears that the step from polymer to plastic often complicates the position to the extent that even simple statements about the properties of plastics cannot be made, it should be remembered that there is the considerable gain that the plastics technologist has more variables at his command, and is therefore able to approach more closely to the requirements of the design engineer in any given problem.

## COMPOUNDING

There are several ways in which additives can be mixed with polymers. Which way is chosen will depend on one or more factors such as, for example, the properties required in the finished article, the equipment available for mixing the raw materials and for processing the final blend, and the relative merits and economics of storing, handling and processing different materials.

It is not possible, nor indeed desirable, in a book of this kind to cover this subject in any detail. Nevertheless, an outline of the various techniques available for compounding will be of interest and of value to anyone concerned with designing and using plastics components.

In any mixing process the aim is to achieve an intimate physical mixture with uniform distribution and dispersion of all the ingredients—except, of course, when special decorative effects such as non-uniform patterns and colours are intended. In certain circumstances this aim can be achieved by carrying out the mixing operation in the cold, but more usually the application of heat and pressure is necessary, and for some materials in some applications the use of 'hot compounding' is essential.

If the mixing has been carried out cold, the compound can be fed directly to the processing equipment, e.g. injection moulding machine, but if heat has been applied, the material will usually be in a form which has to be reduced in some way to particles of a shape and size suitable for feeding to the processing unit. A certain amount of mixing and homogenization will take place in the processing machine itself and modified extruders can, in fact, be used as continuous compounding equipment. The mixing efficiency of conventional extruders and moulding machines will vary from one to another and depend on the material being processed but, nevertheless, some mixing operations, for example the dispersion of pigments, are nearly always completed in the processing machine. Therefore, as already stated, the choice of mixing process is governed to some extent by the processing equipment available. However, there are certain applications, for example the use of black polythene for articles intended for outdoor exposure to sunlight, where hot-compounded material must be used, irrespective of the type of processing equipment available.

### *Hot Compounding*

Hot compounding consists basically of three stages: (1) premixing, (2) gelation and (3) granulation.

1. *Premixing.* In the first stage, the dry ingredients are cold-mixed and any liquids are then added while stirring is continued. This procedure produces a homogeneous mixture known as a premix.

2. *Gelation.* Depending on the type of plastic and scale of operation the premix is transferred either direct to the nip of a hot twin-roll sheeting mill, or first to a heated internal mixer and then either to the mill or to a special type of extruder. The mixer produces a dough-like material and the mill a thin crêpe.

3. *Granulation.* The crêpe produced by the mill can be fed directly to a calender for pressing into foil or sheet, or to an extruder, but more usually it is removed as a strip and fed to a granulating machine for disintegration into randomly or regularly shaped particles.

### *Dry Blending*

In contrast to hot compounding, which relies on mastication and gelation, the dry blend process depends on mixing the ingredients under conditions which result in the production of a dry, free-flowing product. High speed mixers capable of being heated and operating at different speeds are used. Final compounding, particularly the adequate dispersion of additives, depends on efficient working of the material in the processing equipment used, and such equipment must be carefully selected if dry blends are to be used.

### *Dry Tumbling*

Dry tumbling is the name given to that form of dry blending in which the polymer and additives are tumbled together slowly for a period that may vary from a few minutes to a few hours without the application of heat. Additives which can be incorporated in this way include dyes and pigments (dry colouring) and antistatic agents. The degree of distribution and dispersion achieved will depend on the nature of the polymer, the type of tumbling equipment used, and the amount of turbulence occurring in the processing machine. For example, coloured poly(methyl methacrylate), produced by dry colouring finely divided polymer, can be moulded into articles of perfectly uniform colour, whereas other dry coloured plastics, moulded on the same machine, might well give articles of poorer colour uniformity. Two big advantages of dry coloured materials over coloured compounds are that dry coloured materials are cheaper, and that the fabricator himself can carry out the dry colouring and thus avoid the necessity of carrying large stocks of various coloured compounds.

*Masterbatch Technique*

It is often more convenient and sometimes necessary to use the additives in the form of a masterbatch: this is a mixture of polymer and additives in which the additives are present in a concentration up to ten times that required in the final blend. Masterbatches can be made by hot compounding or by tumble blending, and in turn the required amount of masterbatch can be mixed with the polymer by either of these techniques.

# 3

# PROCESSING PLASTICS

The ease with which plastics raw materials can be converted into semi-finished and finished products is one of their distinguishing features and one of their outstanding advantages over many other materials. Also, because plastics are available in such a wide variety of forms, and can be processed and fabricated in so many different ways the design possibilities are practically unlimited.

Plastics can be bought as raw materials in the form of liquids, pastes, powders and granules, and as semi-fabricated products in forms such as sheet, film, foil, tube and rod. Many plastics are available in several of these forms; PVC is available in all.

The processing of all plastics falls into four types of operation:

(a) Conversion—from raw material, or semifabricated form.
(b) Machining—either of semifabricated forms or of components made by conversion.
(c) Jointing.
(d) Decorating.

Some articles require the use of two, three or all four of these methods. Which method or combination of methods is used will depend on many interacting factors, including the type of plastic, nature of the final product, quantity required, and the equipment available. The ultimate aim will be to produce a satisfactory article at minimum cost, and as a help in achieving this an outline knowledge of the main processing techniques, as given below, will be found useful.

## CONVERSION

Although a few plastics materials can be converted into finished articles, or used as coatings, in the cold, it is usual for the many different conversion techniques to include the following operations:

1. Heating the plastic to a temperature at which it becomes soft.
2. Applying pressure to shape the softened material to the required form.
3. Cooling it to retain the shape.
4. Removing it from the mould or tool.

The order in which these operations are performed depends on whether the raw material is thermosetting or thermoplastic, and on whether it is solid or liquid at room temperature.

Solid plastics of both types will soften when heated and will flow under pressure, but whereas thermoplastic materials must be held in shape while they are cooled, thermosetting materials will, provided that the cross-linking process is complete, hold their shape if removed from the mould when hot.

With liquid plastics materials, the shaping operation must be carried out before heating is completed.

*Compression Moulding*

Compression moulding consists of heating and compressing the solid raw material between male and female tools attached to one fixed and one moveable platen. It is usually restricted to thermosetting materials, because they can be removed from the mould while hot and do not require the mould to be alternately heated and cooled, which would be uneconomical. Temperatures of up to 180°C and pressures of up to 4000 lb/in$^2$ are necessary for thermosetting materials. Moulding times depend on the thickness of the part being moulded and can vary from a few seconds for thin-walled components to ten or more minutes for thick-walled mouldings. Mouldings weighing from a fraction of an ounce to 10 lb can be produced.

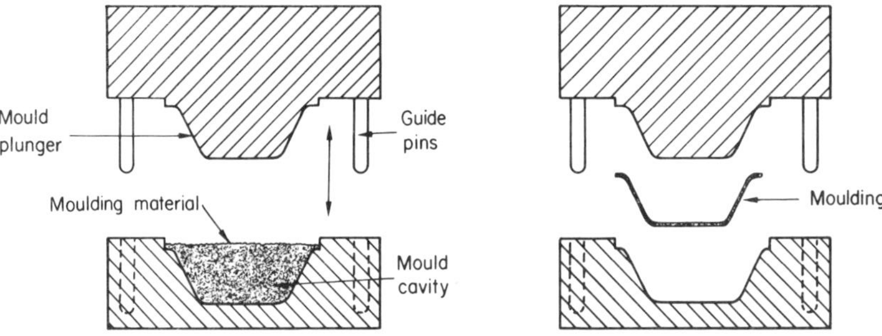


**Figure 3.1.** Compression moulding

*Transfer Moulding*

Transfer moulding is also often used for thermosetting materials. In this process the powder is heated in a separate chamber connected to the mould, instead of in the mould itself. When it is sufficiently molten to flow under pressure, it is forced by a ram out of the chamber and into the mould. Transfer moulding is used mainly for thermoset mouldings that cannot readily be made

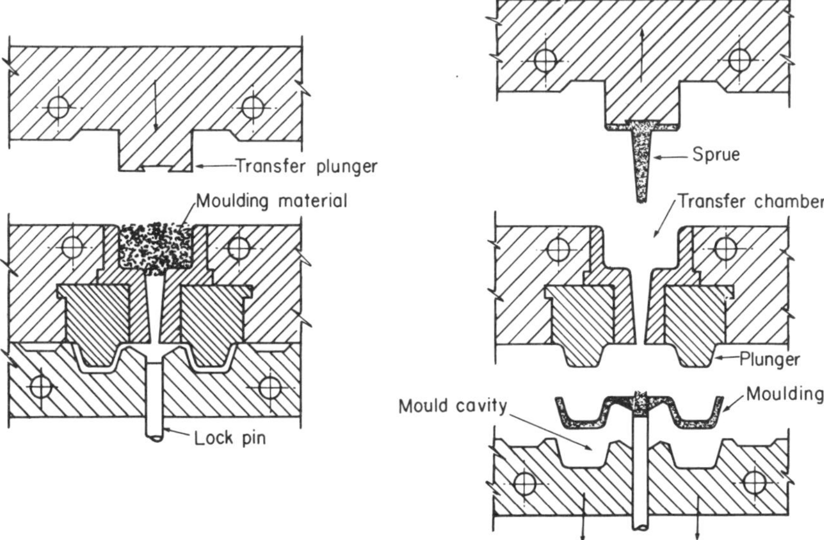


**Figure 3.2.** Transfer moulding

by other methods, i.e. mouldings with delicate inserts, mouldings with very thick sections, and mouldings with internal holes. However, transfer moulding can be used for thermoplastics if the mould is kept cool and the chamber hot, and the plastic is preheated before being loaded into the chamber.

*Injection Moulding*

Injection moulding is used almost entirely for thermoplastic materials which need to be cooled before they can be removed from the mould, although there is growing interest in its development for thermosetting materials.

Injection moulding closely resembles pressure diecasting in that a molten material is forced under pressure through a nozzle into a cool, split mould. An injection moulding machine consists basically of a feed hopper, a heated cylinder in which the plastic is softened, a means of applying pressure to the plastic, a mould, and means of controlling the heating and cycling operations. The two main types differ in the way in which pressure is applied.

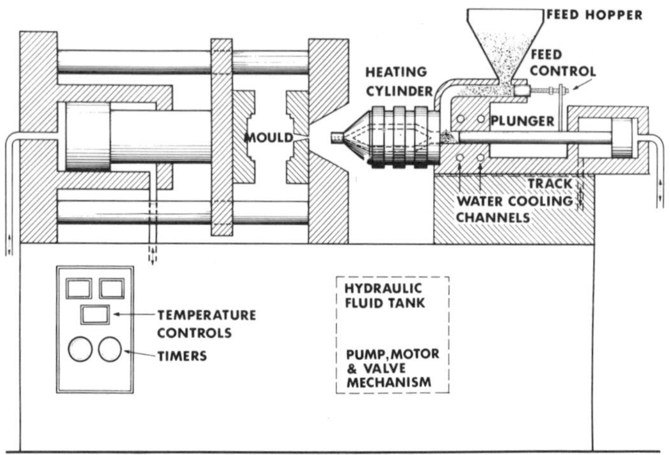


**Figure 3.3.** Plunger injection moulding machine

In a ram or plunger machine, a plunger forces material from a feed mechanism along the cylinder, around the sides of a torpedo or spreader, where it is heated and softened to a fluid state, and then through the nozzle into the mould. When the plunger is withdrawn, material for the next shot is metered into the rear of the cylinder.

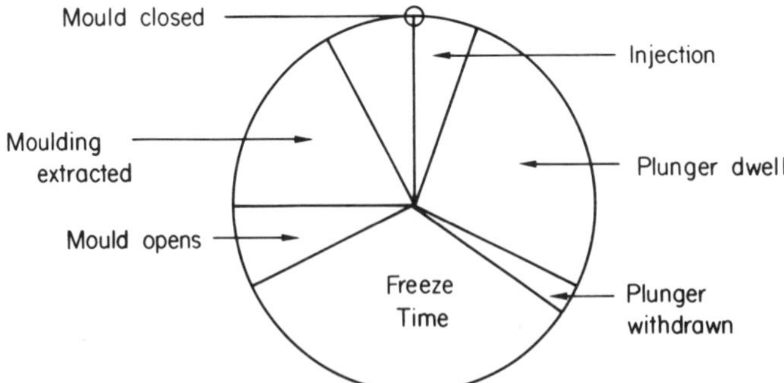


**Figure 3.4.** Moulding cycle for plunger machine

In a screw preplasticizing machine, a screw carries material from a feed hopper along the cylinder, in which it is heated and softened, to the section nearest the nozzle, where it accumulates and forces the screw, which rotates, back against a pressure pad at the rear of the cylinder.

When enough material to fill the mould has accumulated the screw stops rotating. The screw is then pushed forward by hydraulic pressure, forcing the fluid plastic through the nozzle into the mould. Limit switches regulate the distance the screw travels and thus accurately meter the amount of material entering the mould.

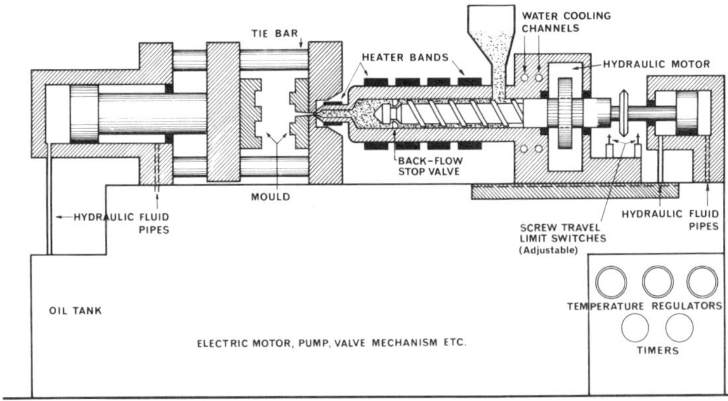


**Figure 3.5.** Single screw injection moulding machine

The pressures and temperatures used depend on the nature of the plastic and lie within the range 5000–20,000 lb/in$^2$ and 150–320°C respectively. The mould, which is kept at a carefully controlled temperature below the fusion point of the plastic, is kept closed by a hydraulic or mechanical locking device of sufficient strength to withstand the hydraulic force developed over the projected area of the moulding. Many modern machines are capable of fully automatic working, and time cycles can be from 5–300 s depending on the nature of the plastic and the size and thickness of the moulding.

In all three moulding processes multi-impression moulds can be used for small parts and metal inserts can be incorporated into the mouldings.

## *Extrusion*

Extrusion is a very important method of converting thermoplastic materials. As in injection moulding, molten material is forced from a heated cylinder into a die, but it differs from injection moulding in being a continuous process, and it can therefore be used to produce profiles of

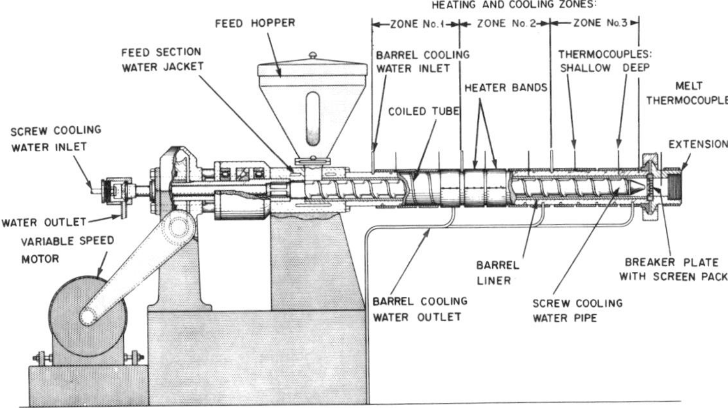


**Figure 3.6.** Single screw extruder

all kinds, including rod, tube, sheet, foil, and flat and tubular film. In an extrusion machine, usually referred to simply as an extruder, an Archimedean screw is used to convey material from a feed system through a heated cylinder or barrel to a die of appropriate shape through which the melt is forced. During its passage through the cylinder the plastic is compacted and mixed by the action of the screw, and softened, partly by contact with the hot barrel walls and partly by the frictional heat arising from the shearing action of the screw. The extrudate is then cooled to retain its shape and at the same time is 'hauled off' by a method appropriate to the material and profile being extruded.

The design of the screw is vital to the efficient operation of an extruder and different screws are used with different plastics. All however can be regarded as consisting of three zones: (1) a feed zone which functions as a screw conveyor, (2) a compression zone in which the plastic is converted into a fully compacted, homogeneous melt at uniform temperature, and (3) a metering zone which delivers this melt at a constant rate and pressure to the die.

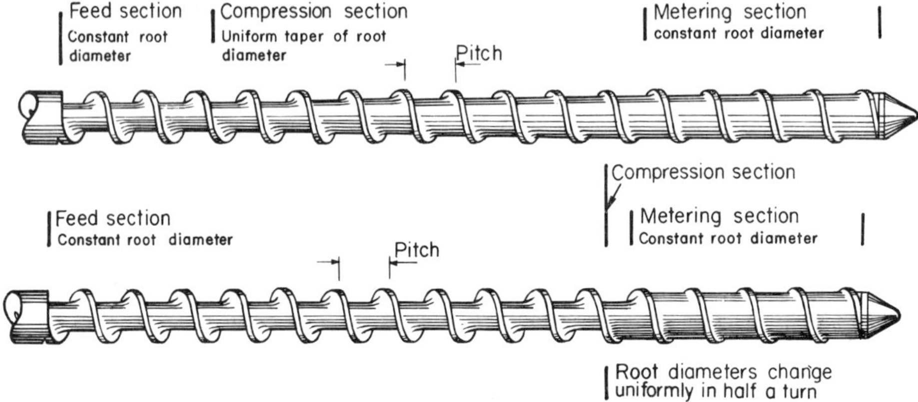


**Figure 3.7.** Typical extruder screws

Extrusion is a versatile process. Not only can it produce a wide range of profiles which, after cutting to size, can be used as such, but these profiles can often be postformed or adapted in some way, often in an integral operation, to yield a more extensive range of products: thin films can be laminated to substrates such as paper and plastics films: wires and cables can be covered; tubes ('parisons') can be fed to split moulds in which they can be blown to form hollow containers, bottles, etc. (the blow-moulding process); and simple sections can be given more intricate shapes, e.g. sheet and foil can be vacuum formed.

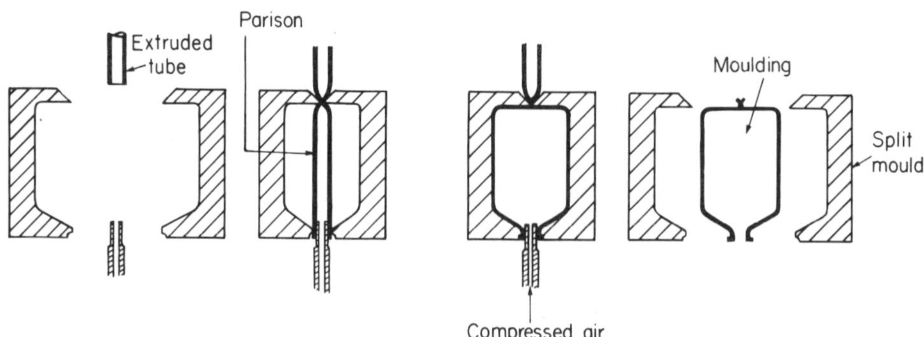


**Figure 3.8.** Extrusion blow moulding

*Other Techniques*

As an alternative to extrusion, sheet, foil and film can be made by a process known as calendering; this consists of compressing the softened thermoplastic between rollers or bowls to give a continuous length of material. Patterns are easily obtained by using a final embossing roller. Thick sheet can be made by laminating plies of thinner calendered sheet, or of plastics-impregnated material, between heated platens.

**Figure 3.9.** Calendering: z-type arrangement of bowls

Many processes have been developed to convert those forms of plastics raw materials, such as pastes, solutions and emulsions, which cannot be injection moulded or extruded. These techniques include dipping, brushing, spraying, knife-coating, impregnation, slush-moulding and rotational moulding. Of particular interest is the rotational moulding of polythene powder to produce small to medium quantities of large, unsupported articles, for example, water tanks and bulk containers.

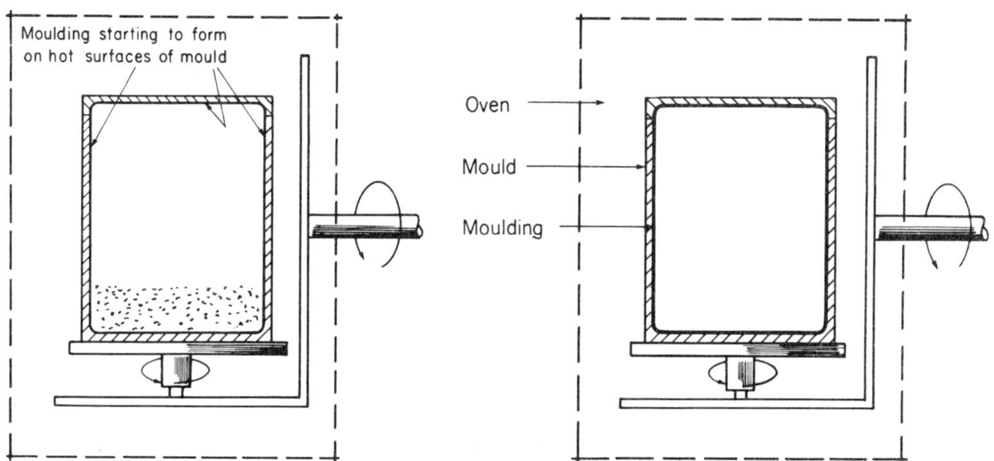


**Figure 3.10.** Rotational moulding

*Thermoforming of Sheet*

Foils and sheets can be shaped in a number of ways, all of which consist essentially of three stages:

1. Heating the sheet until it becomes soft and rubbery.
2. Pushing, blowing or sucking the sheet, thus causing it to conform to a shape.
3. Cooling the shape while it is firmly held.

The necessary shaping pressure can be applied either by using a simple plunger, or by introducing an air pressure difference by means of a vacuum or a compressed air supply. Many shapes are best made without a mould by what is known as free-blowing or free pressing, but moulds are often used to produce complicated or deep-drawn shapes. Moulds are usually made of wood or metal, and because they do not have to withstand high temperatures or pressures are less costly than those used for injection moulding.

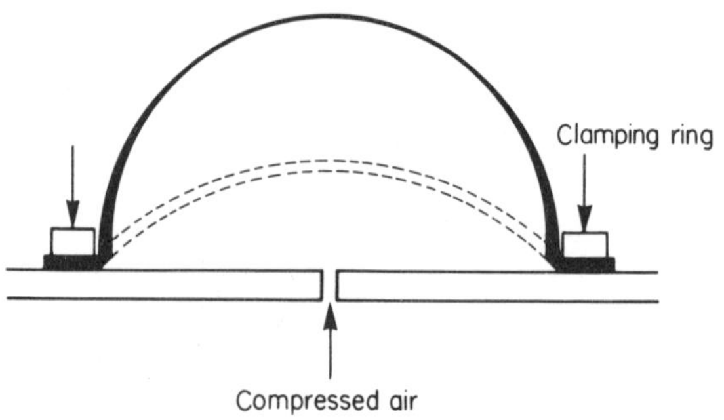


**Figure 3.11a.** Free blowing

The drawing of the sheet naturally leads to its thinning, and a skilful shaper chooses combinations of blowing, vacuum and pressing to obtain the best possible thickness distribution in the final shaping. The choice of mould used may be influenced by the fact that mould markings are usually required to be confined to one particular side of the shaping.

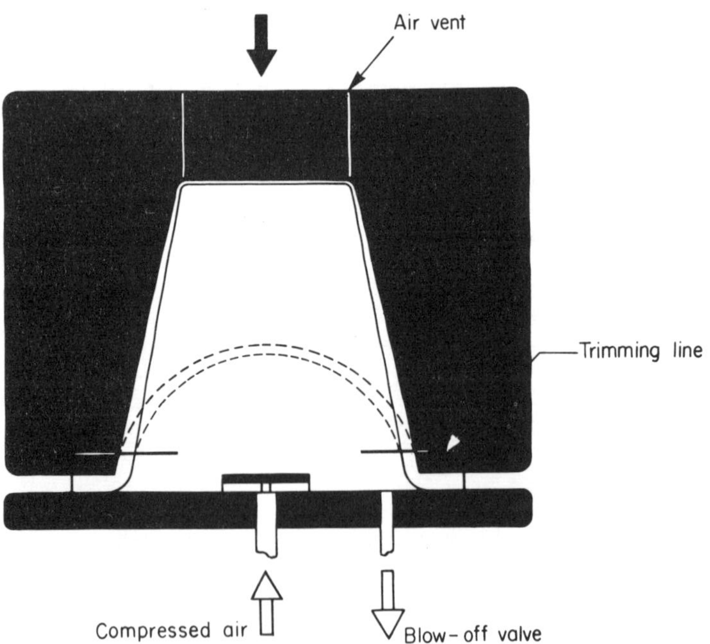


**Figure 3.11b.** Blowing into a female mould

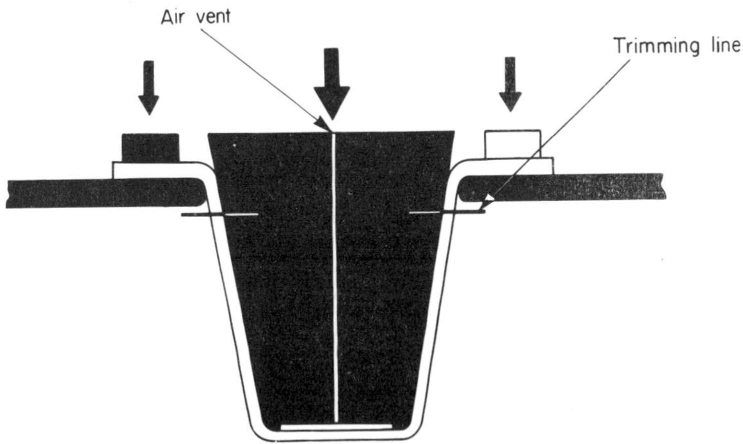


**Figure 3.11c.** Simple pressing

In the technique known as vacuum forming, the sheet is clamped over a mould, heated *in situ*, and then shaped by applying a vacuum in the mould. The process is often modified, for example by using a plunger to produce a deeper draw than would otherwise be obtained. Vacuum forming is sometimes used, rather than injection moulding, as an economical method of producing thin-walled shapings such as, for example, disposable beakers.

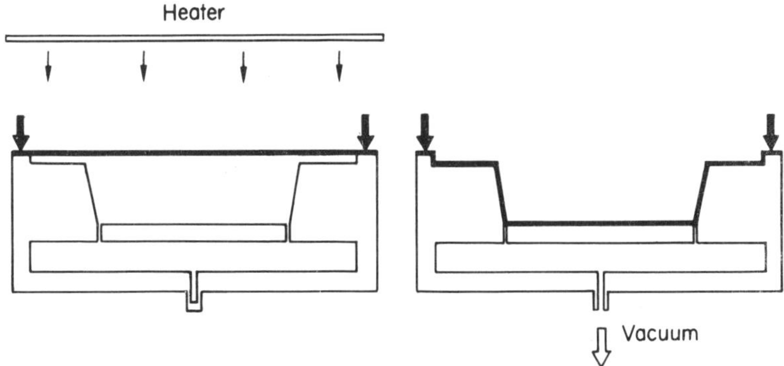


**Figure 3.11d.** Vacuum forming using a female mould

*Reinforced Plastics Techniques*

The term 'reinforced plastics' is used here to denote fibre-reinforced thermosetting resins. In contrast to reinforced thermoplastics, which can be handled by conventional thermoplastic techniques, these materials demand special processing methods, some of which are outlined below.

In the hand lay-up technique, layers of resin and reinforcement are 'laid up' in a mould until the required thickness of section has been achieved. The advantages of this method are inexpensive tooling and lack of restriction on size. The big disadvantage, the large amount of manual effort involved, is partly overcome in the spray-up technique in which a twin-nozzle gun sprays resin and reinforcing fibre simultaneously on to a mould. A further advantage is that complex parts can be formed more easily. In both these methods, curing is effected at room temperature. Components with better physical properties and finer tolerances can be made rapidly by shaping and curing the impregnated material between matched metal dies, in which case the fibrous reinforcement is often preformed to shape. For quick production of complex shapes of less critical dimensions premix moulding is suitable. In this method, resin, fibre and a mineral filler are mixed to a dough, called a dough moulding compound or DMC, and cured in metal dies. Heating is generally used to cure mouldings made in matched metal dies.

*Fabrication of PTFE*

Although polytetrafluoroethylene (PTFE) is a thermoplastic, it has no truly fluid state and cannot, therefore, be processed by conventional methods such as injection moulding and melt extrusion; nor can it be handled by compression and transfer moulding techniques. Special methods have been developed, consisting essentially of compacting the powder and then 'sintering' at a high temperature to produce a continuous structure; both specific shapes and standard forms, such as rod or tube, for subsequent machining, can be produced in this way.

## MACHINING

In general, plastics can be machined on conventional wood-working and metal-working equipment, although methods must be modified to suit the material. Tools must be kept sharp, and because of the low thermal conductivity of plastics with the consequent risk of build-up of frictional heat, some form of cooling must be used to prevent softening and possible decomposition of the plastic; cooling also sweeps away swarf and prevents tools from becoming blunt. Compressed air and liquid coolants are both used.

Because it is possible to introduce thermal and mechanical strains into the material during machining it is necessary to anneal some plastics afterwards, particularly if parts have to be produced to close tolerances.

## JOINTING

Because of the variety and versatility of conversion techniques, plastics components can frequently be made in one piece. However, there will always be some that have to be assembled from two or more parts, involving the use of jointing techniques. Many such techniques are available, including welding, cementing, heat sealing, and the use of snap fits and mechanical joints. Not every technique is suitable for every plastic, although it is true to say that every plastic can be joined either to itself or to other materials by at least one method. The performance of the joint will, of course, vary according to the materials, the type of joint, the efficiency with which the joint was made and the service conditions. Where there is a choice of jointing method, it will usually be found that some factor will dictate a preference for one rather than another.

As an illustration of the above remarks, polystyrene, PVC and acrylics can easily be cemented to themselves; polythene and polypropylene can be cemented with difficulty to produce bonds of limited use; and PTFE, except in very special circumstances, cannot be cemented at all. In the product chapters information is given on suitable methods of jointing.

Both machining and jointing are particularly valuable in the making of prototypes from semi-fabricated forms such as sheet, bar and rod. However, too much significance should not be paid to results obtained from such prototypes, because the properties of any plastics article will depend to some extent on the way in which it was made. For example, an item machined from a compression-moulded block may be entirely free from strain, whereas a similar item injection moulded from powder might contain residual moulded-in strains which could well result in, for instance, a lower resistance to stress cracking.

## DECORATING

Although most plastics articles rely for their sales and aesthetic appeal on their intrinsic colour, there are occasions when it is necessary or desirable to add some further decoration. Various processes can be used, the effectiveness of which depends on the nature of the plastic and also on the effectiveness of any pretreatment necessary. Some operations, such as hot foil stamping and the fixing of labels, can invariably be performed on untreated surfaces, irrespective of the plastic; others, such as painting, plating, vacuum metallizing and printing may well demand a pretreated surface to be at all effective. The type of pretreatment used will depend both on the type of plastic and on the finishing process, but two methods—flaming and acid etching—are most commonly used.

# Part II

## PROPERTIES OF THERMOPLASTICS

# 4

# DEFORMATIONAL BEHAVIOUR

The design of components in plastics, as in other materials, involves two major engineering problems arising out of the loads they are expected to bear. One is that of deformation, which might have to be restricted for functional or aesthetic reasons. The other is the avoidance of failure—either by fracture or by collapse through buckling.

The first problem calls for information on the deformational behaviour of plastics. Deformational behaviour is relevant also to the problem of buckling, which is essentially one of instability and which involves, therefore, the stiffness of plastics components. The remaining aspects of the second problem, that is the avoidance of failure by fracture or rupture, call for information on the strength characteristics of plastics.

These two facets of the mechanical properties of plastics are most conveniently discussed separately: this chapter is confined to a discussion of deformational behaviour and strength characteristics are left to the following chapter. The objects of these chapters, and that of Chapter 6, are however similar. First, the physical properties relevant to the behaviour of plastics as engineering materials are discussed in general terms. Secondly, the methods of characterizing plastics are introduced which are adhered to in the presentations of data in the product chapters; where no data are yet available, these methods indicate how the data should be provided. In this chapter a method is also given for mechanical design; this is referred to as the pseudo-elastic design method and is developed from, and makes use of, the results of creep tests in uniaxial tension.

## DEFORMATIONAL CHARACTERISTICS

The deformational behaviour of engineering materials is generally described in terms of relations between stresses and strains established on the basis of experimental results obtained in tests performed on a macroscopic scale. The most common of these mechanical tests involves uniaxial tension at normal temperature and the results obtained for many metals are well known. In particular, up to a certain strain level, in the range of small strains within which metals are commonly used, the observed strains are directly proportional to the imposed stresses. Consequently, metals are generally represented as linear elastic solids, and the mathematical model of the relation between strain $\varepsilon$ and stress $\sigma$ takes the form of a linear algebraic equation, which corresponds to Hooke's law, i.e.

$$\varepsilon = k\sigma \tag{1}$$

where $k$ is a constant of proportionality, called the compliance. However, in engineering practice this relationship is usually expressed as a ratio of the stress to the strain, which is defined as the modulus of elasticity, or Young's modulus, $E$.

Thus, in uniaxial tension, the extensional behaviour of materials such as metals may be characterized in the region of small strains by a single constant $E$. Moreover, materials for which this is true and which can be regarded as homogeneous, isotropic, linear elastic solids can have their deformational behaviour entirely characterized by only two constants, that additional to $E$ being Poisson's ratio or the modulus of rigidity.

This is obviously convenient and is enshrined in the classical theory of elasticity, well-known to engineers. As a result, there is a tendency to characterize all materials in similar terms. This includes plastics, which are commonly described in terms of $E$, in spite of the fact that they can not be regarded as linear elastic solids, except under special, narrowly defined conditions.

The essential difference between the behaviour of plastics and metals of construction is shown by the fact that when a plastics specimen is loaded, the resulting strain depends not only on the magnitude of the imposed stress, but also on the length of time during which the load is applied. Therefore, even the uniaxial strain in plastics cannot be represented by a single-valued function of stress and, in general, plastics cannot be characterized as simply as can metals.

## CREEP CURVES

The dependence of strain on time, as well as on stress, is conveniently established in tensile creep experiments, where the increase in strain with time is observed at constant values of the imposed uniaxial stress or, usually, at constant load. The results of such experiments are illustrated in diagrammatic form in Figure 4.1, which shows three different creep curves, i.e. curves of strain $\varepsilon$ versus the logarithm of the time $t$ for three different values of the constant imposed stress $\sigma_n$. The necessity for a semilogarithmic plot is dictated by the large time range which is involved, and it is in the form indicated by Figure 4.1 that the basic data on the deformational behaviour of different thermoplastics are presented in subsequent chapters.

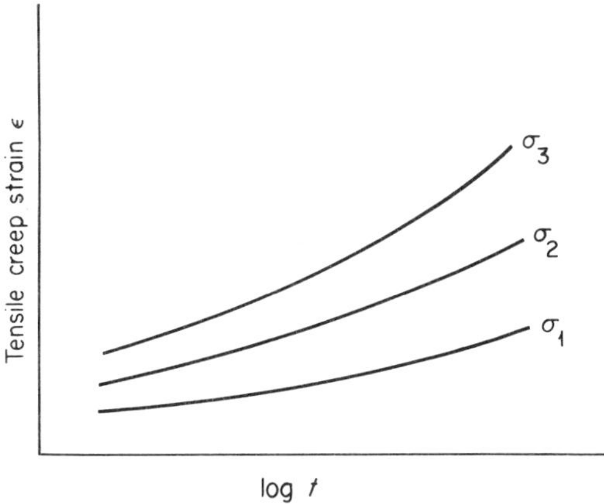


**Figure 4.1.** Curves of strain versus time at constant stress in a typical tensile creep test, where $\sigma_1 < \sigma_2 < \sigma_3$

The results obtained in creep tests with some of the more rigid plastics such as poly(methyl methacrylate), may be expressed, within certain limits of strain and time, by the equation:

$$\varepsilon = \sigma_n f(t) \quad (2)$$

This particular relationship between strain, stress and time implies linear viscoelasticity and the plastics to which it applies are, therefore, classified as linear viscoelastic materials. It should be emphasized that 'linear' means that the strain is equal to a product of stress and a function of time, and not that the stress–strain curve obtained in a conventional tensile test, where the load is progressively increased, is a straight line, as it is for linear elastic solids.

Most of the theoretical work done so far on the deformational behaviour of plastics has been based on the assumption that they are linear viscoelastic materials. However, most plastics cannot be regarded as such, even approximately, except at low strains and in limited ranges of time. For instance, polypropylene exhibits non-linear viscoelasticity for any deformation of practical importance, in which case the general equation for the results of creep experiments takes the form:

$$\varepsilon = g(\sigma, t) \tag{3}$$

Sometimes the effects of stress and time are separable and the experimental results may be expressed by another, more manageable relationship, namely:

$$\varepsilon = h(\sigma) . k(t) \tag{4}$$

but most plastics do not behave in this way for deformations of practical importance.

Consequently, the deformational behaviour of plastics cannot generally be described completely by 'linear' relationships such as those of equations (2) and (4): any relationship between the variables stress, strain, and time is necessarily complex because experience has shown that for plastics the value of any one of these three variables depends not only on a combination of the other two, but on their histories as well.

Faced with the problem of providing data for mechanical design with plastics in a way that takes into account their non-linear viscoelastic behaviour, the solution adopted here has been to provide creep data obtained under uniaxial tension as the data most useful to the designer, and to present them not only in the form of creep curves but also in three other forms which are often more convenient for design purposes. These three additional forms of presentation are:

1. Isometric stress vs. time curves
2. Tensile creep modulus vs. time curves
3. Isochronous stress vs. strain curves

These are described in the following sections: it should be remembered that these curves are appropriate only to constant loading conditions, i.e. creep conditions.

1. *Isometric Stress vs. Time Curves*

The isometric stress vs. time curves, illustrated in Figure 4.2, are obtained by cross plotting from the creep curves at constant strain. The reason for presenting creep information in this form

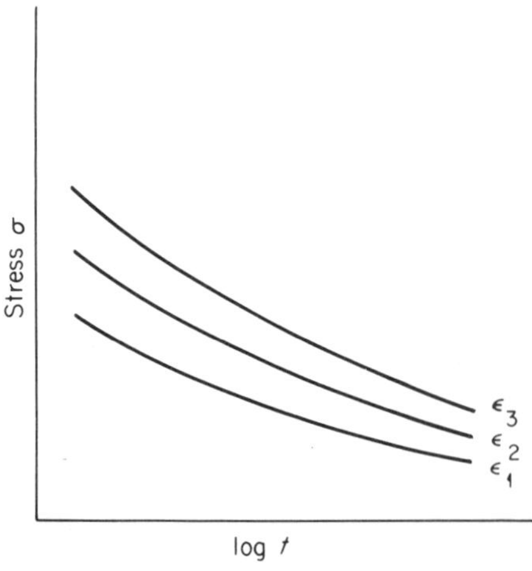


**Figure 4.2.** Alternative presentation of creep results in terms of isometric stress-time curves, that is curves of stress versus time at constant values of strain $\varepsilon$, where $\varepsilon_1 < \varepsilon_2 < \varepsilon_3$

is that in many design calculations strain is the governing parameter. In consequence, it is useful to have curves from which it is possible to read directly that stress which, acting for a specified time, produces a given strain, that strain being chosen for functional or aesthetic reasons.

2. *Tensile Creep Modulus vs. Time Curves*

The creep modulus is defined as:

$$\text{creep modulus} = \frac{\text{stress (constant)}}{\text{strain (time-dependent)}}$$

Unlike Young's modulus it is not constant but varies, for a linear viscoelastic material, with time only; or, for a non-linear viscoelastic material, with time and stress and/or strain level. Moduli often occur in engineering deflection formulae, and for this reason the creep modulus vs. time curves at constant strain are useful. These curves are easily obtained from the isometric stress vs. time curves by dividing the stress at any point by the strain (Figure 4.3).

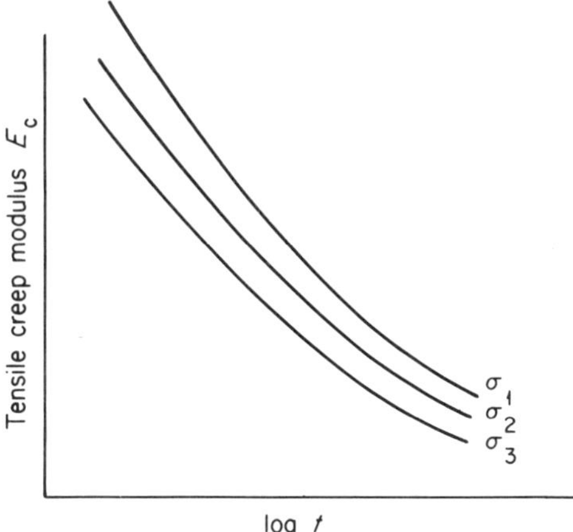


**Figure 4.3.** Results of creep tests presented in terms of tensile creep modulus $E_c$ versus time, when $\sigma_1 < \sigma_2 < \sigma_3$

3. *Isochronous Stress vs. Strain Curves*

Another cross plot of the creep curves gives a series of stress vs. strain curves at constant values of time from the commencement of loading, shown in Figure 4.4. In addition to being obtained from sections taken across families of creep curves, constant time or isochronous stress vs. strain curves can be determined directly, particularly at short time, by repeatedly loading a single specimen to successively higher stress levels, allowing of course, sufficient time for strain recovery between each loading cycle[1].

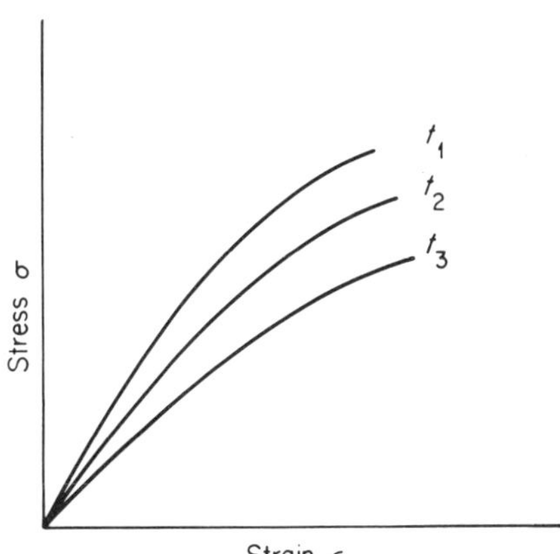


**Figure 4.4.** Constant stress isochronous stress-strain curves at different times from the commencement of loading $t$, where $t_1 < t_2 < t_3$

[1] S. Turner, 'Tensile Creep in Isotactic Polypropylene', *Trans. Plastics Inst.*, **31**, No. 93, June 1963.

Whichever way they are obtained, isochronous stress vs. strain curves provide an immediate indication of non-linear viscoelasticity by their departure from straight lines, and they represent the most useful method of characterizing the strain responses to stresses under different conditions. Stress–strain relationships can be established in other ways, but these cannot provide an alternative to isochronous stress vs. strain curves because the response of plastics to stress is not instantaneous (as is generally assumed with metals), and what happens depends on how the stress is applied. In other words, stress–strain relationships obtained in other ways are bound to differ from isochronous stress vs. strain curves derived from constant stress tests. They are also generally less useful, both because of the limited time scale of the tests on which they are based and because the test conditions under which they are obtained are more difficult to relate to the way in which plastics components are loaded in practice.

For instance, the time from the commencement of loading is far more likely to be known in practice than the rate at which the component is strained; this is particularly so in structural problems. Thus, an isochronous stress vs. strain curve at the appropriate time is more likely to be of use than the results of a constant strain rate test, which has hitherto represented the most common mode of stressing in materials testing.

The difference between constant stress rate, constant strain rate and constant stress or creep tests is shown diagrammatically in Figure 4.5. In constant stress rate tests the stress input takes the form of a straight ramp; in constant strain rate tests it is a curved ramp; and in creep tests it is of the step type. The three different types of strain outputs are shown on the right of Figure 4.5.

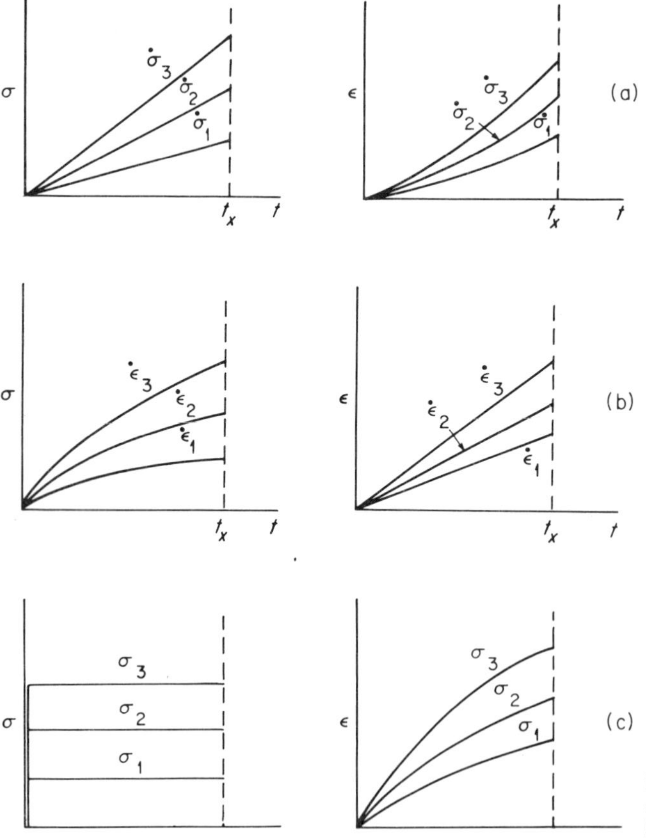


**Figure 4.5.** Comparison of stress inputs and strain outputs in: (a) Constant stress rate tests; (b) Constant strain rate tests; (c) Constant stress (creep) tests

The first of the three sets of results corresponds, in principle, to tests carried out in direct loading machines, where the rate of loading is constant; and the second to the common tests in machines with constant cross-head speeds, which result in specimens being stressed at constant rates of elongation. A special case of the last type of test is the so-called 'static' tensile test which is, in effect, an incremental load test carried out at low rates of strain. It is also the least satisfactory, because of the difficulty of defining the strain–time path illustrated in Figure 4.6.

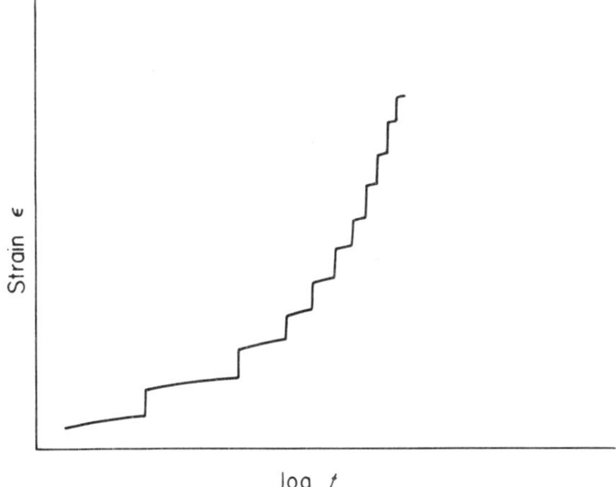


**Figure 4.6.** Typical "static" stress-strain curve plotted on a strain-time graph

If isochronous stress vs. strain curves are to be used as the principal method of characterizing the strain responses to stresses, there remains the question of the times at which they should be determined. In most of the materials chapters it is arbitrarily chosen that the isochronous stress vs. strain data are given at times of 100 seconds, 1 hour, 100 hours and 10,000 hours.

## STRAIN RECOVERY

What has been said so far about the variation of strain with time under constant stress has not brought out the important fact that, provided they have not exceeded the yield point, the creep strains of plastics are recoverable—although this is implied, of course, in the description of plastics behaviour as viscoelastic. In fact, strain recovery needs to be studied separately, as a sequel to creep experiments.

The way in which the strain decreases with time after the removal of stress is shown diagrammatically in Figure 4.7, for three different times of prior stress application. Strain recovery

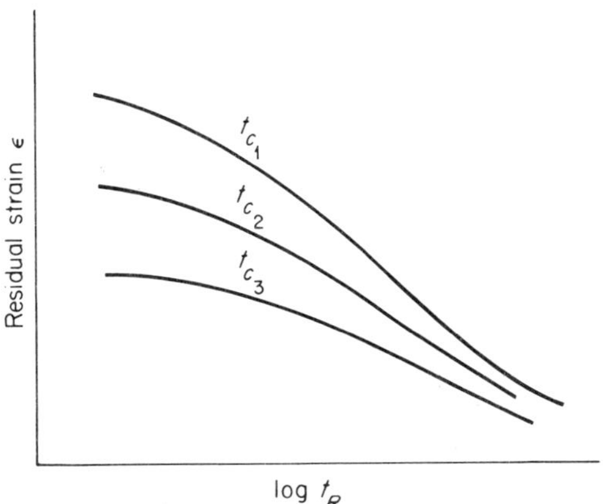


**Figure 4.7.** Curves of residual strain $\varepsilon$ versus recovery time $t_R$ after three different times under constant stress, where $t_{c_1} > t_{c_2} > t_{c_3}$

data may be presented in an alternative and more concise form based on the concepts of 'fractional recovery' of strain and 'reduced time', these being defined as follows:

$$\text{Fractional recovery of strain} = \frac{\text{Strain recovered}}{\text{Total creep strain at the time of load removal}}$$

$$\text{or} = \frac{L_x - L_t}{L_x - L_o}$$

where $L_o$ = original dimension of the specimen, e.g. the gauge length

$L_x$ = dimension when stress is removed at the end of a creep test

$L_t$ = dimension at time $t$ after removal of load

$$\text{Reduced time} = \frac{\text{Recovery time}}{\text{Time under load before recovery measurements began}}$$

Applying this approach to the curves shown in Figure 4.7 produces the results illustrated in Figure 4.8. This indicates the possible economy in the number of recovery tests which need to be performed, because the results generally fall within a band.

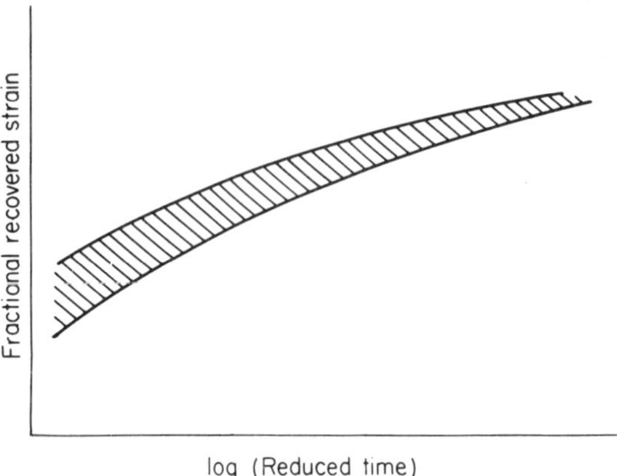


**Figure 4.8.** Recovery after tensile creep presented in terms of fractional recovered strain versus reduced time

Information on recovery after creep is particularly important for plastics components which are loaded intermittently as are, for instance, the blades of axial flow impellers when subjected to periodic use. In all such cases the strain decays during each period of no-load from whatever level it reached during the preceding period under load. The net result is that the cumulative strain produced by an intermittently applied load is considerably less than the strain under the same load applied continuously for the same total time, as shown in Figure 4.9.

Recovery after creep needs to be taken into account, therefore, in all problems involving intermittent loading. Otherwise, the strain accumulated by a plastics component during a given period of service will be overestimated, and if this is done at the design stage the component will be overdesigned.

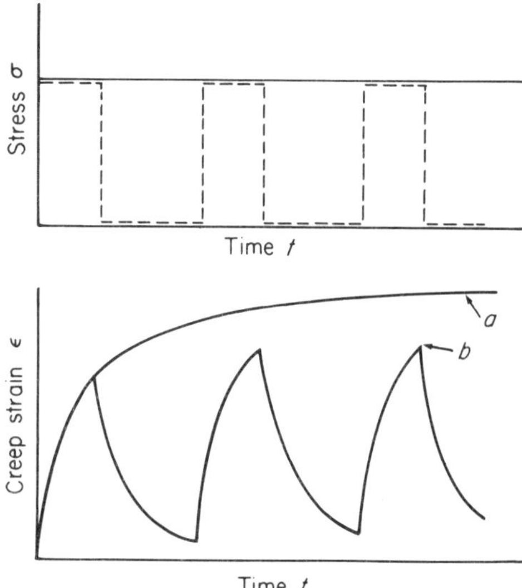


**Figure 4.9.** Diagrammatic comparison of creep strains under the same stress applied continuously (a) and intermittently (b)

## STRESS RELAXATION

Another important aspect of the viscoelastic behaviour of plastics is the decay of stress with time, or stress relaxation, at constant strain. Tests devised to investigate stress relaxation result in families of constant strain curves of stress against time which are similar in shape to, but not identical with, the isometric stress vs. time curves obtained by cross plotting the results of creep tests. The results of stress relaxation can also be expressed in terms of a relaxation modulus, defined as:

$$\text{Relaxation modulus} = \frac{\text{stress (time-dependent)}}{\text{strain (constant)}}$$

in the same way that the results of creep tests can be expressed in terms of a creep modulus, previously defined. Like the creep modulus, the relaxation modulus varies, for a linear viscoelastic material, with time only; or, for a non-linear viscoelastic material, with time and stress and/or strain level.

In the region where a plastic behaves approximately as a linear viscoelastic material, that is at low stresses and strains and short times, the differences in values between the creep modulus and stress relaxation modulus can be fairly simply calculated, and are generally small enough for the two types of moduli to be assumed to be the same for design purposes. However, because plastics are predominantly non-linear viscoelastic materials, under certain conditions the differences between creep modulus and stress relaxation modulus have been observed to be considerable. Therefore, if precise information is required, stress relaxation data need to be obtained separately and in addition to creep and strain recovery data. Priority should, nevertheless, be accorded to creep data because of the paramount importance of creep in many engineering problems, because of the relevance of recovery data and, to some extent, because approximate stress relaxation curves, as well as strain recovery curves, can be derived from creep curves.[2]

[2] S. Turner, 'Creep in thermoplastics—final comments', *Brit. Plastics*, **38** (2), pp. 106–7, February 1965.

## TEMPERATURE DEPENDENCE

Discussion of the deformational behaviour of plastics has been confined so far to the relationships between stress, strain and time at a constant temperature. The behaviour of plastics does, however, vary appreciably with temperature. What is needed, therefore, to describe deformational behaviour more fully are relationships between stress, strain, time and temperature. Such relationships are bound to be complex and even if they could be expressed in terms of mathematical models these would not constitute mechanical equations of state, that is, unique relationships between stress, strain, time and temperature, because the value of any one of the four variables depends not only on the combination of the other three but also on their history. It is better, therefore, to confine attention to the effect of temperature on particular relationships between stress, strain and time.

The stress–strain–time relationship whose temperature dependence is particularly important is that established in creep tests. The effect of temperature on the results of creep tests is best brought out by repeating the tests performed at a normal temperature at higher temperatures. In consequence, creep curves obtained at a higher temperature are included in this book in addition to those obtained at normal temperature. The actual temperatures chosen are 20°C for normal temperature and 60°C for the high temperature.

Comparison of the creep curves obtained at these two temperatures shows up the softening which takes place when plastics are used at temperatures above normal, a feature further illustrated by a comparison of isochronous stress vs. strain curves determined at different temperatures (Figure 4.10). Alternatively, the effect of temperature can be indicated by working out the values of creep modulus obtained at a particular strain from a set of curves of this kind and then plotting them against temperature, thus obtaining a curve of the kind shown in Figure 4.11. This approach has been adopted in the product chapters, the isochronous stress vs. strain curves on which the creep modulus is based being those obtained at 100 seconds, and the strain being 0·2%.

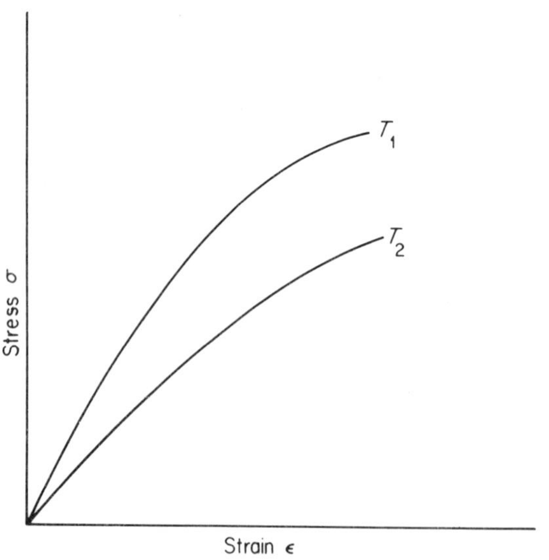


**Figure 4.10.** Constant stress isochronous stress-strain curves at two different temperatures $T$, where $T_1 < T_2$

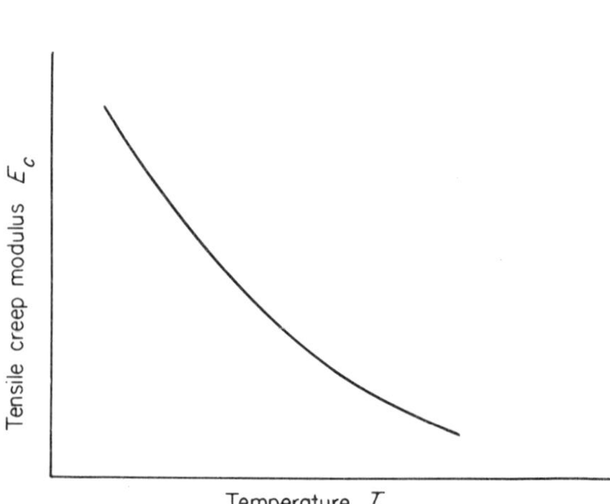


**Figure 4.11.** 100-sec 0·2% strain tensile creep modulus $E_c$ plotted against temperature

## MATERIAL VARIABLES

Other variables such as crystallinity, moisture content and plasticizer content are of less fundamental importance than time and temperature, but they have a significant effect on the deformational behaviour of plastics. Ideally, the stress–strain–time–temperature relationships should be determined for several values of each of these variables, but the amount of work which would be involved in such a procedure would be prohibitive. Only a limited amount of data on the effects of the material variables can, therefore, be expected and the presentation chosen takes the form of short-term isochronous (100 second) stress vs. strain curves for different values of the relevant variables. This procedure is in keeping with the use of isochronous stress vs. strain curves to demonstrate the effects of other variables and enables the magnitude of the effect of the material variables to be gauged. Approximate corrections derived by this method are given in the text of the product chapters whenever they are relevant; for instance, for density changes in crystalline plastics and for variation of moisture content in nylons and polyacetals.

## PSEUDOELASTIC DESIGN METHOD

In the preceding pages it has been indicated that the deformational behaviour of plastics is a complex phenomenon in the sense that plastics are generally non-linear viscoelastic materials with a linear viscoelastic region too small to be of practical importance. Consequently, in this book the deformational behaviour of plastics is characterized by creep data, the presentation of which is chosen as one of the best methods of taking into account non-linear viscoelastic behaviour in the range of time most useful to the designer.

A pseudoelastic design method is now described which makes use of the creep data to provide a method of rational design with plastics. (A worked example is given at the end of this chapter.) Although the method is based on the results of creep tests, i.e. conditions of constant load, which might not be thought to be sufficiently general, the approach is justified for two reasons. First, in practice, constant loading conditions often apply, and secondly, it has been shown that for conditions whereby the load is applied gradually or intermittently the approach provides safe design. Intermittent loading has already been discussed in this chapter in the section on Strain Recovery, and quantitative data for polypropylene are given in Chapter 7 (Figure 7.8), showing that creep under intermittent loading is less severe than creep under continuous loading.

With the pseudoelastic design method a designer can do one of two types of calculation. Knowing the appropriate elastic deflection formula and creep data, he can either calculate the deflection due to a particular constant stress acting for a known time, or he can determine the constant stress such that the deflection, or deformation, does not exceed a certain value in a known time. The latter type of calculation is more often used, because in many applications it is necessary to limit the deflection, or deformation, for functional or aesthetic reasons. Also, when the application of a component is limited only by the ultimate properties of the material of which it is made, it has been found by experience that a more rational design method is provided by adopting a criterion of failure related to a maximum strain, rather than by adopting the traditional approach of applying an approximate safety factor to the ultimate tensile stress. Assuming that a known constant tensile load is applied to a component for a specified time $t$, that the maximum temperature in service is $T$, and that there is a limit set on the allowed deformation or strain, the stress for the prescribed limiting strain appropriate to $(t, T)$ can be taken from any one of the following three families of curves:

1. Tensile creep curves.
2. Isometric stress vs. time curves.
3. Isochronous stress vs. strain curves.

Because these curves each contain the same information it is a matter of convenience which representation is chosen. All three representations are given for most of the plastics described in the engineering data chapters.

In most cases of engineering interest the stress is related to strain by deflection formulae derived on the basis of the classical theory of elasticity and conveniently tabulated in engineering textbooks. Advantage is taken of these in the pseudoelastic method, which makes use of the formulae but replaces the single-valued modulus of elasticity, or Young's modulus, of the classical theory of elasticity by the creep modulus determined at the appropriate time, strain level, and temperature; the tensile creep modulus vs. time curves at constant strain are the most convenient curves for this purpose, and these are given for most of the plastics in the engineering data chapters.

Another function which often occurs in deflection formulae is Poisson's ratio, a function describing the bulk behaviour of a material. For plastics, Poisson's ratio is generally a function of time and stress and/or strain level just as is the tensile creep modulus. Fortunately, in most cases, the variation in the value of Poisson's ratio can be sensibly considered negligible compared with that in tensile creep modulus. Also, Poisson's ratio often appears in deflection formulae only as a weak factor, so that its value can be assumed to a good approximation: from general considerations it is about 0·35 for a rigid plastic and nearly 0·5 for a rubbery one. When it is an important factor in deflection formulae it is advisable to substitute both the maximum and minimum values which are thought possible into the formulae, and compare the results. For instance, if the plastic were to change from rigid to rubbery during the deformation, values of 0·35 and 0·5 should be tried, but this is an extreme case and better estimates of Poisson's ratio are normally possible.

In classical elasticity theory, on which the deflection formulae are based, it is also assumed that the moduli in tension and in compression are equal. All the evidence so far indicates that the creep moduli of plastics in tension and in compression are not greatly different: if anything the creep modulus in compression is slightly greater than that in tension, and so the use of tensile creep data for designs in compression errs on the side of safety.

## EXAMPLE

Components are to be fabricated from the material defined by Figures 4.12 and 4.13 and the accompanying correction factors, to withstand a constant uniaxial compressive load for three years at a maximum temperature of 40°C. Deformations must not exceed 2%. What is the maximum allowable stress if the density of the component is 0·905 g/cm$^3$?

Isometric stress vs. time curves are available at 20°C and 60°C; at 40°C design stresses may be calculated by linear interpolation. The limiting strain is 2%; 3 years $\approx 10^8$ seconds.

At 20°C density 0·909 g/cm$^3$, the design stress is 850 lb/in$^2$.
At 60°C density 0·909 g/cm$^3$, the design stress is 550 lb/in$^2$.
At 40°C density 0·909 g/cm$^3$, the design stress is 700 lb/in$^2$.

For every 0·001 g/cm$^3$ change in density the design stress should be changed by 4% in the same sense. In this case the density change is $-0{\cdot}004$ g/cm$^3$, i.e. the design stress must be reduced by 16%.

$\therefore$ At 40°C, density 0·905 g/cm$^3$,

$$\text{design stress} = \frac{100-16}{100} \times 700 = 588 \text{ lb/in}^2$$

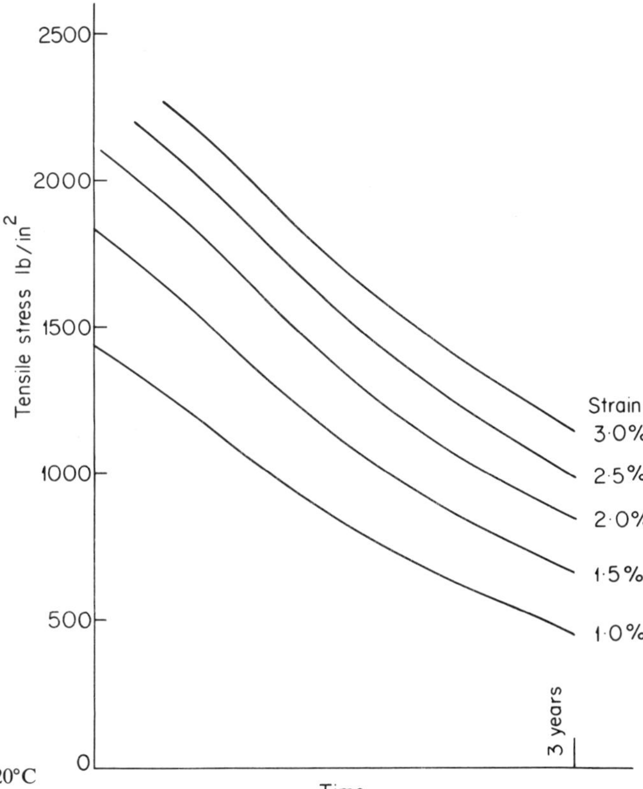


**Figure 4.12.** Isometric stress versus time curves 20°C

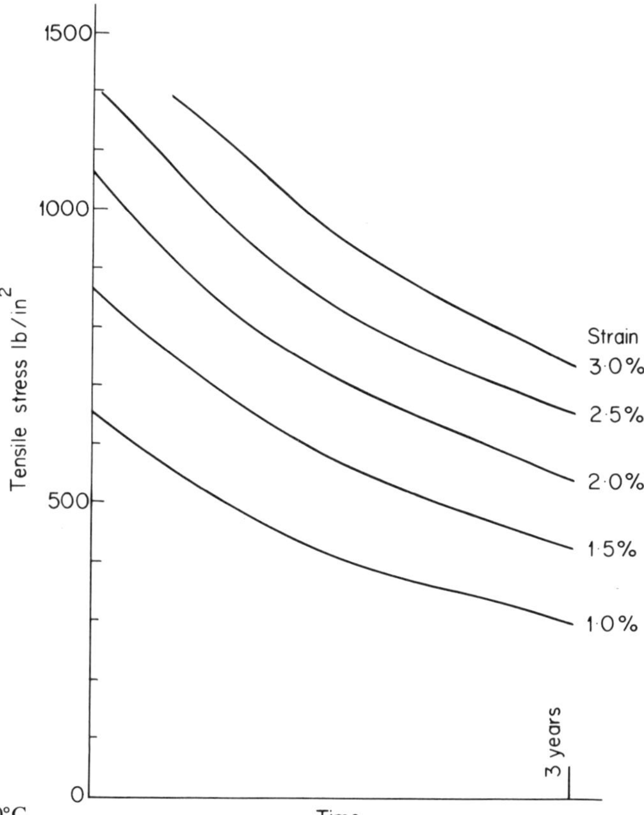


**Figure 4.13.** Isometric stress versus time curves 60°C

# 5

# STRENGTH CHARACTERISTICS

Because strains in plastics are large in relation to the imposed stresses, engineering applications of plastics are governed by strain considerations to a much greater extent than are those of other materials. Often, however, there are no functional or aesthetic limits to the deformation of plastics components and their design is therefore governed by the magnitude of the maximum permissible stresses. In many other instances plastics components have to withstand occasional impact loads or stresses under particular environmental conditions, in addition to meeting deformation requirements under normal long-term service loads. Thus, problems other than that of deformation as dealt with in Chapter 4 have to be considered.

In particular, this chapter will be concerned with the problems of short term (impact) failure and longer term failure because of, for example, yielding, brittle fracture or fatigue. The dependence of failure characteristics on time, temperature and other relevant variables will also be discussed. Like the previous one, this chapter is intended to provide an introduction to the relevant sections of later chapters dealing with individual materials, and to introduce the strength data contained in these chapters.

The usual method of characterizing the strength of a metal at room temperature is in terms of its 'tensile strength'. This quantity is, for all practical purposes, a constant for the material, independent of time and temperature under normal service conditions. It is not surprising, therefore, that the most common method of characterizing the strength of plastics is also in terms of uniaxial 'tensile strength'. This is generally interpreted as the maximum tensile stress which a plastic can develop, but it does not necessarily represent the limit to which plastics can be stressed.

For plastics which, under given conditions, behave in a brittle manner and whose uniaxial tensile stress–strain curves are of the kind shown in Figure 5.1, tensile strength is synonymous

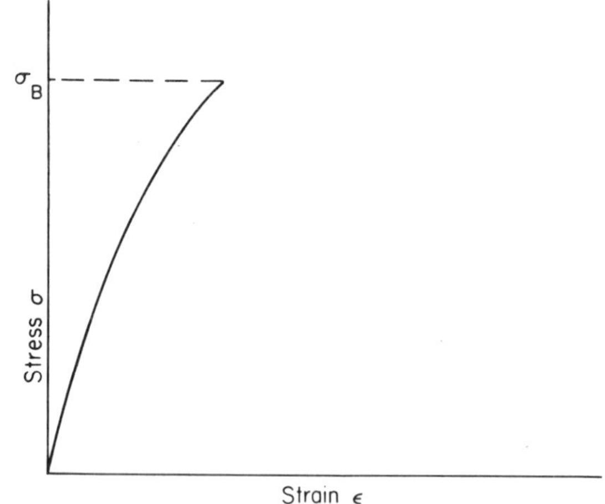


**Figure 5.1.** Stress-strain curve at constant strain-rate for a 'brittle' plastic

with the stress at fracture, or rupture, $\sigma_B$, and this represents unambiguously the limit to the tensile stresses, under similar conditions. But for plastics which behave in a ductile manner it is necessary to recognize the existence of another important stress. This is the yield stress, $\sigma_Y$, which corresponds to the onset of large permanent deformations, and which is usually defined by the point where the rate of change of the nominal stress with strain first reaches zero, as shown in Figure 5.2. The value of the yield stress $\sigma_Y$, may be lower or higher than that of the stress at break, $\sigma_B$. If lower, it is more likely to represent the limit to which plastics may be stressed in practice.

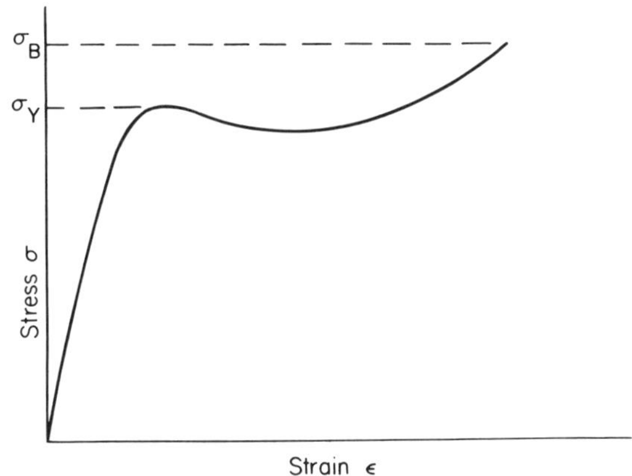


**Figure 5.2.** Stress-strain curve at constant strain-rate for a 'tough' plastic

## TIME DEPENDENCE

Moreover, not only is it necessary to recognize the existence of more than one type of limiting stress, but also that the limiting stresses vary with the rate at which plastics are strained. In fact the yield stress in ductile behaviour, and the fracture stress in brittle failure, increase as the strain rate increases, as shown in Figures 5.3(a) and 5.3(b). Thus the strength of a plastics material cannot be characterized properly by the value of either brittle or yield stresses at one particular strain rate, which is all that is usually quoted.

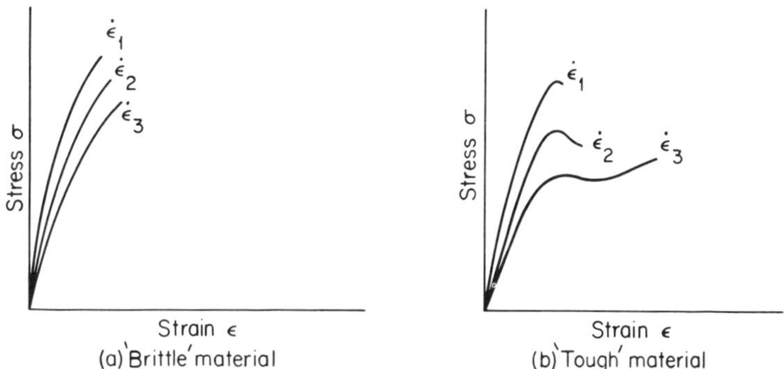


**Figure 5.3.** Stress–strain curves at different strain rates $\dot{\epsilon}_1$, $\dot{\epsilon}_2$, $\dot{\epsilon}_3$, where $\dot{\epsilon}_1 > \dot{\epsilon}_2 > \dot{\epsilon}_3$

Curves of the kind indicated in Figures 5.3(a) and 5.3(b), would represent an advance on the commonly quoted single-value yield or fracture stresses, but they would still be inadequate because they do not give a useful indication of the time dependence of long term strength. The most useful presentation of data will be in terms of applied stress vs. time under load to failure. To obtain the required information it is necessary to resort again to tensile creep tests, but this time carried to the point of specimen rupture. A set of the resulting tensile creep curves is shown diagrammatically in Figure 5.4 and from curves of this kind it is possible to determine the time

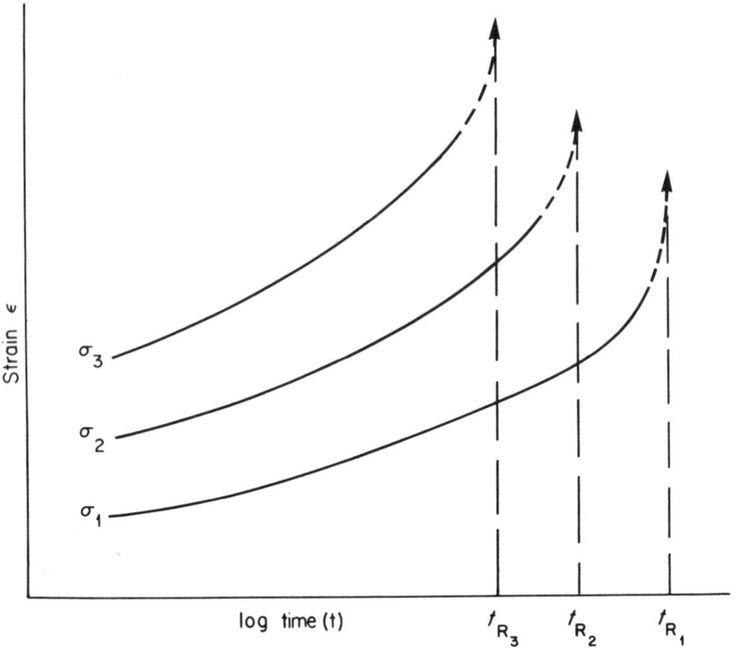


**Figure 5.4.** Curves of strain at constant stress $\sigma_1, \sigma_2, \sigma_3$, versus time under load where $\sigma_1 < \sigma_2 < \sigma_3$

to rupture at a given constant load. This leads to curves of creep rupture stress vs. logarithmic time, an example of which is illustrated in Figure 5.5. It is more difficult to define in these tests behaviour corresponding unambiguously to the yield point in conventional tensile tests, but the phenomenon of run-away creep is obviously connected.

Curves similar to that of Figure 5.5 show clearly that the creep rupture stress decreases with time under constant load. This has an obvious bearing on permissible design stresses, but the amount of relevant data is limited, and curves of the kind shown in Figure 5.5 will be found in only some of the chapters dealing with individual materials. Where these data are not available, it is recommended that a maximum strain of 3% be imposed upon the component design, that is, the isometric stress vs. log time curves appropriate to 3% strain should be employed.

Even more data might be required in some cases, because rupture and yield stresses are not the only factors which set a limit to the loads which can be imposed on plastics. The limit depends on the criterion of failure for the given component: the component might be thought to have failed because of the onset of crazing or microvoiding, and the stresses corresponding to these events also vary with time under load. The onset of these phenomena is detailed in the stress vs. time to failure plots.

In addition to their dependence on time, the mechanical properties of plastics are also markedly dependent on temperature. To cover this effect in terms of stress vs. time to failure it would be necessary to generate a family of curves such as Figure 5.5 for a range of temperatures. The

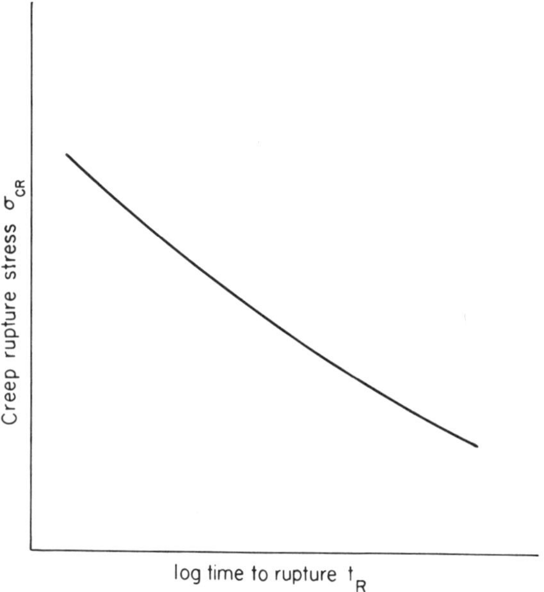


**Figure 5.5.** Curve of creep rupture stress versus time to rupture

amount of work involved in such an approach is, unfortunately, prohibitive. However, some indication of the expected effect within the time scale of the constant strain rate tests can be gauged from yield or fracture stress vs. temperature characteristics at one or more constant strain rates, as shown in Figure 5.6.

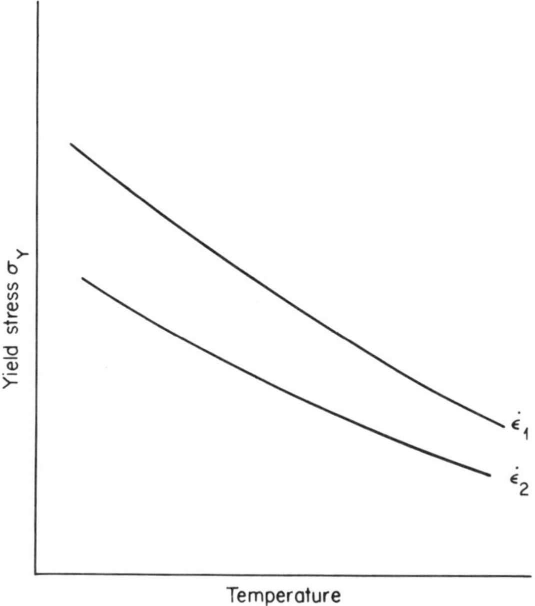


**Figure 5.6.** Curves of yield stress versus temperature at two different strain rates, $\dot{\varepsilon}_1$ and $\dot{\varepsilon}_2$, where $\dot{\varepsilon}_1 > \dot{\varepsilon}_2$

## MATERIAL VARIABLES AND ENVIRONMENTAL EFFECTS

Although loading time and temperature are, in principle, the two most important variables affecting the mechanical properties of plastics, they are by no means the only ones which have to be considered. The others are material and fabrication variables which, in particular circumstances, can be at least as important as time or temperature. In consequence it is necessary to consider the variation of the failure stresses with these variables.

For instance, what might ostensibly be one type of material is, in fact, a group consisting of several grades which can be characterized by their molecular weight. As a result it is necessary to consider the variation of the failure stress with molecular weight.

Processing might produce molecular orientation so that the material can no longer be regarded as isotropic. In consequence the difference between the strength in the direction of orientation and that in a direction perpendicular to it, which is lower, needs to be considered. Processing might also produce varying degrees of crystallinity, and the effects of this on strength have to be considered.

The presence of flaws—either naturally existing in the material because of sample preparation (e.g. voids) or contamination in the material, or deliberately introduced in the form of notches, or inherent in the design of a component—will adversely affect the strength of the test piece or component.

The nature of the ambient environment is also important: chemical attack on the plastic, without any applied stress being present, can seriously reduce mechanical strength. The combined action of constant or varying stress and 'active' environments inevitably produces a much more complicated pattern of behaviour, ranging from solvent stress cracking in acrylic materials to environmental stress cracking in polyolefines. The results of such tests may be presented as curves of failure stress vs. log time to failure, with an explanatory note on the nature of the environment and of the applied stress.

The strength of some plastics is also affected to a significant extent by the amounts of additives which they contain: that of others is strongly affected by humidity. Information on the effect of the material or environmental variables is, therefore, included where appropriate in subsequent chapters.

## LOADING HISTORY

The long term behaviour will depend on whether the load is maintained constant or is allowed to vary with time. Furthermore, in any particular application or test the behaviour may well be influenced differently, depending on whether the situation is primarily one involving a constant load, or constant deformation.

The situation where the load is maintained constant has already been discussed but it is now necessary to consider the consequences of varying the load. This gives rise to 'fatigue' which is traditionally used to describe failure caused by a repeatedly applied load but which for plastics is more precisely called 'dynamic fatigue'. The usual way of presenting the fatigue characteristics of metals is in terms of the maximum stress vs. number of cycles to failure, that is in terms of an S–N curve. For plastics the shape of the S–N curve will depend on the test temperature and, furthermore, because plastics have a low thermal conductivity and exhibit high mechanical hysteresis, the energy which is dissipated as heat will cause a rise in temperature, with a consequent decrease in the resistance to deformation in the specimen. If the test frequency is reduced, the problem of temperature rise may be overcome.

If the load is applied in a cyclic manner, the shape of the loading waveform and the distribution of the loading with respect to zero load may also be important. The effect of cyclic loading is presently being studied by applying to a waisted specimen clamped at one end, a square wave stress input having a frequency of 30 cycles/min, and observing the time at which the specimen fails because of the applied cyclic stress. The results of such tests in air can be presented as curves of failure stress vs. log (time under load). This also approximates to conventional S–N representation.

## IMPACT STRENGTH

The dependence of failure stress on time extends to the very short times involved when plastics are subjected to impact loads. In other words, there is a continuous spectrum of behaviour as a function of time or straining rate.

However, strain rate is only one aspect of impact behaviour because the normal concern in impact behaviour is the incidence of brittle failure. In this sense, it is not so much continuity of behaviour as a function of testing speed that is important, but the appearance of brittle failures as the severity of the test is increased by higher testing speeds, lower temperature or stress concentrations (notches). Thus, although tensile impact testing provides a logical extension of tensile stress vs. strain measurements to impact speeds, it cannot be claimed to be a complete measurement of impact resistance, because it ignores the practically important effect of stress concentrations. It is also too sophisticated a technique for routine use. Experience has shown that when plastics break in service under impact conditions, the failures are nearly always brittle: the strains are small and signs of gross yielding are rare. Satisfactory impact performance necessitates the avoidance of low energy brittle failures and the most satisfactory material is one which will behave in a ductile manner over the widest range of conditions. The tendency to change from tough to brittle behaviour is enhanced by increasing the severity of the test conditions, e.g. by increasing the speed of impact, by decreasing the temperature or by introducing stress concentrations (notches). Thus, definition of impact strength devolves into defining the importance of these factors in promoting brittle failure by affecting the tough–brittle transition. It should be stressed, however, that these are not the only variables which affect behaviour and that materials and processing variables must also be considered.

At present it is not possible to develop formulae for impact behaviour which could be used in the rational design of components in plastics materials, any more than it is in metals. Nevertheless, the logical selection of materials for specific applications is greatly facilitated by the presentation of data in a uniform manner. It would be convenient if the impact behaviour of a material could be characterized completely, but because of the difficulty of relating service conditions to impact tests an empirical approach has had to be adopted to characterize the impact behaviour of plastics; this is described later. In general there are four principal factors which affect the impact behaviour of thermoplastics: some of these are also relevant to thermosetting plastics and other materials.

1. A decrease in service temperature may cause an otherwise ductile material to behave in a brittle manner. In practice the transition may take place over a wide or narrow temperature range and may or may not take place in the range of practical importance.

2. Orientation resulting from fabrication can lead to an increased tendency to cracking parallel to the orientation direction, particularly for uniaxial orientation. This is especially true of injection moulding where the hot plastic melt is forced at high speed along a long flow path in a narrow cavity: the material is weakest when stressed across the direction of flow.

3. For practical reasons it is very often necessary to design into an article corners and changes in cross-section; these configurations give rise to stress concentration effects and an increase in stress concentration increases the likelihood of low energy brittle failure on impact.

4. An increase in straining rate has a similar effect to a decrease in temperature, in that it may cause a transition from tough to brittle behaviour. Over a wide range of practical interest, an increase in straining rate by a factor of ten is approximately equivalent to a decrease in temperature of 8–10°C. Thus practical decreases in service temperature are more likely to cause brittle behaviour than practical increases in straining rate. Doubling the straining rate will not usually be significant unless the component is to operate very close to the tough–brittle transition on the impact strength vs. temperature plot, which is not a desirable operating condition.

A second group of factors includes those which are inherent in particular compositions, such as molecular weight, molecular weight distribution, branching and the nature of additives. Additionally, processing variables such as melt and mould temperatures and the rate of cooling affect the crystallinity level.

### *Test Method*

The influence of these factors on the impact strength of plastics is being satisfactorily investigated by using a Charpy type impact test. This test is formally equivalent to a flexural stress–strain test carried out at high rates of straining, but exact equivalence of results should not be expected as the striker decelerates as it breaks the specimen.

The basic method of test employs specimens 2 in long, $\frac{1}{4}$ in wide, and about $\frac{1}{8}$ in thick, usually cut across the flow direction of an injection moulded disc or plaque (for moulding grades) or across the extrusion direction (for extrusion grades). This ensures that the results obtained are for the weakest direction in the sample, because this is likely to be the direction of failure in practical applications. A 45° notch is machined at the centre of the long side, thus reducing the width of the specimen. The tip of the notch may be of radius 0·010 in or larger, and the notch depth is usually 0·110 in but can be varied. The notched specimen is supported on two anvils defining a test span of $1\frac{1}{2}$ in with the notch at the centre opposite the point of impact: this constitutes essentially a 3-point loading system. The energy to break the specimen in a brittle manner is determined by striking the specimen with a pendulum weight at 8 ft/s, and measuring the energy lost by the pendulum in breaking the specimen.

### *Presentation of Data*

In order to rationalize the results obtained from tests made on specimens of different dimensions, impact values are now quoted in terms of energy to break per unit area of residual cross-section of the specimen.

$$\text{Impact Energy} = \frac{E}{t(w-c)}$$

where $E$ = energy required to break the specimen

$t$ = specimen thickness

$w$ = width of specimen before notching

$c$ = notch depth

To make tests of different severity, specimens are prepared with different notch tip radii and/or depth, usually the former. Tests are made over the range of temperatures of practical interest to investigate the impact strength of specimens of different notch geometry.

The results of these tests are presented in terms of median impact strength[1] (energy per unit area) vs. temperature, under standard notch conditions: i.e. notch depth 0·110 in and notch tip radius 0·010 in or 0·080 in, as shown in Figure 5.7. Where appropriate, supplementary curves at

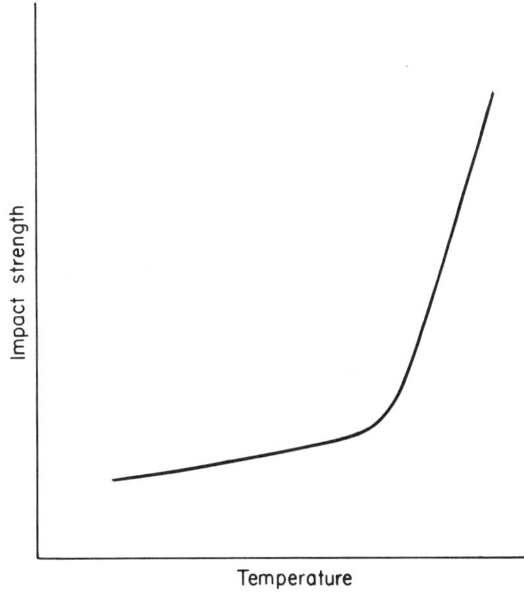


**Figure 5.7.** Impact strength versus temperature

one or more temperatures are also presented to show the variation in impact strength with notch depth; where practicable these results are unified into a plot of impact strength vs. the notch tip radius on a logarithmic scale, Figure 5.8.

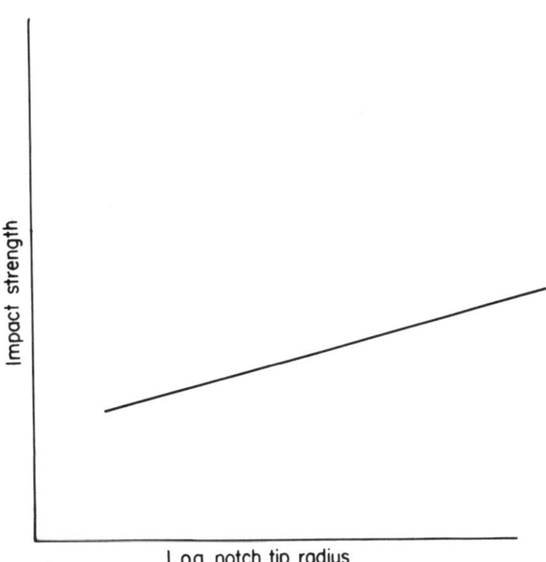


**Figure 5.8.** Impact strength versus notch tip radius

[1] i.e. the median of the test results on at least six specimens.

*General Classification*

From the point of view of impact at room temperature plastics may be grouped into three main classes:

1. Essentially brittle when unnotched.
2a. Tough unnotched and brittle when bluntly notched.
2b. Tough unnotched and brittle when sharply notched.
3. Essentially tough under all normal conditions.

Several well-known materials are grouped in Table 5.1 according to the proposed general classification at room temperature.

TABLE 5.1

| Class | Material |
|---|---|
| 1 | Poly(methyl methacrylate), poly(4-methylpent-1-ene), dry glass-filled 66 nylon, polystyrene |
| 2a | Propylene homopolymer, acetal copolymer, polyvinyl chloride |
| 2b | High density polythene, polycarbonate, dry nylon 66 |
| 3 | Low density polythene, PTFE, propylene–ethylene copolymer, ABS, wet nylon 66 |

It may be concluded from this table that all class 3 are better in respect of toughness than all class 2, and all class 2 are better than all class 1. Furthermore, conditions have to be very mild indeed to cause class 1 materials to break in a tough manner, and very severe indeed to cause class 3 materials to break in a brittle manner—a low temperature is required to enhance the severity of a very sharp notch. Thus under normal conditions the performances of class 1 and class 3 materials are reasonably predictable.

What is not so clear cut is the prediction of the behaviour of class 2 materials (the notch is only a convenient means of increasing the severity of the performance conditions); i.e. under what cumulative conditions of severity is such a material likely to behave in a brittle way and therefore fail under small impact loads? The answer to this question is left to the discussion of the individual classes of thermoplastics.

*Comparison of Results*

A meaningful comparison of the impact behaviour of two different materials can only be made by inspecting test results obtained under similar conditions and observing where the tough–brittle transition occurs. As a rough working rule a material may be described as 'satisfactory' from the point of view of impact strength, when a sharply notched specimen fractures at an impact energy of 5 ft-lb/in$^2$ or greater. Test results do vary, however, and it must be noted that, in the brittle region and under given conditions, differences between materials of 40% are not necessarily significant from a test point of view. Thus although the absolute value of impact energy may vary to some extent, the position of the transition region on a temperature plot is not greatly affected thereby. It is the temperature of the transition which is significant.

The interpretation of results may be illustrated, in principle, by considering the idealized curves obtained under the same conditions of specimen geometry, etc. in Figures 5.9(a), (b), (c), and (d).

(a) Between $T_A$ and $T_B$ X is better than Y, and below $T_A$ both are brittle.
(b) Below $T_B$ X and Y are both brittle, with X rather better than Y.
(c) Between $T_A$ and $T_B$ Y is better than X, and below $T_A$ both are brittle, with X rather better than Y.
(d) Between $T_A$ and $T_B$ X is better than Y, and below $T_A$ both are brittle, with X rather better than Y.

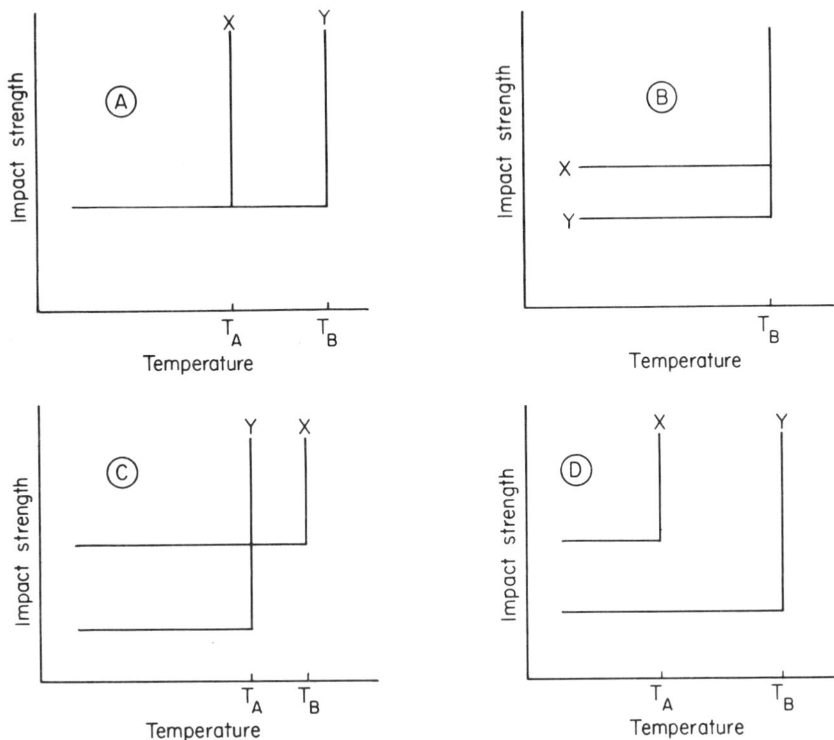


**Figure 5.9.** Idealized curves of temperature impact strength versus temperature

In practice, of course, impact curves are seldom as straightforward as these idealized curves; but examples in the comparison of real materials, which follow the basic patterns, may be given as follows, in Figures 5.10(a), (b), (c), and (d).

(a) sharp notch, cast acrylic sheet ('Perspex') and propylene homopolymer ('Propathene' GPE 33).
(b) blunt notch, PVC ('Welvic' RI7/551) and propylene homopolymer ('Propathene' GPE 33).
(c) blunt notch, acetal copolymer ('Kematal' M90) and propylene homopolymer ('Propathene' GPE 33).
(d) sharp notch, polythene ('Alkathene' XRM 21) and propylene/ethylene copolymer ('Propathene' GWM 201).

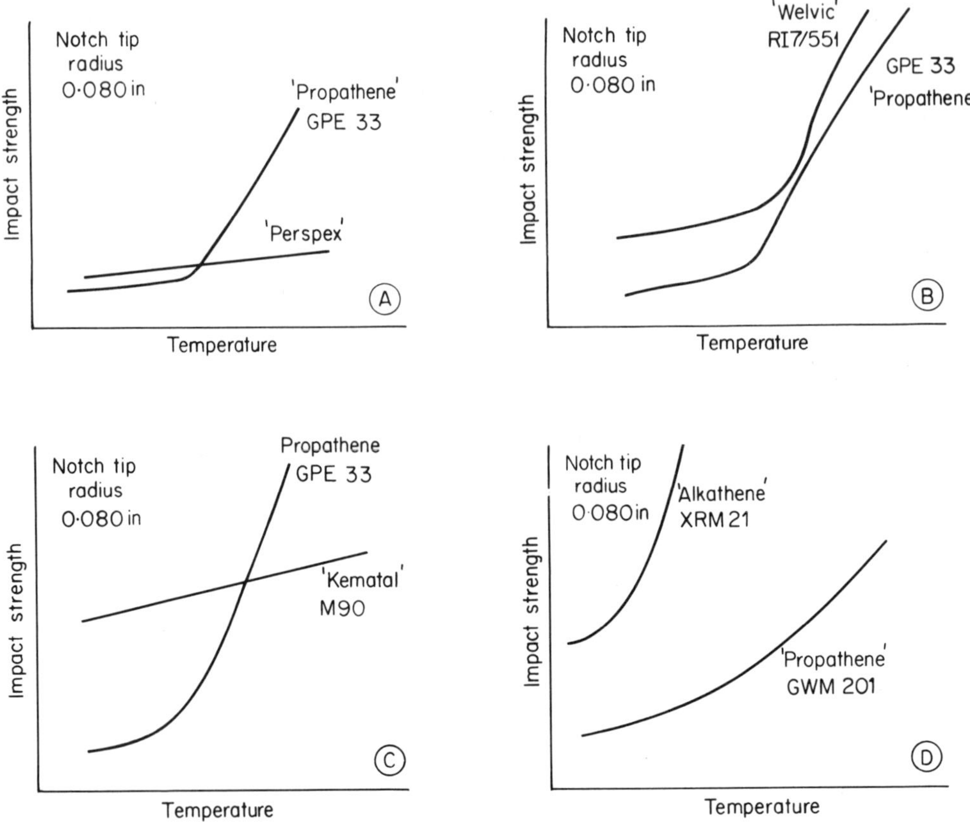


**Figure 5.10.** Comparison of materials under impact conditions

# 6

# OTHER PHYSICAL PROPERTIES

## ELECTRICAL PROPERTIES

Like many organic compounds, plastics are non-conductors of electricity; in fact, some are amongst the best insulating materials and all are normally adequate insulators in general applications. Thus it frequently happens that the choice of a plastic for a particular application depends primarily on factors other than the electrical properties, for example, on the mechanical properties, corrosion resistance or thermal stability. However, it is appropriate to consider briefly in this book the response of plastics to electrical stresses.

Broadly, there are two classes of phenomena observed when an electrical potential is applied across a sample of a plastics material. For high potential gradients, discharge currents can flow continuously or discontinuously through the sample, sometimes forming an electrically conducting path of carbon and at other times eroding an air path through the material. For low applied voltages, plastics in general behave as insulators and here some standards terminology can be misleading because an important standard criterion of performance is in terms of the one-minute value of volume resistivity. This is derived from the relationship between the applied voltage and the transient current observed one minute after electrification. One might infer from this quantity that equilibrium resistivity (or conductivity) is established within one minute of applying the potential across the specimen but, as will be shown below, this is usually very far from the truth. Other properties of some importance are permittivity (dielectric constant) and loss tangent: usually however, applications requiring such information are within the province of the electrical engineer and not the mechanical engineer for whom this book is primarily written.

### *Discharge Phenomena*

As noted above, electrical discharges through a plastic can result in gross mechanical damage. It may also happen that a discharge in the vicinity of a component may erode some of the plastic and, in extreme cases, lead to premature failure of the component.

The inherent dielectric strengths of plastics are high and of some plastics very high, compared with the dielectric strength of air or other gases. In fact, only plastics decidedly deficient in electrical properties are comparable with air in this respect, so that it is not surprising that any residual pockets of air or 'voids' in a sample subjected to high electrical stresses can result in failure at stresses very much lower than the plastic itself would stand. A full discussion of the design factors leading to the highest discharge resistance is not possible here, but it must be noted that most electrical failures in practice arise from gas inclusions or particles of conducting impurity in the plastic.

Although many test data are relevant to room temperature, it must be realized that discharge behaviour, in common with other electrical and mechanical properties, does depend on factors such as temperature and additives in the material.

*Resistivity*

A warning on the interpretation of the term 'volume resistivity' was sounded earlier. Formally, the situation is closely similar to that for deformation, considered in detail in Chapter 4: for both properties, values are quoted which are frequently interpreted as material constants, independent of time and other parameters, and this is at least approximately true for conventional materials such as metals. However, it most definitely is not true for plastics.

In the standard experimental procedure for measuring volume resistivity, a potential is applied across the specimen and the resulting current actuates a galvanometer: instruments to do this are available commercially. A complete analysis of the current response shows, however, that after application of the potential the measured current is not independent of time, as would be required of an ideal resistor, but gradually decays with increasing time over many decades of time extending into days and weeks. This behaviour, which is typical of many plastics at room temperature, clearly cannot lead to a unique value of resistivity, that is, a constant ratio between applied voltage and resulting current. In consequence, in the product Chapters 7 to 14 values of apparent volume resistivity will be given as functions of time.

There are circumstances under which plastics do behave like conventional materials, exhibiting a resistivity independent of time. As might be expected, plastics materials differ from one to another in this respect, but it is generally true that the chance of conduction occurring is increased at elevated temperatures and by the presence of impurities, particularly ionic components and impurities which are water-attractive. As a corollary to this, it is found not infrequently that the onset of conduction is enhanced by increases in humidity.

*Permittivity and Loss Tangent*

Generally, permittivity and loss tangent are of little concern except to the specialist electrical engineer. However, there are circumstances where loss tangent can be important and can lead to phenomena affecting other than the electrical properties; for instance, materials of high loss tangent transform electrical energy into heat, a factor of particular importance at high frequencies. Under these conditions, very considerable heat may be generated leading frequently to softening of the plastic and this is the basis of dielectric heating used commercially.

Many non-polar plastics, including polythene and polypropylene, have permittivities closely similar to those calculated from the Clausius–Mosotti equation relating permittivity with density:

$$\frac{\varepsilon - 1}{\varepsilon + 2} = kd$$

$\varepsilon$ = permittivity
$d$ = density
$k$ is constant

For such materials the permittivity is essentially constant with frequency and changes with temperature in accordance with density change. For polar materials these considerations do not apply, and the temperature and frequency dependence of permittivity are given in graphical form in later chapters.

Although the permittivity is sometimes calculable from density and can be essentially independent of frequency, the loss tangent has maxima which are frequency and temperature-dependent, approximately obeying activated rate process theory. The levels of dielectric loss are very variable ranging from a few units of $10^{-6}$ in $\tan \delta$ to values as high as 0·3. In view of such variability, tan $\delta$ results are given graphically as functions of temperature and frequency. For low values of loss tangent, since tan $\delta \rightarrow \delta$ (in radians), it is usual to modify tan $\delta$ to the angular notation in microradians. 1 µrad $\equiv$ 0·000001 in tan $\delta$ units.

## OPTICAL PROPERTIES

Although the theme of this book is mechanical engineering with plastics, these materials do have optical properties which can be of direct benefit in some applications. Unlike metals, most plastics are not inherently opaque and may vary from transparent to translucent. Plastics can be used for their optical properties in areas such as the optical industry, building, lighting and packaging. Each type of application requires the plastic to have a different combination of optical properties; these properties fall into one or more of the following categories:

*Refraction*, which characterizes the properties of a plastic used in the optical industry.

*Transparency*, which describes the quality of an image of an object viewed through the plastic in terms of contrast and resolution.

*Light Transfer*, which is self-explanatory and covers the characteristics of plastics materials of interest mainly in lighting and advertizing.

Each of these properties can be expressed basically by optical quantities which are now examined in greater detail.

### *Refraction*

(a) The *refractive index* of a material in a particular direction, which must be specified for anisotropic materials, is defined by the ratio of the velocity of light in vacuum and the velocity in the material along the specified direction.

(b) The variation of refractive index with the wavelength of light in vacuum is described by the *dispersion curve*. The refractive index should be specified for some, or all, of the following wavelengths, and this practice will be followed in this book for the appropriate materials.

7800 Å (Rb) refractive index denoted $n_A$,
7682 Å (K)
6563 Å (H) refractive index denoted $n_C$,
6438 Å (Cd) refractive index denoted $n_{C'}$,
5893 Å (Na) refractive index denoted $n_D$,
5876 Å (He) refractive index denoted $n_d$,
5461 Å (Hg) refractive index denoted $n_e$,
4861 Å (H) refractive index denoted $n_F$,
4800 Å (Cd) refractive index denoted $n_{F'}$,
4358 Å (Hg) refractive index denoted $n_g$,
4047 Å (Hg) refractive index denoted $n_h$.

Data for refractive index are assumed to be at 20°C unless otherwise stated.

(c) The *critical angle* $\alpha_d$ for total internal reflection of light for the helium yellow line is defined by the equation:

$$\alpha_d = \sin^{-1}\left(\frac{1}{n_d}\right)$$

and similarly for other spectral lines.

The variation of refractive index with temperature is given either in terms of a temperature coefficient or as a curve of refractive index ($n_d$) vs. temperature. Usually the temperature range $-20°C$ to $+60°C$ is covered.

*Transparency*

Transparency is a general quality of plastics materials, depending on their light-absorbing and light-scattering properties. The former depends largely on the molecular structure, and therefore on the family of plastics under consideration, whereas the latter results from local variations in refractive index in the material. This variation in refractive index has a double effect: it affects both the amplitude and phase of a passing wave; and it produces scattering of a certain amount of the energy. The first effect is responsible for the astigmatism of the image formed by any optical system, be it film, sheet or lens. The confusion, in the plane of the image, of two distinct points in the object plane is commonly known as 'loss of clarity'. The second effect results in a deterioration of contrast and is associated with 'haze' and milkiness in the sample. These are two aspects of what is generally understood by 'transparency'.

Both the loss in clarity and loss in contrast are determined by the variation of refractive index in the material, the average value of this variation and the average extent. For partially crystalline polymers, the situation is complex and is not amenable to easy theoretical treatment. However, in all cases, the transparency can be described by measured characteristics which are discussed below.

If a parallel beam of monochromatic light, amplitude $\phi$, falls perpendicularly on a translucent sheet, a certain amount $\phi_{sc}$ will be scattered, a certain amount $\psi_a$ absorbed (ignoring fluorescence —reemission at a changed frequency), and the remainder $\phi_{undev}$ will be transmitted without deviation through the material. Hence,

$$\phi - \phi_a - \phi_{sc} = \phi_{undev}$$

The direct transmission factor, $T$, is defined by the ratio

$$T = \frac{\phi_{undev}}{\phi}$$

and is a dimensionless number, usually expressed as a percentage. The direct transmission factor is a function of the wavelength of the light, which must be specified, and is dependent on the thickness of the sample.

For weak scattering the connection between the direct transmission factor and the thickness $l$ is approximately given by

$$T = e^{-(\sigma + K)l}$$

$\sigma$ is a measure of the amount of light scattered and is called the *scattering coefficient* (sometimes turbidity). It represents the fraction of the light scattered in passing through unit thickness of the material and has the dimension $(\text{length})^{-1}$: it is usually expressed in $cm^{-1}$. The wavelength in vacuum of the light must be reported. It is perhaps the best and only single-valued quantity of the overall phenomena of transparency. $K$ is a measure of the amount of light absorbed and is called the *absorption coefficient* (or absorptivity). It represents the fraction of light absorbed in passing through unit thickness of the material and has the dimensions of $(\text{length})^{-1}$, usually $cm^{-1}$. Again the wavelength of the light in vacuum must be reported. The sum $(\sigma + K)$ is known as the *attenuation coefficient*.

The total amount of scattered light may be thought of as involving two contributions, the forward scattered part and backward scattering. The 'milkiness' of translucent samples is caused largely by the backward scattering.

The deterioration of contrast (and colour saturation) in the object results mainly from the light scattered forward at high angles to the undeviated transmitted beam and is usually expressed through the forward scattered fraction. In some cases, it is possible to characterize the contrast deterioration by the proportion of light scattered forward between two appropriate angles, compared with the total transmitted flux. This quantity is usually (and improperly) known as 'haze'. Although it is sometimes found to be a useful concept, one must always remember its arbitrary definition and its restriction within a particular system, in other words to a certain angular distribution of the scattered light.

The *clarity* is associated with the capacity of the sample for allowing details in the object to be resolved in the image which it forms. The clarity is perfect only when no light is scattered by the sample and it is strongly dependent on the angular distribution of the scattering intensity and on the distance between the object and the sample.

An assessment of the clarity can be made by indicating the loss in angular separation of two points which can just be resolved in the image in comparison with the object (for an angular magnification of the optical system = 1). The thickness of the sample must be stated.

The *direct reflection factor* is defined by the ratio of light intensity reflected at an angle equal to the incidence angle and the intensity of the incident beam (assumed parallel). The incidence angle, as well as the wavelength in vacuum of the light, must be specified. More information is given by a plot of direct reflection factor against the angle of incidence.

It is possible to characterize the appearance of a surface by recording the ratio between the flux scattered within a certain solid angle around the reflection direction and the flux incident, assumed concentrated in a parallel beam. By reflection direction is meant that given by the laws of geometrical optics.

This ratio, measured for different angles of incidence (which must be indicated) and various solid angles (also indicated) is known as 'gloss'. Although sometimes a useful concept, its definition is arbitrary and therefore it is restricted within a particular system, in other words to a certain angular distribution of the scattered light.

All the quantities which characterize the transparency of a sample are defined for unpolarized light.

### *Light Transfer*

The *total transmission factor* is defined by the ratio between the total transmitted light flux and the incident flux, assumed concentrated in a parallel beam perpendicular to the sample.

Similarly, the *total reflection factor* is defined by the ratio between the total reflected light flux and the incident flux, assumed concentrated in a parallel beam perpendicular to the sample.

## HARDNESS

Hardness is a complex engineering property, and the stresses and strains in the various hardness tests which result in permanent deformation are difficult to analyse. The general opinion appears to be that, in the absence of work hardening, the quantity measured in a hardness test is directly proportional to the tensile yield stress. With work hardening present, which is generally true for plastics materials, it has been stated that hardness is 'some complicated function of the whole stress vs. strain curve in the relevant range of strain'. In this situation, it is not easy to see the value of a hardness test; where a measure of hardness is needed, the value of the yield stress and its variation with fabrication and other factors would appear to be a more than satisfactory substitute.

The case for adopting this approach is strengthened by considering hardness tests currently available. For metals, hardness is an important control quantity and there are three principal methods of measurement: Brinell, Vickers, essentially British in origin, and Rockwell, of American origin. The indenters used in these three methods differ in shape but all the methods have this in common, that the hardness is derived from the extent of permanent deformation or damage after the load has been removed. For rubbers, on the other hand, hardness is defined in terms of the penetration of the indenter whilst still under load, because there is no permanent deformation. Plastics are phenomenologically more like metals in that they yield and suffer permanent damage at comparatively low strains, i.e. strains of the order of 3–10%. Because permanent damage is likely to be the more important facet of behaviour in design, it is appropriate to work from the basis of the yield stress as suggested above.

A rough comparison can be made between plastics and metals on the basis of scratch hardness data for which Mohs' scale is perhaps appropriate. On this scale, most plastics are in the range 2–3 whereas lead has a value of 1·5, aluminium is in the range 2–2·9, iron 4–5 and steel 5–8·5. The relative scratch resistance of plastics materials is of some importance, for example in the selection of materials for domestic applications.

## FRICTION

When two bodies are in contact and one of them moves or tends to move in relation to the other, an opposing force develops in a direction tangential to the surfaces in contact. This opposing force is termed friction. The tangential force required to start motion is called static friction and that required to maintain motion is known as kinetic friction. It is further necessary to distinguish between sliding friction, as of a sledge, and rolling friction, as of a wheel. The two fundamental laws of friction, that:

1. the frictional force is proportional to the load, and,
2. the frictional force is independent of the area of contact

were first stated by Amontons but were later verified and clarified by Coulomb, after whom they are generally known. They lead to the definition of a coefficient of friction $\mu$, the ratio of the tangential to the normal force between the two surfaces.

The coefficient of friction is not a material constant. It depends on the nature of the two surfaces in contact and is affected also by load, speed of movement and temperature. The static coefficient of friction depends on the time for which the surfaces have been in contact; the dynamic coefficient depends on the extent to which material has been transferred from one surface to the other.

Quite simple equipment can be used to measure the frictional force, although the number of variables affecting the measurement virtually prohibit complete coverage. Such data as are immediately available will be given in later chapters. The main requirements of the equipment are:

1. low inertia so that rapid changes in the force can be recorded,
2. a driving system with considerable spare power, so that the sliding speed is not affected by changes in the frictional force, and
3. a wide range of speed and temperature.

For reproducible results, great care must be taken in the preparation of the sample surface. Even the smoothest surfaces have projecting asperities about 100 Å, (100 million Ångström units = 1 cm) or 0·0000004 in high. When two surfaces appear to touch, the interatomic forces, which are large but which have a range of only a few Ångström units, operate only on a fraction of the nominal area. This fraction is commonly about 0·01 % but can be as large as 1 %.

When sliding occurs between two surfaces, the frictional force is made up of two parts. The first is a result of adhesion at the points of true contact where strong junctions are formed by plastic flow, and the shearing of these junctions is the main cause of frictional resistance. The second arises when the asperities of the harder material plough through the surface of the softer material; this can often be neglected.

A simple outline theory of the first effect can explain the basic laws of friction for metals and other materials which deform plastically under the high local pressures.

$$\text{True area of contact, } A = \frac{\text{Load } W}{\text{Mean yield strength of asperities, } P}$$

$$\text{Frictional force } F = AS = \frac{WS}{P}$$

where $S$ is the mean shear strength of asperities. Hence

$$\text{Coefficient of friction} = \frac{F}{W} = \frac{S}{P}$$

This implies the basic laws that the frictional force is proportional to load and independent of the area of contact. This same approximate theory can also be applied to brittle solids and to plastics. Such differences as exist between metallic and polymeric friction are due to the difference between the plastic response of metals and the viscoelastic response of polymers. Thus, as load is reduced, plastics deform in an increasingly elastic manner and the coefficient of friction increases steadily. Time-dependent effects have been observed with plastics and there is a more marked sensitivity to temperature because plastics are working relatively near to their softening points. The softness of plastics often allows a transfer of material to the mating surface: in such cases, the coefficient of friction tends towards the value for plastic on plastic, even where the surface in contact with the plastic is metallic. Typical values of the friction coefficient for surfaces involving a plastic are in the range 0·3 to 0·6, but PTFE is an exceptional material giving values of 0·05 to 0·1 against most other materials. A suggested explanation is that the large, negatively charged fluorine atoms screen the positive carbon chain atoms, resulting in low interchain forces and a low value of $S$, the mean shear strength. However, the molecular structure is such as to make polymer chains rigid, giving a fairly high value of $P$, the mean yield strength. The combination of these two factors leads to an abnormally low value for the coefficient of friction.

This combination of rigidity and small area of contact with low shear strength can also be achieved in two-component systems. Thus plastics are often reinforced with fillers to reduce friction and wear: in bearings it is common practice to use a thin surface layer of a soft alloy or plastic on a rigid steel backing. These systems can give coefficients of friction similar to those of PTFE. In all these systems, the abnormally low coefficient of friction increases with sliding speed, a characteristic which prevents stick-slip.

One of the most important applications of the frictional properties of plastics is in journal bearings where the use of plastics avoids the catastrophic failure characteristics of metal bearings which, in the absence of lubrication, will seize and cause considerable damage to the shaft.

Plastics behave well in highly abrasive conditions, where their softness enables them to absorb abrasive particles and thus avoid damage. Plastic components also have the advantage of quietness in operation, and good impact resistance. The main disadvantages are poor dimensional stability, low thermal conductivity and low strength. All these disadvantages are to some extent overcome by the use of plastics as thin surface films, supported by a strong metal backing.

In designing conventional metallic bearings, engineers often use the $PV$ factor, the product of the pressure $P$, defined as the load divided by the projected area of the bearing, and relative velocity of the bearing surfaces $V$. Thus it has the dimensions of a rate of energy dissipation—indeed, there is a direct proportionality if the coefficient of friction is assumed to be constant. As a rule of thumb, it is possible to quote a maximum permissible $PV$ value beyond which a given bearing will fail. This concept can also be applied to plastics bearings. By calculating the $PV$ factor for an intended application, it can quickly be decided whether such a use is feasible and whether to proceed with prototype trials, provided the limiting $PV$ values are known for the materials under consideration. In practice, the $PV$ limit is dependent on velocity and critically dependent on the degree of lubrication, but information on $PV$ values will be presented where it is available.

The choice of a plastic for a given bearing application depends primarily on the amount of lubrication to be provided. Thus PTFE with various fillers is excellent for dry bearings. When hydrodynamic lubrication is provided, the surfaces are separated by a relatively thick layer of lubricant: the frictional properties are unimportant and a plastic will be chosen, if at all, for softness, low wear, compatibility, and cost. The intermediate case is that of boundary lubrication where the surfaces are separated by a layer of lubricant a few molecules thick. Nylon and polyacetals are commonly used in light duty applications because of their ability to work well with only initial or infrequent lubrication. In such applications molybdenum disulphide-filled nylon is often preferred because of better self-lubrication, lower friction and better abrasion resistance.

*Lubrication*

The improvement obtained by lubrication of plastics is less marked than with metals. However, heavier loadings can often be used and the $PV$ limit may be increased by as much as a factor of ten. With nylon bearings, slight initial lubrication has a most marked effect and, in these circumstances, nylon can easily out-perform metal. Water can be used to lubricate plastics in such applications as pumps. Naturally, the extreme pressure lubricants, which rely for their effect upon a chemical reaction with metals, are of no use with plastics. Occasionally, friction and wear are increased by lubrication: in such cases, it seems likely that the plastic is softened and swollen by penetration of the lubricant. This can be avoided by care in the selection of a lubricant.

## WEAR

Wear is usually considered under four main headings:

1. Adhesive wear.
2. Abrasive wear.
3. Corrosive wear.
4. Surface fatigue.

Adhesive wear is a natural consequence of the shearing of the junctions between surfaces and is always present when surfaces rub. By using plastics rather than metals, adhesive forces and, therefore, adhesive wear, are reduced, and the disastrous effects of seizure can be completely avoided. Abrasive wear is caused by abrasive particles which may be introduced from outside or which may result from wear. The softness of plastics is an advantage here because such particles may sink beneath the surface and do no further harm. Corrosive wear, the result of an interaction between corrosion and mechanical forces, is not a serious problem with plastics; on the other hand, surface fatigue is important, particularly in such applications as gear teeth.

There is no direct relationship between friction and wear although factors which lead to an increase in friction will usually cause increased wear. Performance in a standard test for abrasion resistance is of little design value for predicting wear and satisfactory predictions of wear resistance can only be made after long-term testing in a rig which approximates to the working conditions.

## THERMAL PROPERTIES

The classification 'thermal properties' is often understood to include information on the effect of temperature on mechanical properties with respect to test procedures such as deflection temperature under load and brittleness temperature. In this book, phenomena such as these are dealt with as appropriate temperature effects on the mechanical properties concerned, and the term 'thermal properties' is restricted to crystalline melting point, specific heat, thermal expansion and thermal conductivity. These are essentially equilibrium properties, which means that time is not an important variable affecting them. However, they are not material constants because they are dependent to some extent on temperature; the specific heat of partially crystalline plastics near the crystalline melting point is particularly affected by temperature.

### *Crystalline Melting Point*

It has been noted earlier (Chapter 1) that chains of certain polymers can form into ordered regions, or crystallites, and that these crystallites have a marked influence on material properties. It is not unexpected, therefore, that the melting of the crystallites changes the properties abruptly and that the temperature at which this occurs sets an upper limit to the working range of most crystalline plastics.

The crystallites of plastics materials vary in size and perfection, depending on local variations in molecular structure such as chain branching and on the previous thermal history of the sample. For these reasons, the melting of such crystallites is not a single catastrophic phenomenon occurring over a very narrow temperature range, as is the case for crystals of low molecular weight organic compounds, but a process that takes place progressively with increasing temperature. The melting range may extend over many tens of degrees: typically a range of some 75°C is frequently quoted for low density polythene whilst that of polypropylene is usually 10–20°C. The properties are likely to be affected by the melting process over similar temperature ranges; however, the temperature at which the last traces of crystallinity disappear is usually the absolute upper working limit for form stability and this is the temperature usually quoted.

Measurements of crystalline melting point can be made by a variety of techniques, the most direct of which is monitoring the disappearance of crystallinity by x-ray diffraction as the temperature of the sample is increased. Other more rapid techniques are also frequently employed such as the birefringence disappearance method and differential thermal analysis[1].

### *Specific Heat*

In general, the specific heats of plastics are higher than those of metals, although the difference between the two classes of material is not so marked as for other thermal properties. Strictly, it

[1] 'Newer Methods of Polymer Characterization' (Ed. Bacon, Ke.), *Polymer Rev.*, **6**, 1964.

is wrong to attribute to crystalline plastics a specific heat in the conventional sense of the term because the crystallinity changes with temperature and some of the measured heat change should be considered as latent. However, for the designer, main interest will be in the behaviour in the temperature region around 20°C and brief data at this condition only are given in terms of this wider interpretation.

*Thermal Expansion*

When materials are heated, they usually expand. The temperature coefficient of expansion for most plastics is some ten times greater than that of most metals, so that a knowledge of the expansion behaviour is essential in many applications. The reversible change in dimensions with temperature is the information usually required: changes in dimensions caused by changes in water content are considered separately in the appropriate product chapters, and irreversible changes resulting from chemical reaction are not considered.

The expansion of plastics is not usually linear with temperature. The expansion coefficient increases with increasing temperature, so that the best presentation of the data is in the form of a curve of expansion or contraction plotted against temperature. However, for many purposes, less accurate data suffice and an average value for the temperature range −30°C to +30°C is given in later chapters. The expansion is determined simply by observing the change in length of a test rod as its temperature is changed. Non-isotropic plastics expand differently in different directions, and it is then relevant to measure the expansion in these directions.

The large thermal expansion of plastics has two important practical consequences. Firstly, because plastics are moulded at elevated temperatures, the finished article will be somewhat smaller than the mould and allowance for this has to be made in mould design. Secondly, a finished component will change in size with temperature.

There is a growing demand for plastics with thermal expansions of the same order as those of metals, and some success has been achieved in reducing the thermal expansion of plastics by incorporating fillers in them. The incorporation of fillers, especially if they are fibrous, may give anisotropic mouldings, the properties of which, including thermal expansion, will be different in the flow direction and the direction perpendicular to it.

*Thermal Conductivity*

Plastics are essentially thermal insulators; they have conductivities some hundreds of times lower than those of most metals. This fact can have important practical consequences. For instance, the freezing of water in polythene pipes is far less rapid than in metal pipes, and plastics, particularly foams, are used as thermal insulators.

Thermal conductivity is usually measured by a variant of the Lee's disc method of classical physics; as for specific heat measurements, care is required to ensure that true equilibrium has been established before measurements are made. The values obtained are usually temperature-dependent, the temperature effect being more marked for partially crystalline plastics. This is again shown in the variation of conductivity at room temperature for materials of different crystallinity: for example, crystalline polythene is four times more conducting than the amorphous component.

## PRACTICAL ASSESSMENTS OF MOLECULAR WEIGHT

It was stated in earlier chapters that certain properties, such as impact strength, are markedly dependent on molecular weight. A further property, melt viscosity, which is important in the fabrication of thermoplastics, is also extremely dependent on molecular weight. For a complete description of the properties of plastics, therefore, a knowledge of the dependence of some properties on molecular weight, and hence processability, is necessary.

The determination of absolute molecular weight is a very difficult and lengthy process, and it has become the custom to employ indices of molecular weight rather than absolute values. There are two commonly accepted methods:

1. *Solution viscosity.* The viscosity of a dilute solution of the plastic is related to the molecular weight of the solute: the higher the solution viscosity, the higher the molecular weight. Sometimes the solution viscosity obtained under specific conditions has been adopted as a measure of molecular weight in a materials' technology—for instance, the *K*-value of PVC (see Chapter 13).

2. *Melt viscosity.* The viscosity of a plastics melt is directly related to the ease of fabrication, particularly in injection moulding and extrusion processes, and is again related to molecular weight: the greater the molecular weight, the higher the melt viscosity. A particularly convenient measurement of melt viscosity can be made in the apparatus shown schematically in Figure 6.1. The rate of extrusion of a plastics melt held in the cylinder of the instrument is determined by weighing the amount of material extruded under predetermined load and temperature conditions in 10 minutes. This weight, expressed in grams, is known as the Melt Flow Index (MFI) and the instrument as the Melt Flow Indexer. Various combinations of temperature and load are defined for the various plastics materials: there is no easy method of comparing one material with another where the melt flow indices have been obtained under different conditions. Occasionally, for very

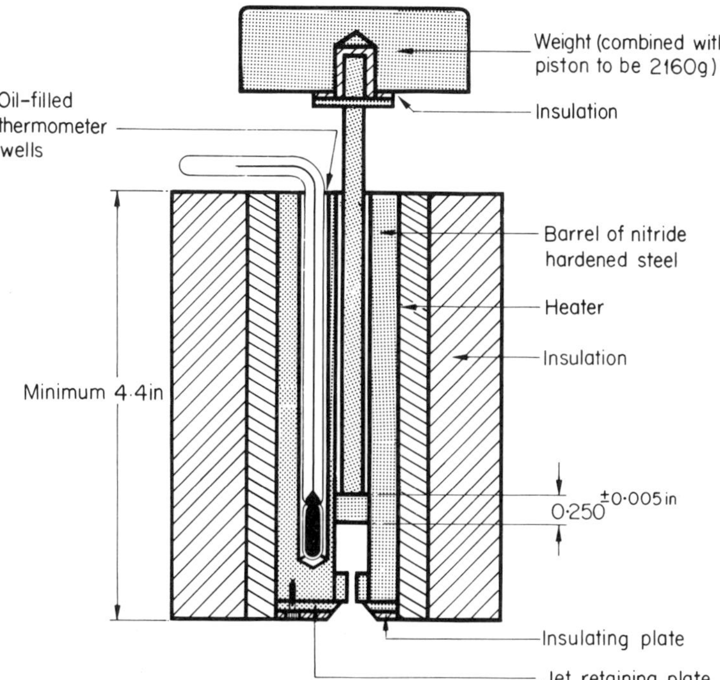


**Figure 6.1.** Apparatus for determining melt flow index using large external diameter cylinder, jet retaining plate and insulating plate

fluid melts, an extrusion die of much smaller size is employed but this variant will not be found for the materials grades discussed in Chapters 7 to 14. It must be stressed that the melt flow index is an inverse measure of molecular weight, that is, the higher the melt flow index, the lower the molecular weight.

## CHEMICAL PROPERTIES

As was seen in Chapters 1 and 2, few of the plastics offered commercially are pure polymers. Most contain other additives such as plasticizers, heat stabilizers, antioxidants or pigments and some consist of blends of polymers. Therefore, the chemical resistance of any material will depend on the nature and amount of each of its ingredients and detailed information on the likely behaviour of a given material in a given environment is best obtained from the raw material manufacturer.

In particular, great care should be exercised in using tables of chemical resistance of plastics. Tests on the resistance of plastics to chemical attack are difficult to interpret because, firstly, the conditions of exposure may well have an important bearing on the result and, secondly, plastics may be attacked in various ways.

All such tables should be taken only as a guide and on no account should they be used to deduce results, e.g. the effect of mixed acids may not be the sum of the effects of the acids taken separately, or to compare the behaviour of different plastics, except in the most general way.

Nevertheless, there are some broad generalizations that can be made, as indicated at the end of Chapter 1, and the following will give an indication of the probable usefulness of plastics in a given situation and within their working temperature ranges.

(a) Plastics are generally resistant to weak acids, weak alkalis, and salt solutions.
(b) Strong oxidizing acids may attack plastics causing discolouration and embrittlement.
(c) Fuels, fats, oils and organic solvents may attack plastics, causing swelling, softening and ultimately dissolution. The degree of resistance to these agents depends on the composition of the plastic, the attacking agent and the temperature, and can vary from complete to very poor.
(d) Exposure to ultraviolet (u.v.) radiation tends to cause deterioration of the properties of plastics, although some show excellent resistance to exposure and others can have their useful lives outdoors extended by a factor of up to a 100 by appropriate compounding.
(e) Most plastics will degrade with consequent loss of usefulness, if kept for long times at elevated temperatures, particularly in the presence of air.

# Part III

## ENGINEERING DATA

# 7

# POLYPROPYLENE

## INTRODUCTION

Since its introduction in the mid-1950's, polypropylene has established itself in many engineering applications through its combination of strength, rigidity, temperature resistance, chemical inertness, low price and ease of fabrication.

Polypropylene is a member of the polyolefine group of polymers. It is related structurally to polythene and is made by polymerizing propylene, the next member in the olefine series after ethylene. Until the early 1950's all attempts to polymerize propylene into solid, useful thermoplastics had failed, only syrupy liquids being obtained, but in 1953 Professor K. Ziegler in Mülheim discovered that certain combinations of metal alkyl complexes and transition metal halides catalyse the polymerization of ethylene so that the reaction can be carried out at temperatures and pressures a little above atmospheric. Previously, polymers of ethylene had been prepared only by using high temperatures and very high pressures. Ziegler's work provided the first example of a new type of polymerization reaction. The following year Professor G. Natta of Milan, working in collaboration with the Montecatini Chemical Company, prepared solid, high-molecular weight polymers of propylene using catalysts of the Ziegler type. He further developed these catalysts so that they became stereospecific. These discoveries have led to the industrial production of stereoregular polypropylene, a rigid, easily moulded material with the excellent chemical resistance and electrical insulation properties normally expected of polyolefines. In contrast, polypropylene in a randomly polymerized form is of little commercial interest.

## NATURE OF POLYPROPYLENE

Polypropylene molecules can assume any of the three structures already referred to in Chapter 1, that is atactic, isotactic or syndiotactic. It is doubtful if molecules consisting wholly of any one structure ever exist and molecules of commercial polypropylene are predominantly isotactic, but with short lengths of atactic and syndiotactic material incorporated in them.

Two main types of polypropylene are available: homopolymers of propylene, and ethylene/propylene copolymers in which propylene is the major constituent; the latter may contain various amounts of ethylene and a subdivision into low ethylene ($\sim 7\%$ by weight) and high ethylene ($\sim 15\%$ by weight) copolymers is appropriate. However, this characterization of copolymers must not be regarded as unique because much depends on the manner of incorporation of the ethylene. It should be added that random polymers are not considered here.

## GRADES AND FORMS

Both homopolymers and copolymers can be obtained in various grades, all of which have virtually the same density, but which differ in melt viscosity and in the degree to which they are stabilized against oxidation; they thus meet the requirements of specific end-uses and processing techniques. Many special grades have also been developed: these include glass-filled and flame-retardant materials, and grades to produce articles with reduced dust pick-up and improved weathering performance.

Polypropylene is usually sold as regularly cut granules, but some grades are also offered in the form of free-flowing powders. In its natural form polypropylene is translucent and white, but compounds in a wide range of colours are available, and the natural material itself can easily be coloured by the fabricator if so desired.

Semifabricated forms such as film, foil, sheet, block, rod and tube are all produced commercially and are readily available. These are convenient starting materials in the making of prototypes.

## PROCESSING

Polypropylene is easily processed by conventional techniques. It can be moulded on all modern injection moulding machines with adequate heating capacity and good temperature control, and readily extruded into film, pipe, wire and cable covering, sheet, monofilament and fibre, and tubes for subsequent blowing into bottles and other hollow articles. Compression moulding, although not economic on a commercial scale, can be used to make blocks and sheets for prototype work.

Polypropylene sheet is readily bent and shaped into simple curves when heated, and it can also be vacuum formed.

The usual machining techniques for thermoplastics are applicable to polypropylene. Jointing techniques such as welding and cementing can also be used, although the strength of cemented joints is not adequate for every application because of the chemical inertness of the material, and whenever possible it is desirable to design articles so that adhesives are not required.

## PROPERTIES

Some physical properties of polypropylene are given in Table 7.1. As with all crystalline polymers, they depend on the amount of crystallinity; in polypropylene, the proportion of stereoregular polymer represents the amount of potentially crystallizable material, and it thus determines the softening temperature, stiffness, hardness and tensile strength; increasing the amount of amorphous atactic material reduces the total crystallinity and hence leads to a reduction in value for all these properties. However, the actual crystallinity of a finished article depends also on the rate at which the molten material is cooled and solidified. On crystallization, the crystallites tend to arrange themselves in ordered aggregates known as spherulites. On slow

TABLE 7.1

| Property | Units | Typical Value |
|---|---|---|
| Density | $g/cm^3$ | 0·90–0·91 |
| Crystalline melting point | °C | 165–175 |
| Coefficient of linear thermal expansion | | |
| −20 to 20°C | $°C^{-1}$ | $9 \times 10^{-5}$ |
| | $°F^{-1}$ | $5 \times 10^{-5}$ |
| 20 to 80°C | $°C^{-1}$ | $1{\cdot}35 \times 10^{-4}$ |
| | $°F^{-1}$ | $0{\cdot}75 \times 10^{-4}$ |
| Thermal conductivity, 20 to 150°C | cal/cm s °C | $3{\cdot}5$ to $5 \times 10^{-4}$ |
| | B ThU in/ft$^2$ h °F | 1·0 to 1·45 |
| Specific heat, 20 to 80°C | cal/g °C | 0·46 |
| | B ThU/lb °F | 0·46 |
| Flammability | | burns slowly with drips |

cooling from the melt few crystal nuclei are formed, and this results in a small number of large spherulites. On fast cooling (quenching) the rate of nuclei formation is very much higher and a large number of small spherulites are formed. Slow cooling produces a higher overall crystallinity. This dependence of crystallinity and texture on cooling rate must be borne in mind during processing; in general, slowly cooled articles are rather more rigid and temperature resistant than those which are rapidly quenched, but these properties are obtained at the expense of a slight decrease in transparency and toughness.

A further variable which affects the physical properties of polypropylene and its flow properties at high temperatures is the molecular weight. As the molecular weight increases, the impact strength increases. Molecular weight also has a marginal effect upon stiffness and hardness because crystallinity, which determines stiffness, is slightly dependent upon molecular weight. Low molecular weight tends to be associated with higher crystallinity, other things being equal. The average molecular weight of polypropylene is not normally measured and in practice polymers are classified by the melt flow index (MFI) at 230°C and 2·16 kg.

For ethylene/propylene copolymers the properties are determined by the proportion of ethylene present and by the way in which it is introduced into the chain structure, as well as by the factors already mentioned above.

### *Deformation*

The deformational behaviour at 20°C of polypropylene homopolymer of density 0·909 g/cm$^3$ is given by the tensile creep curves in Figure 7.1. The corresponding isochronous stress vs. strain, isometric stress vs. time, and tensile creep modulus vs. time curves are given in Figures 7.2, 7.5 and 7.6. Because polypropylene is a partially crystalline material, the deformational properties depend quite markedly on the degree of crystallinity in a particular sample. Crystallinity affects the density and thus the density can be a convenient, but not always unambiguous, index of the crystallinity of a sample. The influence of changes in crystallinity, brought about by different cooling conditions, on the 100 second isochronous stress vs. strain curve is given in Figure 7.3: also included are results obtained on propylene/ethylene copolymers and it can be seen that to a good approximation the isochronous stress vs. strain curves of Figure 7.3 can be indexed by density. When the curves of Figure 7.3 are plotted against logarithmic axes simplifying regularities appear[1] as shown in Figure 7.4. The spacings of these curves are such that an increment in density of 0·001 g/cm$^3$ corresponds to an increment in stress of 4%. This numerical correction factor for density can be applied to any of the data in Figures 7.2, 7.5 and 7.6.

Recovery data at 20°C for propylene homopolymer of density 0·909 g/cm$^3$ are given in Figure 7.7, and show the high degree of recovery characteristic of thermoplastics even from long term and high strain creep. Data are available also on another aspect of recovery. As described in Chapter 4, stresses applied to a specimen intermittently result in a strain considerably lower that that which would be attained for the same stress acting continuously. This is shown in Figure 7.8 where two stress levels are examined and ratios of $\frac{\text{time at constant load}}{\text{time at zero load}}$ of $\frac{1}{5}$ and $\frac{1}{11}$ are used. The maximum strain reached in any one of the loading cycles (it will be at the end of that particular loading cycle) is plotted against the total creep time (i.e. the sum of the times under load) to the end of that cycle. The creep strain resulting from continuous loading is included for comparison. An important result is that the horizontal line condition, i.e. constant strain, is approached as the total time for which the specimen is subjected to intermittent loading increases.

[1] Turner, 'Creep in Thermoplastics—Polypropylene', *Brit. Plastics*, **37** (8), pp. 440–4, August 1964.

The deformational behaviour at 60°C of propylene homopolymer of density 0·909 g/cm$^3$ (at 20°C) is given by the tensile creep curves of Figure 7.9. The corresponding isochronous stress vs. strain, isometric stress vs. time, and tensile creep modulus vs. time curves are given in Figures 7.10, 7.11 and 7.6. Again the density correction is to change the stress by 4% for each 0·001 g/cm$^3$ change in density (at 20°C).

The deformational behaviour between 20°C and 60°C can be obtained to a good approximation by linear interpolation between the 20°C and 60°C tensile creep data. For temperatures outside this range estimates can be made from the 100 second creep modulus (at 0·2% strain) vs. temperature curves given in Figure 7.12 for a homopolymer and high ethylene copolymer. For instance, for temperatures in the range 60°C to 100°C the stress should be reduced by approximately 1·3% for every 1°C rise in temperature.

### *Limiting Stresses*

Too little is known of the failure characteristics of polypropylenes to enable design on the basis of failure to be a reliable method and, as was concluded in Chapter 5, data for design stress at some limiting strain, such as 3%, should be used. However, some applied stress vs. time to failure data are available for homopolymer and low and high ethylene content copolymers and are presented in Figure 7.13. Because polypropylene under such long time stresses behaves in a ductile manner, the failures plotted in Figure 7.13 are for 'necking' ruptures at short times and 'necking' and cold drawing at long times. The data are for 20°C and for polymers and copolymers of the densities given. In Figure 7.14 the failure curve for homopolymer is compared with the appropriate isometric stress vs. time curves, corrected to a density of 0·906 g/cm$^3$ from the curves of Figure 7.5.

The yield stress obtained in a conventional tensile test at a straining rate of about 50% per minute is plotted as a function of temperature in Figure 7.15 and the effect of density on the yield stress of a variety of polypropylenes is indicated in Figure 7.16. These two figures are included to illustrate the effect of temperature and density on the yield stress; they should not be used for design purposes except for short times under load, that is, times of the order of 1 minute.

### *Dynamic Fatigue and Environmental Effects*

There are certain service conditions such as intermittent stressing or stressing in an active environment, which may promote premature failure. There are insufficient data yet available for such effects to be discussed in detail and current investigations aim merely to indicate whether the limiting stresses or strains indicated in Figures 7.13 and 7.14 are unrealistically high. Dynamic fatigue tests in flexure using a low frequency square wave suggest that polypropylenes have a high fatigue resistance. This high fatigue resistance is utilized in a particularly important practical application, the 'integral hinge', which is described later in the section on 'Applications'.

The effect of an 'active' environment on stressed polypropylenes is different from that observed in polythenes. Whereas in polythenes cracks initiate at stress raisers and propagate in an apparently brittle fashion, in polypropylenes, although cracks initiate, subsequent growth can be impeded. Experience of practical applications also suggests that polypropylenes are less susceptible to detrimental environmental effects than are polythenes. However, certain agents do appear to increase the chance of brittle failure: these are mentioned in the section on 'Chemical Properties'.

*Impact Behaviour*

Figure 7.17 shows the impact strength of a moulding grade of polypropylene vs. temperature for un-notched and bluntly and sharply notched specimens. The homopolymer is a fairly brittle material, particularly if notched. It may be considerably improved in impact behaviour by incorporating small amounts of ethylene into the polypropylene. This is clearly shown in Figures 7.18 and 7.19 where moulding grades of the homopolymer, low ethylene copolymer and higher ethylene copolymer are compared for 0·010 in and 0·080 in notches respectively. A parallel toughening by incorporation of ethylene occurs in the extrusion grades; compare the homopolymer and low ethylene copolymer for sharply and bluntly notched specimens taken across the machine direction for extruded sheet in Figures 7.20 and 7.21.

The effect of molecular weight on impact behaviour is not so marked as is the effect of copolymerizing with ethylene. Nevertheless there is a considerable improvement in impact strength as the molecular weight is increased: compare the behaviour of moulding and extrusion grades for sharp notches (Figure 7.22) and blunt notches (Figure 7.23) where both homopolymer and low ethylene copolymer grades are considered. However, the situation is somewhat complicated by the observation that the residual orientation in extrudates is somewhat less than that in injection mouldings. Thus the former have the better impact behaviour even if there is no effect of molecular weight; this is demonstrated in Figure 7.24.

Figure 7.25 shows the effect of changing the cylinder temperature during the injection moulding process on the impact strength at room temperature of homopolymer, and demonstrates the desirability of moulding in the optimum range 260–290°C for this particular moulding. A low melt temperature results in a moulding with high residual orientation and consequent low impact strength across the flow direction. The impact strength along the flow direction is enhanced by the orientation. At high cylinder temperatures degradation of the material is promoted and there is a progressive decrease in impact strength for both the across-flow and along-flow directions. At high temperatures there is little orientation and the distinction between the two directions is much reduced.

Following fabrication, certain structural changes are known to take place over time scales extending into many months. Such changes are manifested in after-moulding changes in dimensions and are ascribed to molecular processes leading to relaxation of moulding stresses and crystallite reorganization. The importance of storage at 20°C on the impact strength at that temperature has been assessed by obtaining data similar to the 'across-flow' results in Figure 7.25 for various storage times before test. It can be seen in Figure 7.26 that there is a progressive improvement in the impact behaviour with storage, a result contrary to that found in similar tests on compression moulded samples.

A component may be in service for some part of its lifetime at a high temperature. Tests made at room temperature on specimens of homopolymer after treatment for various times at 130°C give results which show a significant increase in impact strength for short treatments (these are conditioning treatments similar to that recorded in Figure 7.26), a steady impact strength over long periods of exposure at 130°C, and a final catastrophic decrease in impact strength. This last is to be associated with oxidative degradation of the polypropylene and its onset will depend on the efficiency of the stabilization system incorporated in the plastic.

*Friction and Wear*

Coefficients of friction between unlubricated polypropylene/polypropylene and polypropylene/steel surfaces are low, and comparable with those for nylon, but the effect of lubrication is much less marked for polypropylene as illustrated in Table 7.2 where the data were obtained by measuring the frictional resistance of a loaded hemisphere sliding over a flat surface at various speeds at 20°C.

The abrasion resistance of polypropylene is good in practice as can be judged from the list of satisfactory applications at the end of this chapter. Laboratory tests for abrasion resistance are notoriously unreliable but even here, Tabor abrasion results (ASTM 1044–56) show a satisfactorily high abrasion resistance of 18–28 mg per 1000 revolutions of a C5–17 wheel.

TABLE 7.2 Comparison of the Effect of Surface Finish and Lubricant on the Coefficients of Friction of Polypropylene (PP) and Nylon 66 at 20°C.

| Specimen | | Sliding Speed (cm/s) | | | | | |
|---|---|---|---|---|---|---|---|
| Slider | Plate | 0·003 | 0·01 | 0·04 | 0·08 | 0·30 | 1·06 |
| | *Unlubricated tests* | | | | | | |
| PP (as moulded) | PP (as moulded) | 0·54 | 0·65 | 0·71 | 0·77 | 0·77 | 0·71 |
| Nylon (as moulded) | Nylon (as moulded) | 0·63 | — | 0·69 | 0·70 | 0·70 | 0·65 |
| PP (abraded) | PP (abraded) | 0·26 | 0·29 | 0·22 | 0·21 | 0·31 | 0·27 |
| Nylon (machined) | Nylon (machined) | 0·42 | — | 0·44 | 0·46 | 0·46 | 0·47 |
| Mild steel | PP (abraded) | 0·24 | 0·26 | 0·27 | 0·29 | 0·30 | 0·31 |
| Mild steel | Nylon (machined) | 0·33 | — | 0·33 | 0·33 | 0·30 | 0·30 |
| PP (abraded) | Mild steel | 0·33 | 0·34 | 0·37 | 0·37 | 0·38 | 0·38 |
| Nylon (machined) | Mild steel | 0·39 | — | 0·41 | 0·41 | 0·40 | 0·40 |
| | *Water-lubricated tests* | | | | | | |
| PP (abraded) | PP (abraded) | 0·25 | 0·26 | 0·29 | 0·30 | 0·28 | 0·31 |
| Nylon (machined) | Nylon (machined) | 0·27 | — | 0·24 | 0·22 | 0·21 | 0·19 |
| Mild steel | PP (abraded) | 0·23 | 0·25 | 0·26 | 0·26 | 0·26 | 0·22 |
| Mild steel | Nylon (machined) | 0·23 | — | 0·20 | 0·20 | 0·19 | 0·17 |
| PP (abraded) | Mild steel | 0·25 | 0·25 | 0·26 | 0·26 | 0·25 | 0·25 |
| Nylon (machined) | Mild steel | 0·20 | — | 0·23 | 0·23 | 0·22 | 0·18 |
| | *Liquid paraffin-lubricated tests* | | | | | | |
| PP (abraded) | PP (abraded) | 0·29 | 0·26 | 0·24 | 0·25 | 0·22 | 0·21 |
| Nylon (machined) | Nylon (machined) | 0·22 | — | 0·15 | 0·13 | 0·11 | 0·08 |
| Mild steel | PP (abraded) | 0·17 | 0·17 | 0·16 | 0·16 | 0·14 | 0·14 |
| Mild steel | Nylon (machined) | 0·16 | — | 0·11 | 0·09 | 0·08 | 0·08 |
| PP (abraded) | Mild steel | 0·31 | 0·30 | 0·30 | 0·29 | 0·27 | 0·25 |
| Nylon (machined) | Mild steel | 0·26 | — | 0·15 | 0·12 | 0·07 | 0·04 |

### *Electrical Properties*

Polypropylene and ethylene/propylene copolymers are entirely hydrocarbon in character, so that they are basically non-polar with permittivities corresponding to values obtained from the Clausius–Mosotti relationship of around 2·25. In common with other hydrocarbon polymers, dielectric losses can be extremely low but sometimes, somewhat higher values are met in practice because of the inclusion of process impurities or compounding ingredients. Even the most pure form of polypropylene (dielectrically speaking) shows a spectrum of loss with frequency or temperature where 'relaxation processes' can be discerned; this loss pattern is shown in Figure 7.27. With such low losses the permittivity is essentially independent of frequency whilst the temperature dependence is related to the density changes.

The interpretation of current vs. time curves resulting from a step-wise application of a voltage has been considered in Chapter 6; for polypropylene such experiments indicate that a conduction process becomes of importance by ~ 120°C, sometimes at lower temperatures: many data are covered by the band in Figure 7.28. The apparent volume resistivity at 20°C for propylene homopolymer is plotted in Figure 7.29 as a function of the time of electrification.

Polypropylene film is finding application in capacitors and other electrical components. Here its durability in the presence of stray discharges may be of some importance. In fact, so far as such applications are concerned polypropylene, together with the other polyolefines, appears to be more durable than other polymer systems. Similarly, polypropylene is of some interest in high voltage insulation.

### *Optical Properties*

Polypropylene is essentially translucent and in bulk form is not used directly in optical components. It does, however, find some application in lighting fittings because of its light transfer properties and its 'see-through' properties can be important in packaging. Film can be obtained in strikingly transparent form.

### *Refraction*

Polypropylenes, even grades nucleated to give fair transparency, are in no sense materials for the optical industry and the data presented under this heading are limited in consequence: they are for homopolymer of density 0·907 g/cm³.

Refractive Index $n_e$ (Hg 5461 Å) 1·505
$n_D$ (Na 5893 Å) 1·504
Critical Angle $\alpha_D$ (Na 5893 Å) 41°/48′

### *Transparency*

Although polypropylene is not transparent its clarity may be sufficient to allow its use in liquid measuring devices, e.g. hypodermic syringes. Modified materials in which the crystallization has been nucleated by 'seeds' do have a degree of transparency although they are decidedly 'milky'. Data for the two types of material are given in Table 7.3 where the results have been corrected for reflection and surface defects.

TABLE 7.3

| Property | Value unmodified polypropylene | Value nucleated polypropylene |
|---|---|---|
| Thickness (mm) | 1·418 | 1·46 |
| Direct transmission factor (%) | 5 | 32 |
| Scattering coefficient ($cm^{-1}$) | 21·5 | 7·7 |
| Forward scattered fraction (ASTM D1003) (%) | 80 | 44 |
| Corrected direct transmission factor for 1 mm thickness (%) | 11 | 46 |

*Chemical Properties*

Polypropylene shows the very high resistance to chemical attack typical of polyolefines. At ambient temperatures it has extremely high resistance to water, water vapour, aqueous solutions of inorganic salts, most mineral acids and bases, and most organic chemicals. There is no solvent for the material at room temperature. At higher temperatures polypropylene still has good resistance to a wide variety of chemicals. However, there are some important exceptions to the above general rules—for example, at 20°C concentrated sulphuric acid can cause deterioration in properties and at 100°C 30% hydrochloric acid attacks it.

TABLE 7.4

| | |
|---|---|
| Equilibrium water content (immersed) | 0·03% |
| Gas permeabilities (cm³ cm/cm² s cm Hg) × 10⁹ at 20°C and 0% rh | |
| Carbon dioxide | 0·3 |
| Oxygen | 0·1 |
| Nitrogen | 0·04 |
| Water vapour permeability (g/m² 24 h) at 38°C and 90% rh for 0·001 in thickness | 11 |
| Resistance to | |
| Mineral acids (dilute) | Excellent |
| Mineral acids (conc.) | Good, except with oxidizing acids which may cause cracking |
| Alkalis | Excellent |
| Solvents | |
| alcohols | Excellent |
| ketones | Good at room temperature, but poor with higher molecular weight ketones at 60°C and above |
| aromatic hydrocarbons | Fair (or poor) |
| chlorinated hydrocarbons | Fair (or poor) |
| Detergents | Excellent |
| Greases and oils | Good at room temperature, poor at 60°C and above |

## APPLICATIONS

Apart from the properties already mentioned, polypropylene possesses three other characteristics which can be exploited to considerable advantage: these are the ability to form an integral hinge, the ability to accept a textured finish, and the ability to provide snap fits.

Polypropylene excels in its ability to form an integral hinge, which is simply a very thin web of material in which orientation of the long chain molecules has produced a great increase in strength in a direction at right angles to the hinge length. This orientation is initiated during the moulding process (although hinges can also be produced by extrusion, machining or forming) and completed by subsequent flexing of the hinge, at first slowly and then at normal speed. When correctly designed and moulded, polypropylene hinges will withstand flexing indefinitely and in addition have remarkably high tensile and tear strengths.

Mouldings with textured finishes can readily be produced from moulds on to which the appropriate pattern has been engraved photochemically. Because such moulds do not require polishing, cast moulds can be used with a considerable saving in cost.

The possibility of using snap-fit assemblies is a consequence of the toughness and resilience of polypropylene, and the unique combination of this feature with the integral hinge and textured finish often leads to a reduction in the number of components and assembly operations required, as well as providing aesthetic advantages. Many of the successful applications listed below incorporate one or more of these features.

*Typical examples*

Mouldings:
- Car components, e.g. accelerator pedals, fascia panels, glove boxes, trim panels, radiator fans and heater ducting.
- Washing machine parts, e.g. agitators, tops, tubs and lint filters.
- Radio and T.V. parts, e.g. housings and cabinets, capacitors, coil formers and control knobs.
- Crates, tote boxes, bottles and closures.
- Chair shells.
- Tool handles, e.g. saws, spades and garden forks.
- Hospital equipment, e.g. kidney dishes, bedpans and tweezers.
- Valves and pipe fittings.
- Toys, shoe heels, hinged containers, safety helmets, tableware.

Pipe and sheet:
- Waste systems, tanks, reactors, fans, cyclones, pumps and filters.

Wire covering:
- Cellular distribution cable.

Cast film:
- Garment display bags, drum liners and paper coating.

Oriented film:
- Overwrapping of food products.

Fibre:
- Ropes, netting, textiles and carpets.

Film yarn:
- Woven sacks, carpet backing, baler twine and industrial cordage.

Monofilament:
- Road-sweeping brushes and industrial brooms.

## LIST OF FIGURES

**7.26.** Impact strength vs injection moulding cylinder temperature: 20°C, 0·080 in notch tip radius. Effect of time of storage at 20°C on impact strength. Propylene homopolymer ('Propathene' GWM 22)

**7.27.** Loss angle vs frequency and temperature. Propylene homopolymer, dielectric grade

**7.28.** Apparent volume resistivity vs temperature. Propylene homo- and co-polymers. (Various 'Propathene' grades)

**7.29.** Apparent volume resistivity vs time of electrification: 20°C. Propylene homopolymer ('Propathene' HWE 24)

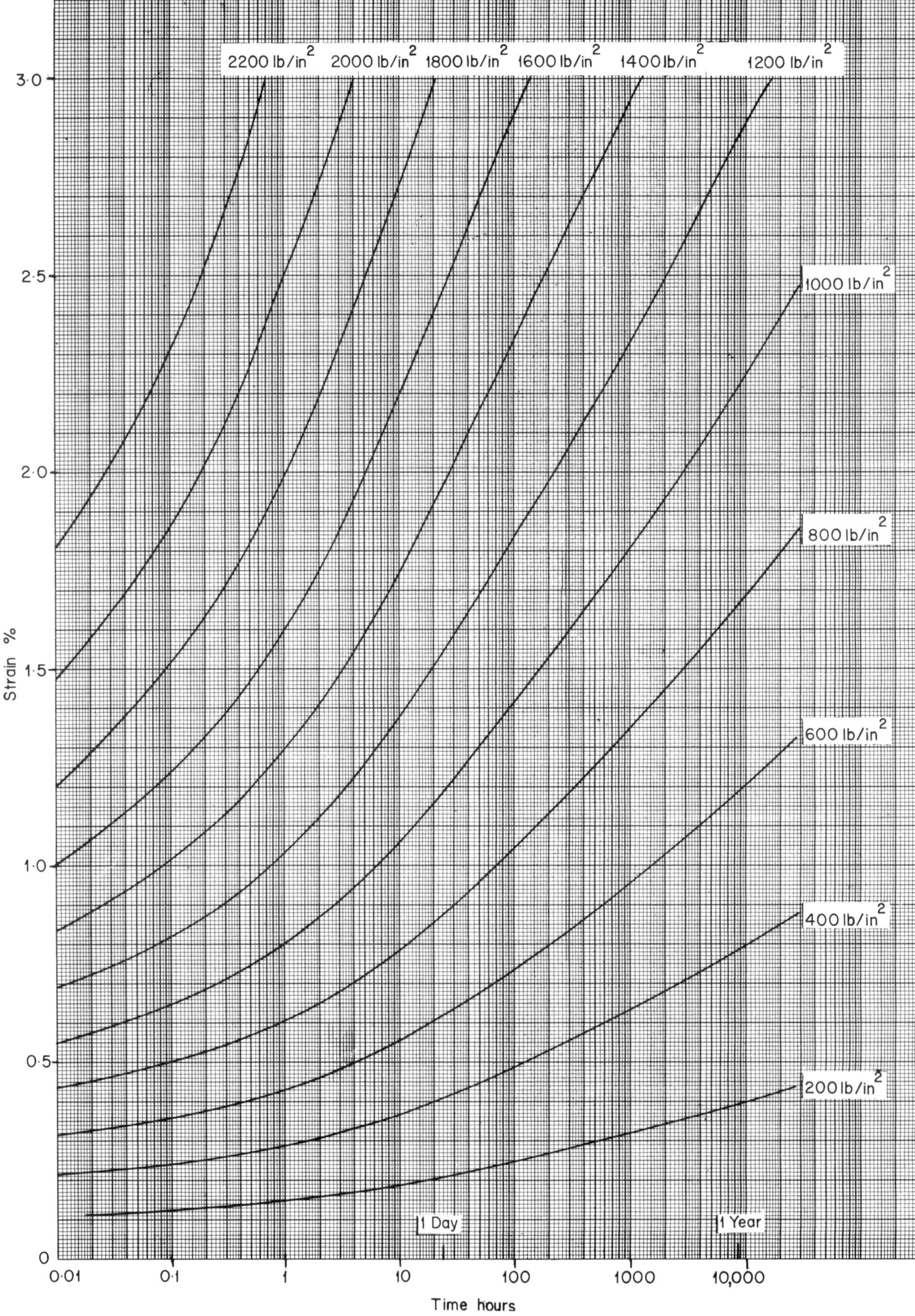


**Figure 7.1.** Creep curves in tension: 20°C. **Propylene homopolymer, density 0·909 g/cm$^3$** ('Propathene' GSM 34)

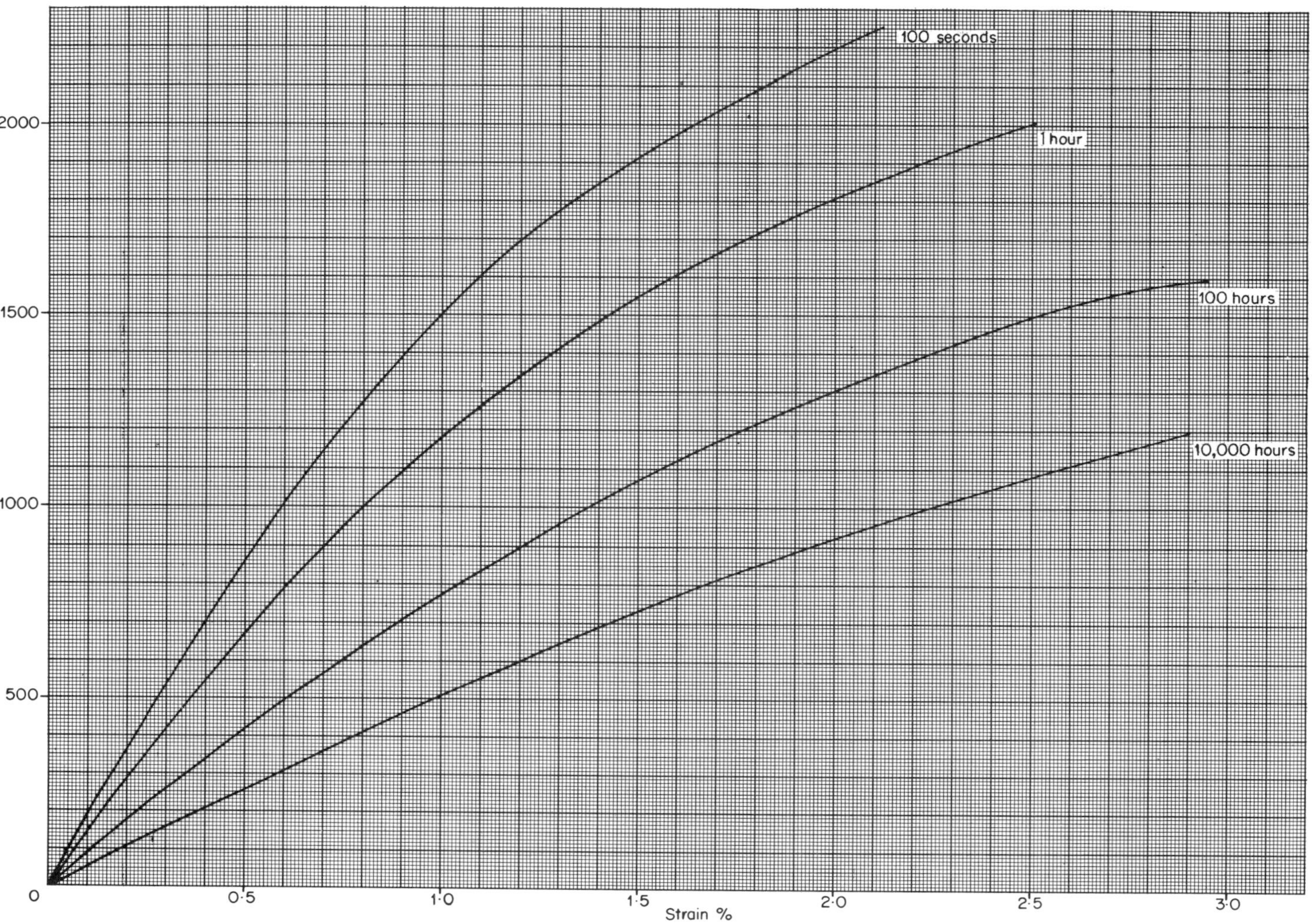


**Figure 7.2.** Isochronous stress vs strain curves: 20°C. Propylene homopolymer, density 0·909 g/cm$^3$ ('Propathene' GSM 34)

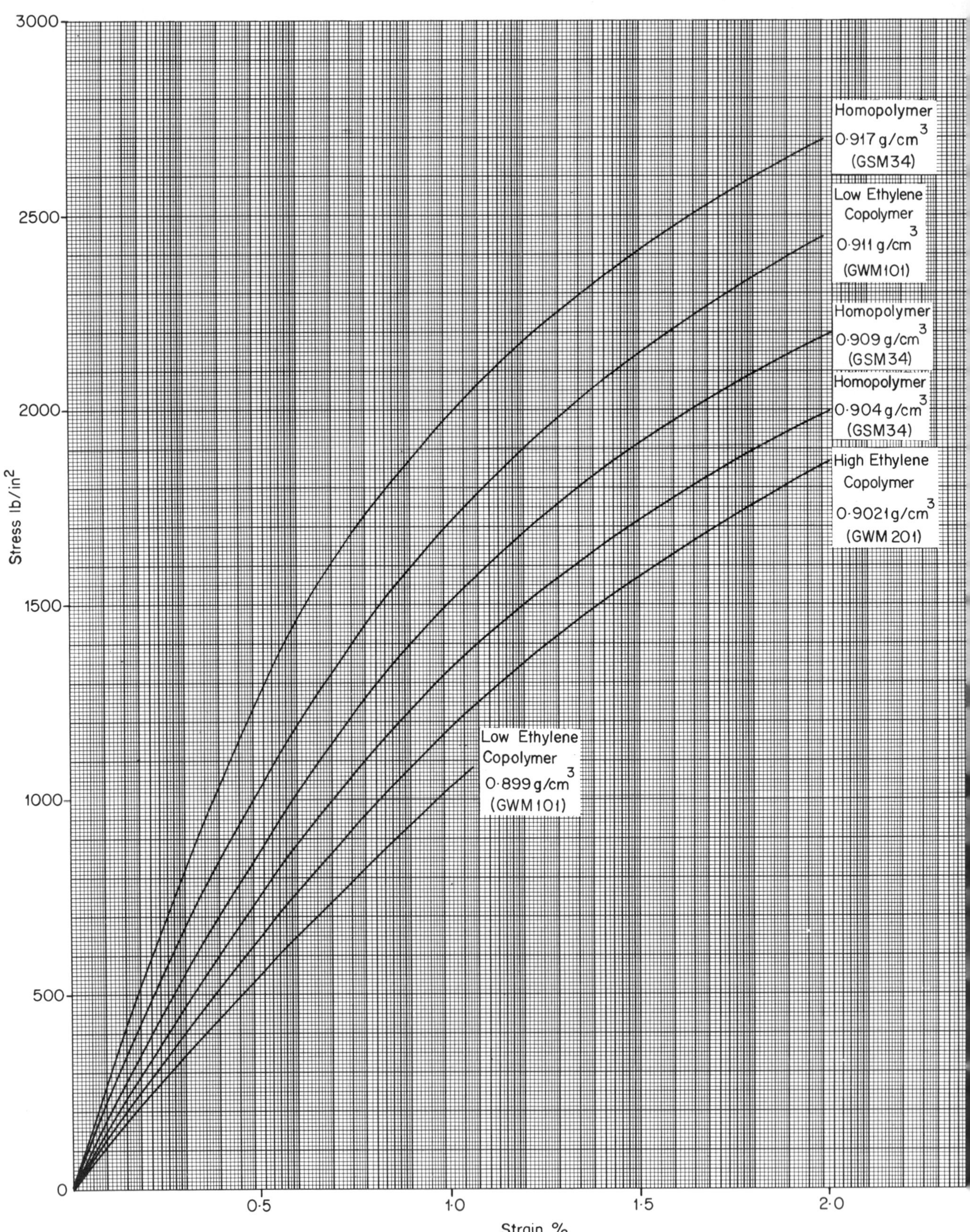


**Figure 7.3.** Isochronous stress vs strain curves: 20°C, 100 sec. Effect of variation in density. Propylene homo- and co-polymer ('Propathene' GSM 34, GWM 101, GWM 201)

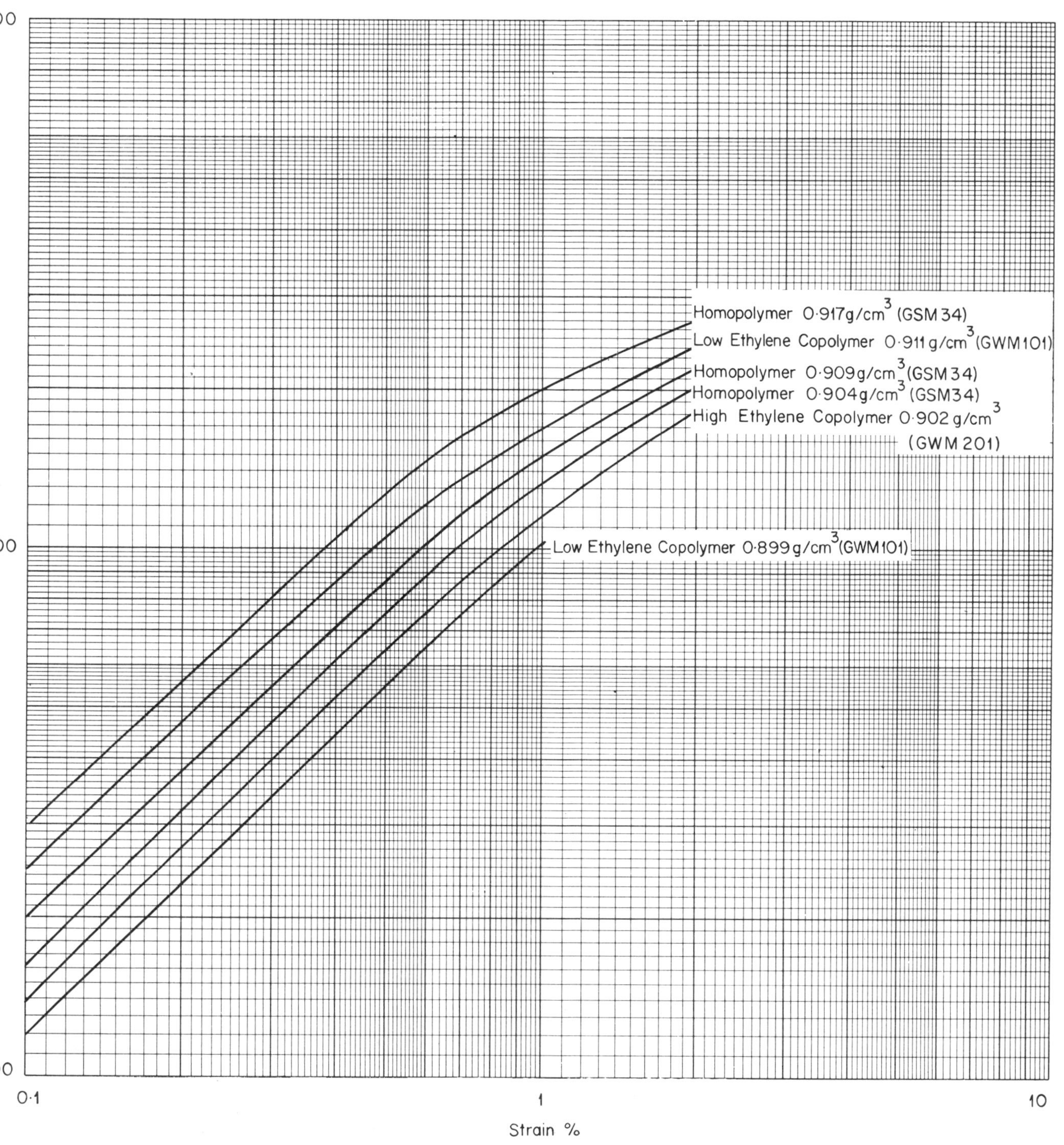


**Figure 7.4.** Isochronous stress vs strain curves: 20°C, 100 sec: logarithmic axes. Effect of variation in density. Propylene homo- and co-polymers ('Propathene' GSM 34, GWM 101, GWM 201)

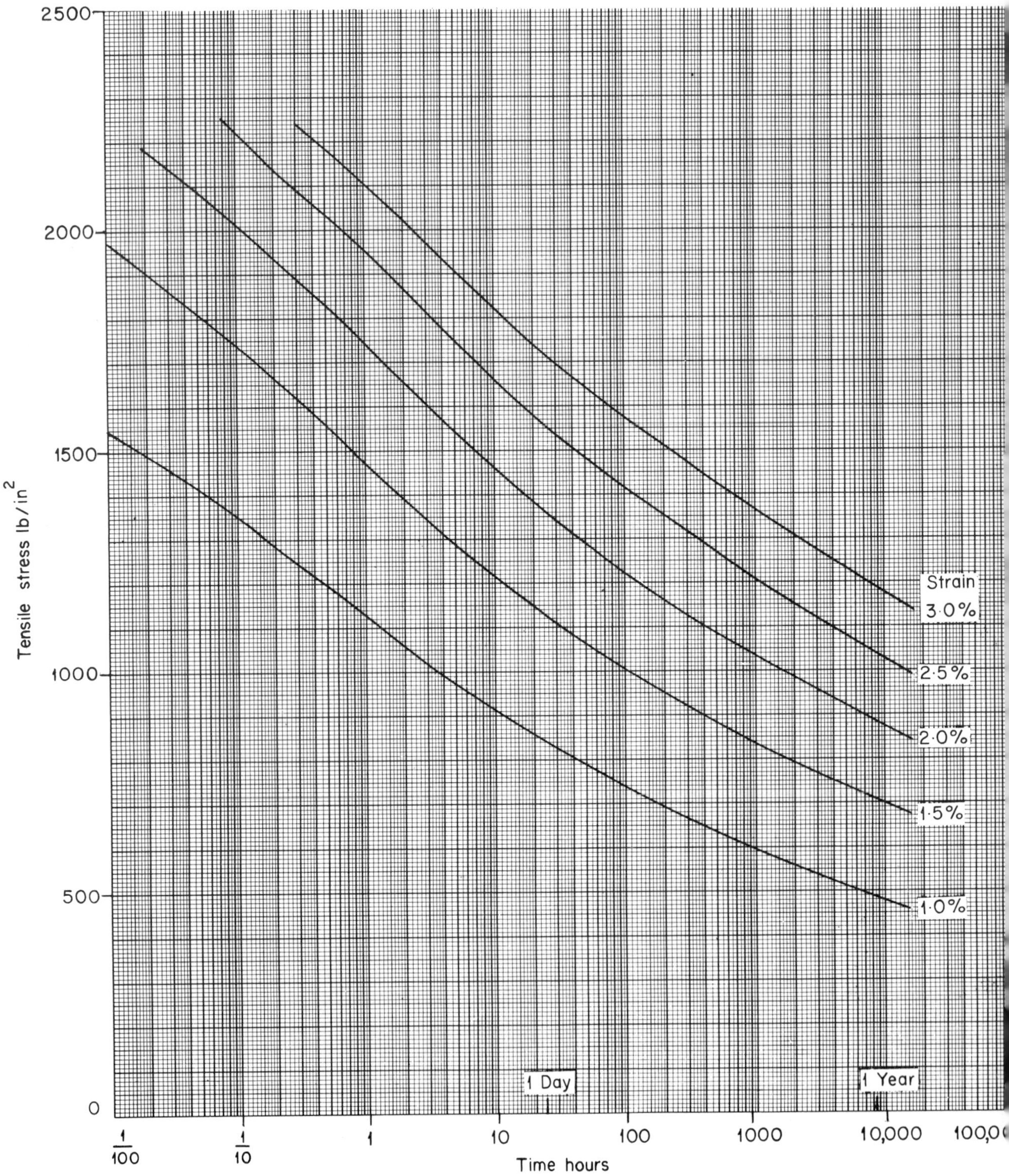


**Figure 7.5.** Isometric stress vs time curves: 20°C. Propylene homopolymer, density 0·909 g/cm$^3$ ('Propathene' GSM 34)

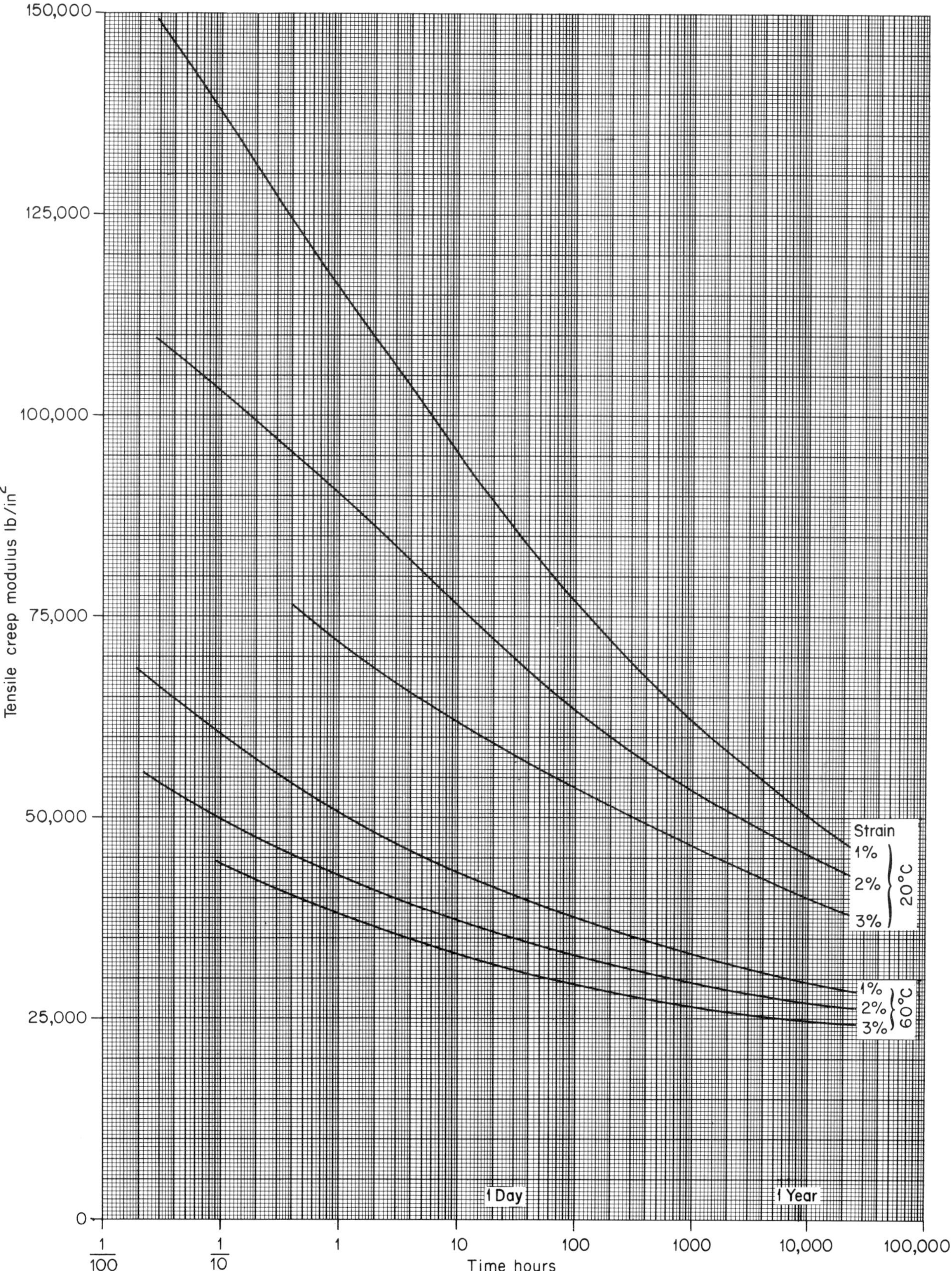


**Figure 7.6.** Tensile creep modulus vs time curves: 20°C and 60°C. Propylene homopolymer, density 0·909 g/cm$^3$ at 20°C ('Propathene' GSM 34)

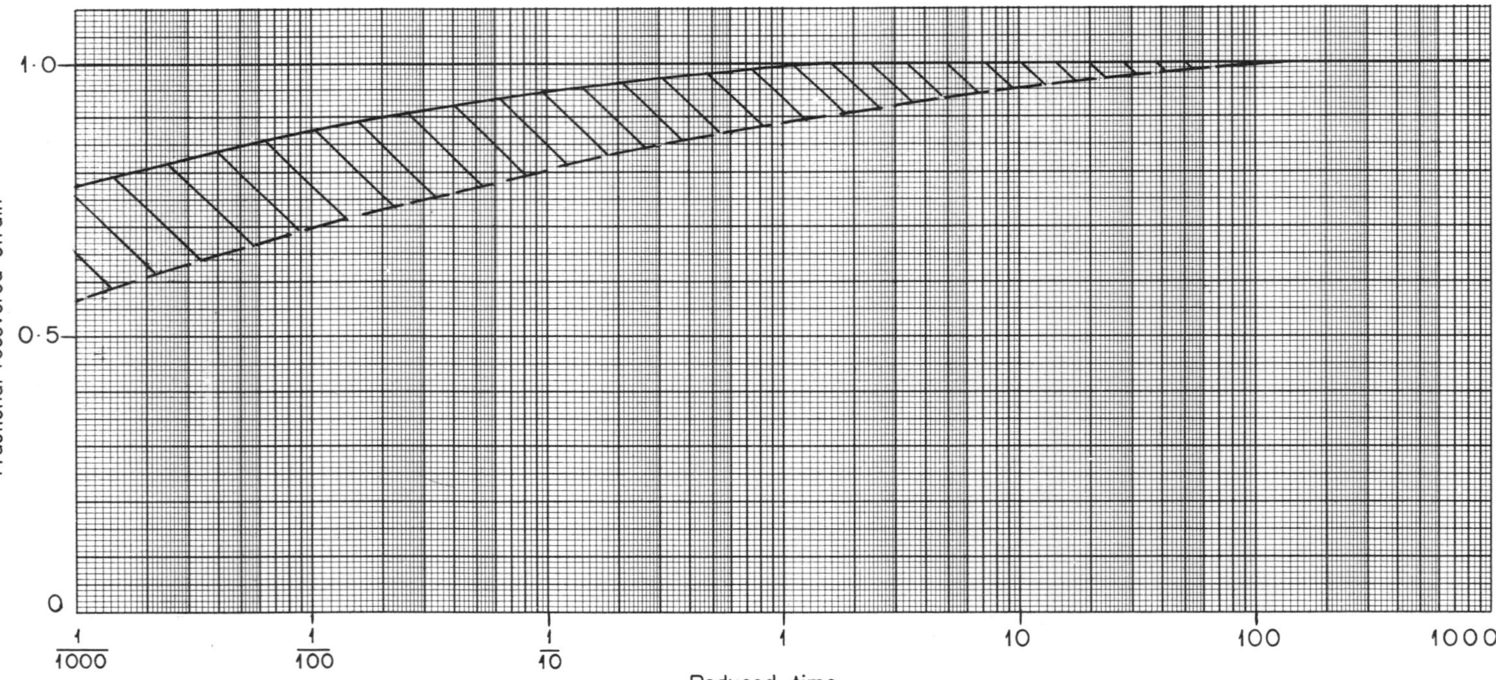


**Figure 7.7.** Recovery from creep in tension: 20°C. Propylene homopolymer, density 0·909 g/cm$^3$ ('Propathene' GSM 34)

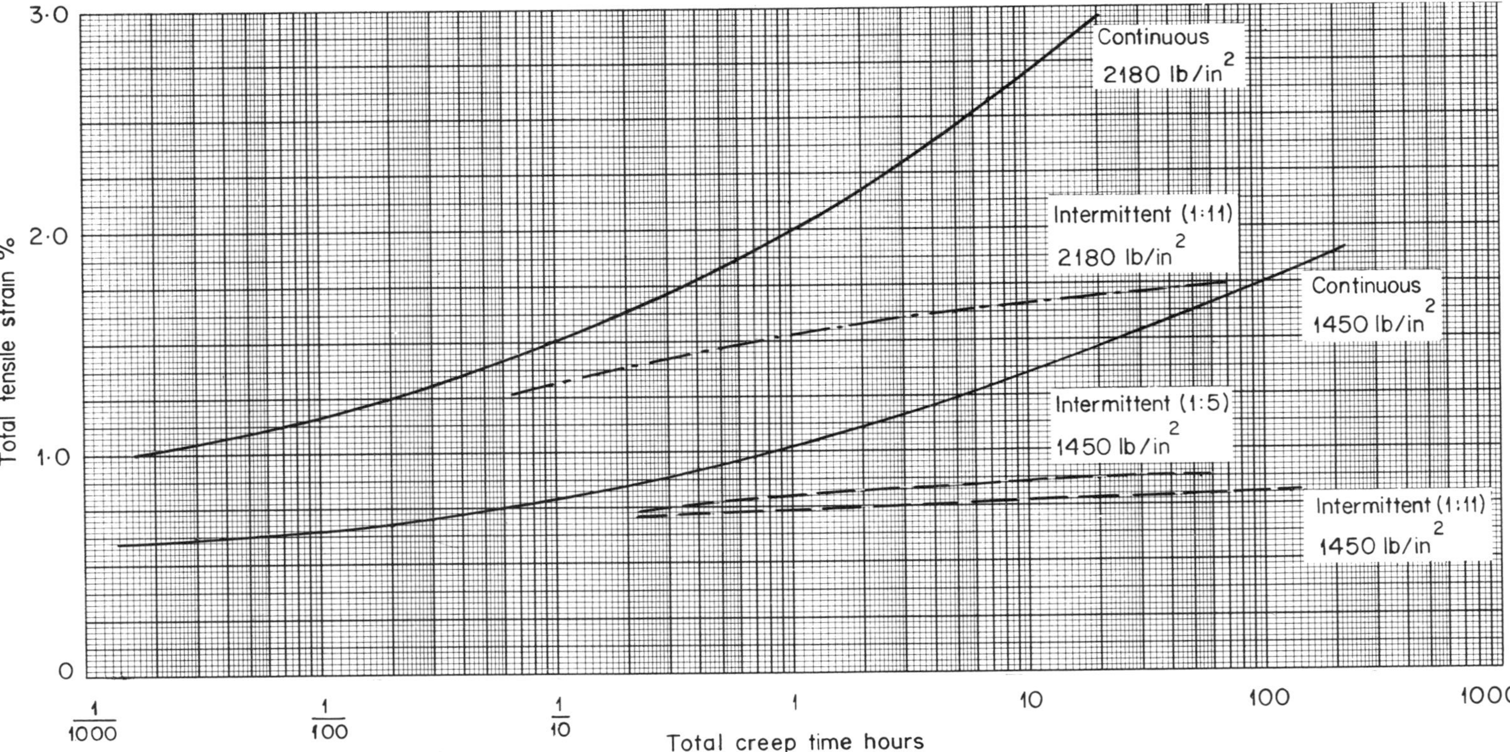


**Figure 7.8.** Creep curves in tension: 20°C. Effect of intermittent loading. Low ethylene copolymer, density 0·912 g/cm$^3$ ('Propathene' GWM 101)

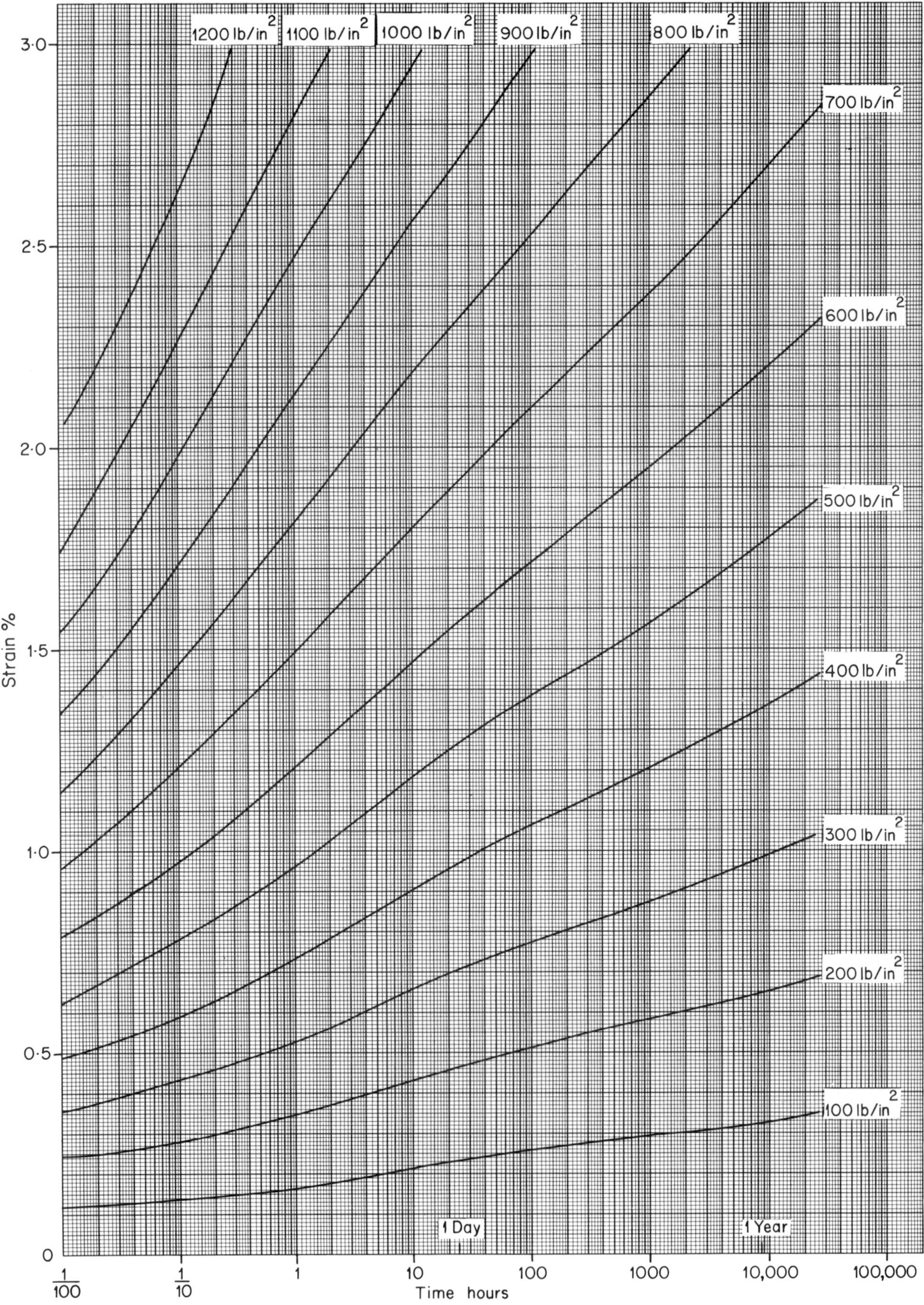


**Figure 7.9.** Creep curves in tension: 60°C. Propylene homopolymer, density 0·909 g/cm$^3$ at 20°C ('Propathene' GSM 34)

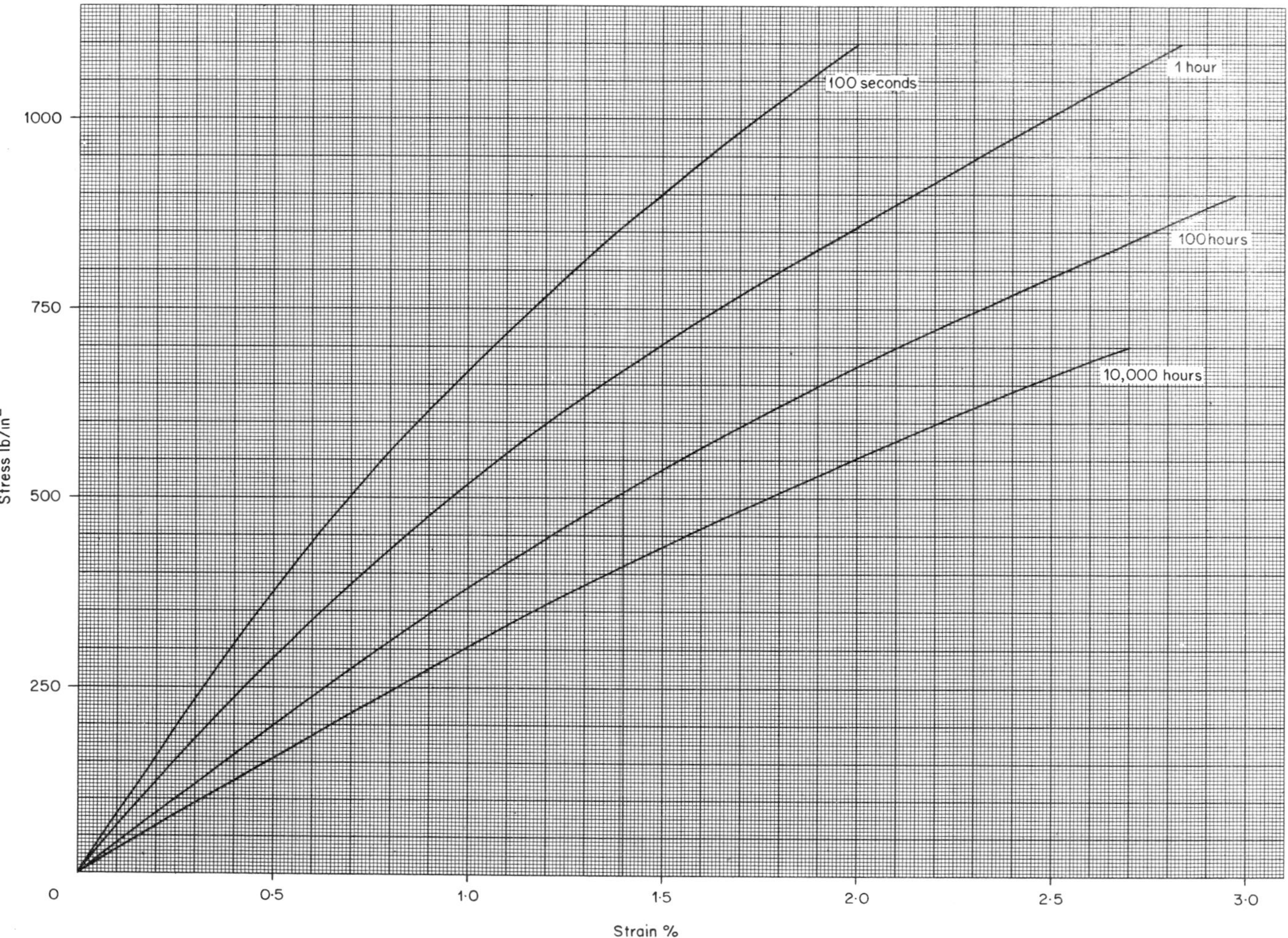


**Figure 7.10.** Isochronous stress vs strain curves: 60°C. Propylene homopolymer, density 0·909 g/cm$^3$ at 20°C ('Propathene' GSM 34)

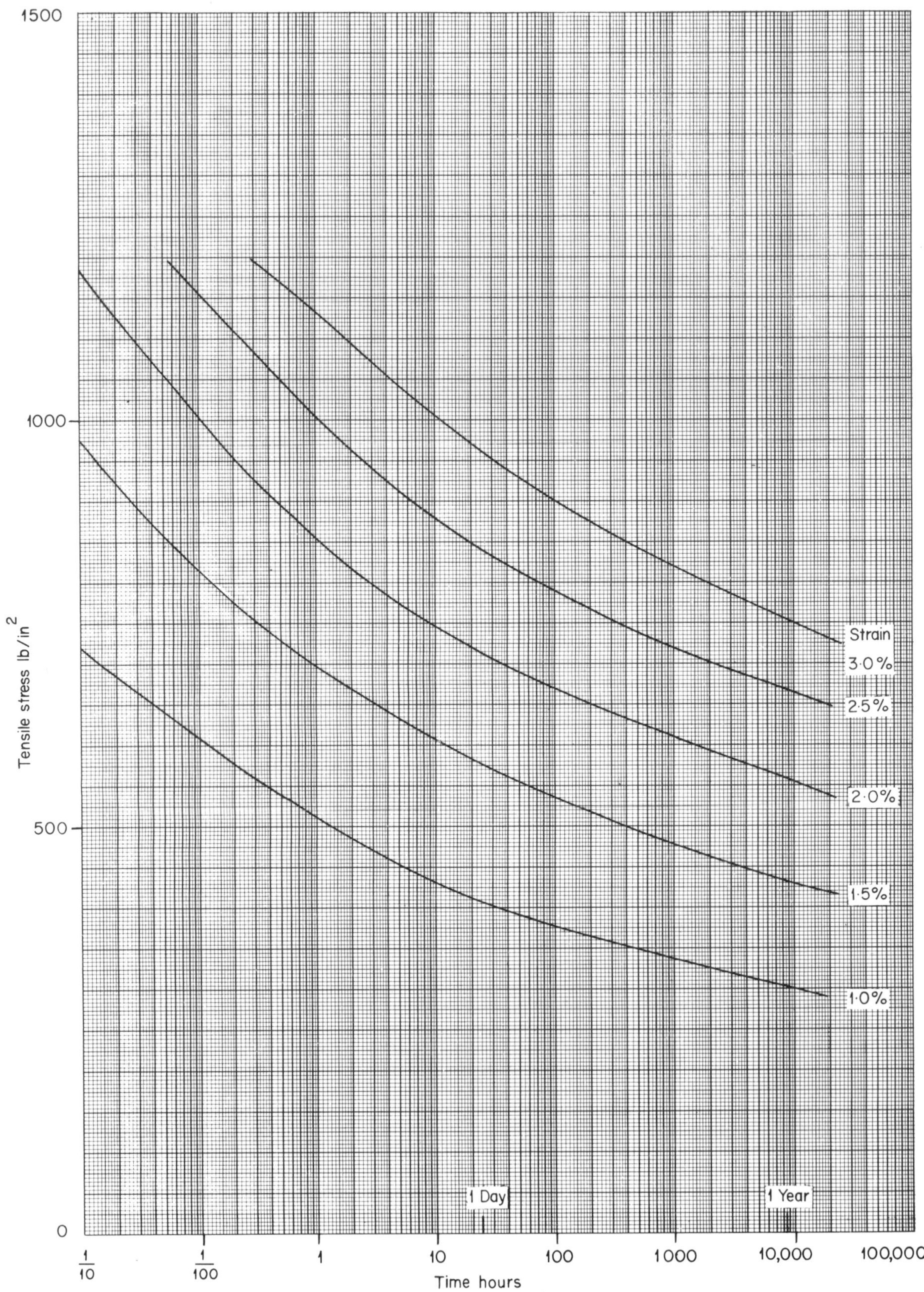


**Figure 7.11.** Isometric stress vs time curves: 60°C. Propylene homopolymer, density 0·909 g/cm$^3$ at 20°C ('Propathene' GSM 34)

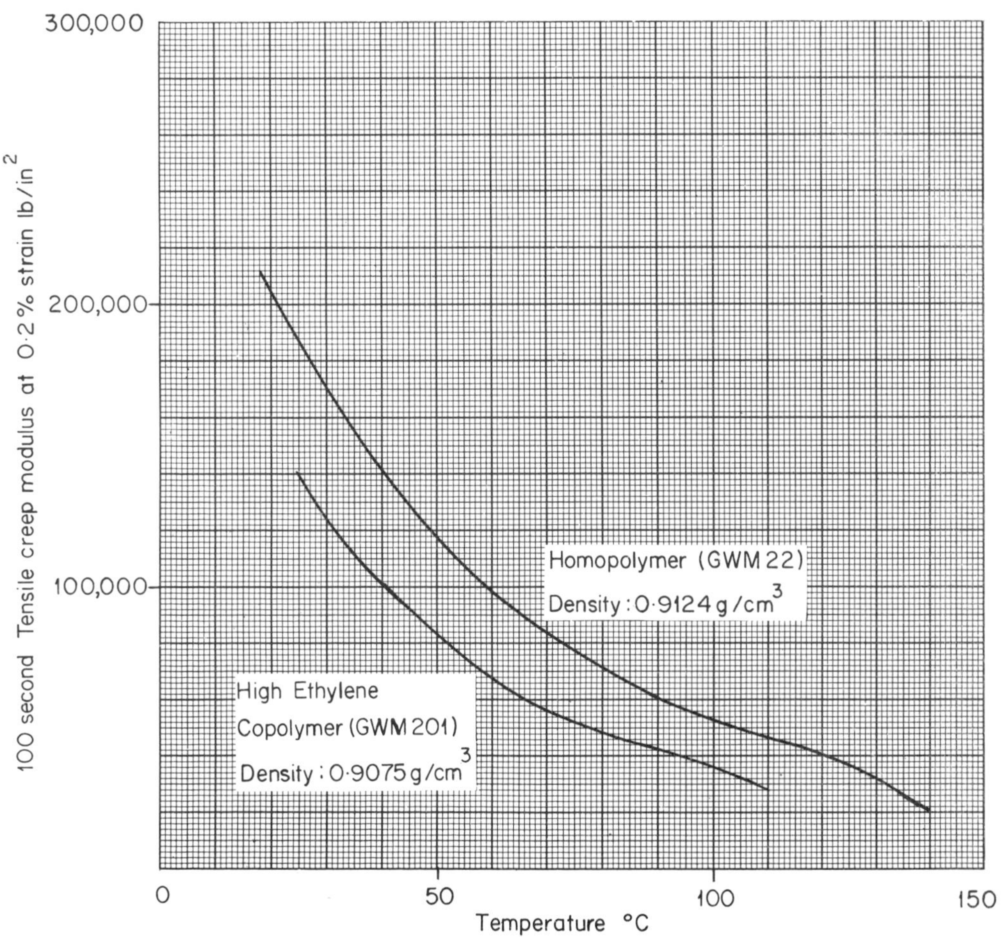


**Figure 7.12.** Tensile creep modulus (100 sec, 0·2% strain) vs temperature. Propylene homo- and high ethylene copolymer ('Propathene' GWM 22, GWM 201)

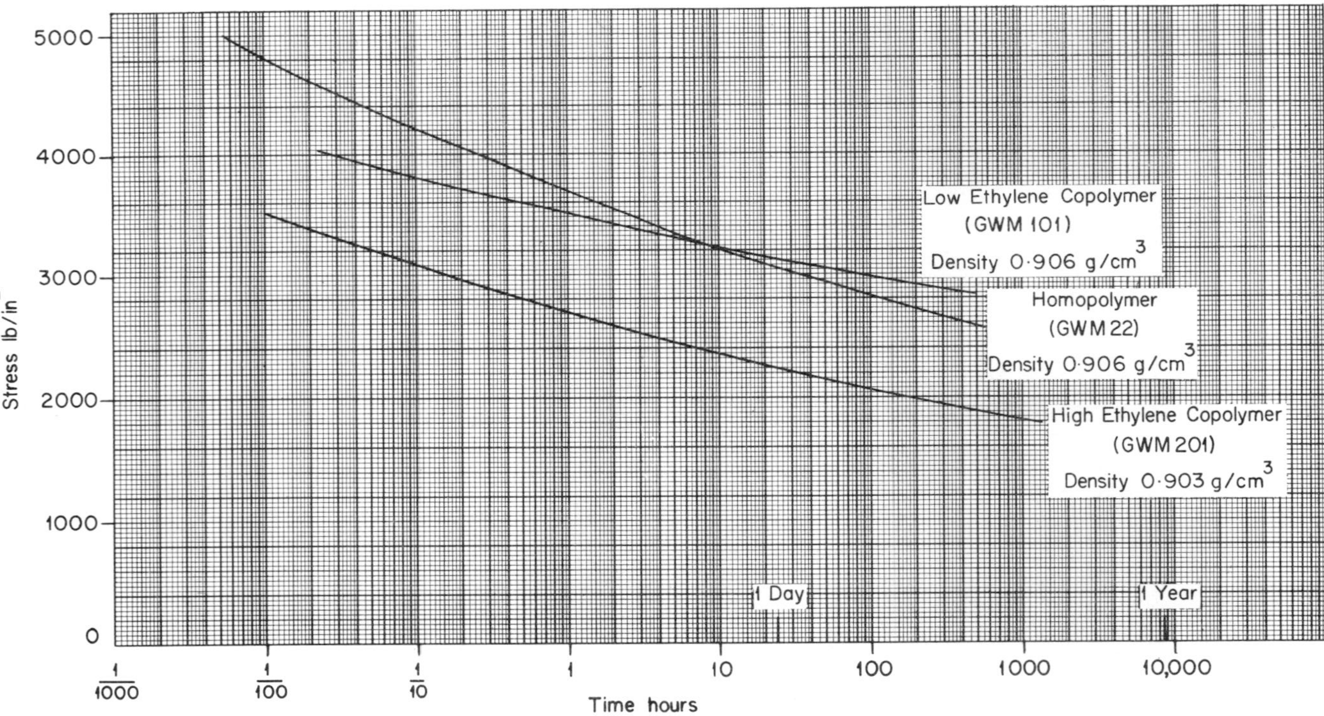


**Figure 7.13.** Creep rupture stress in tension vs time to failure: 20°C. Propylene homo- and co-polymers ('Propathene' GWM 22, GWM 101, GWM 201)

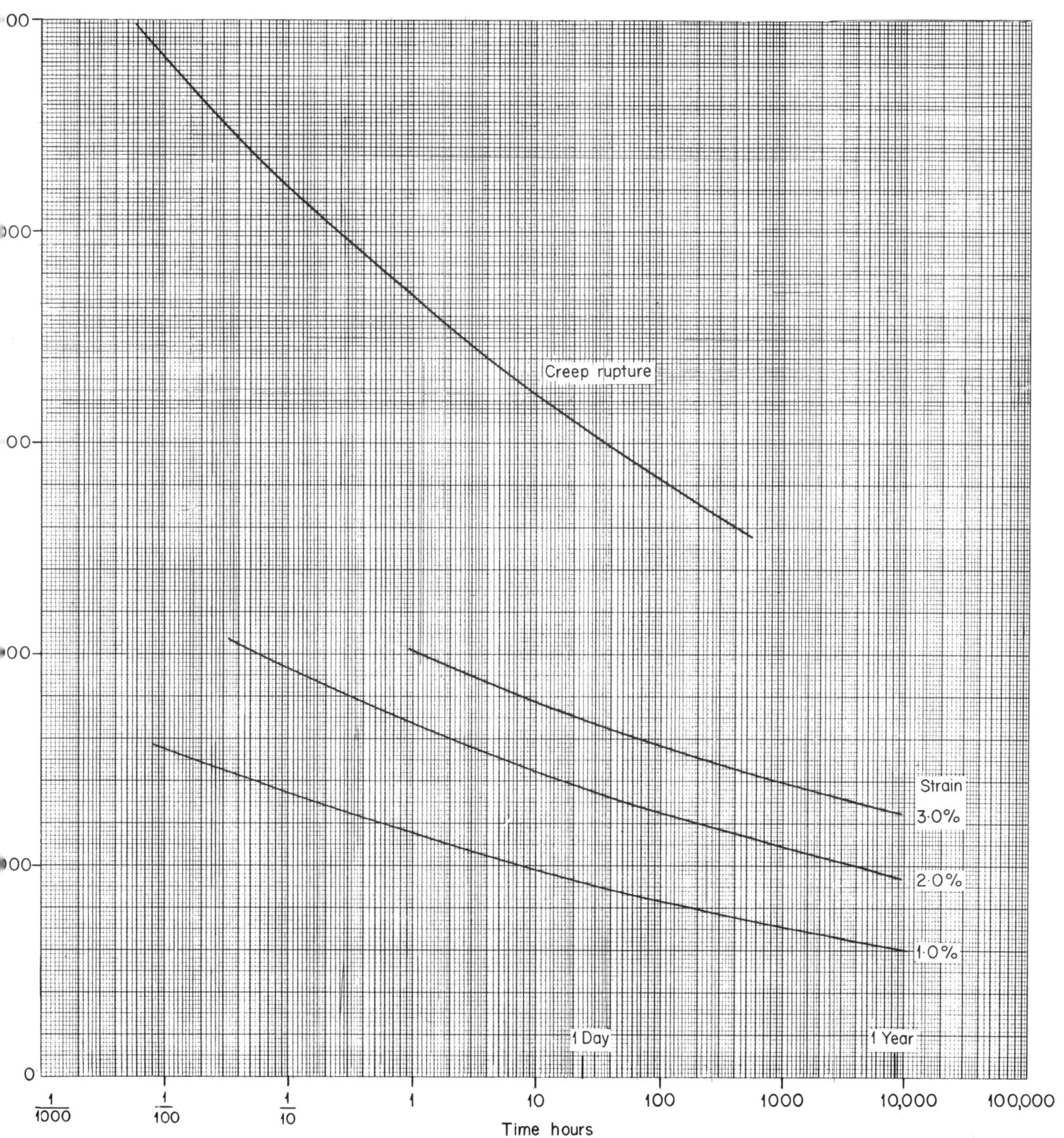


**Figure 7.14.** Comparison of creep rupture curve with isometric stress vs time curves: 20°C. Propylene homopolymer: density 0·906 g/cm³ ('Propathene' GWM 22)

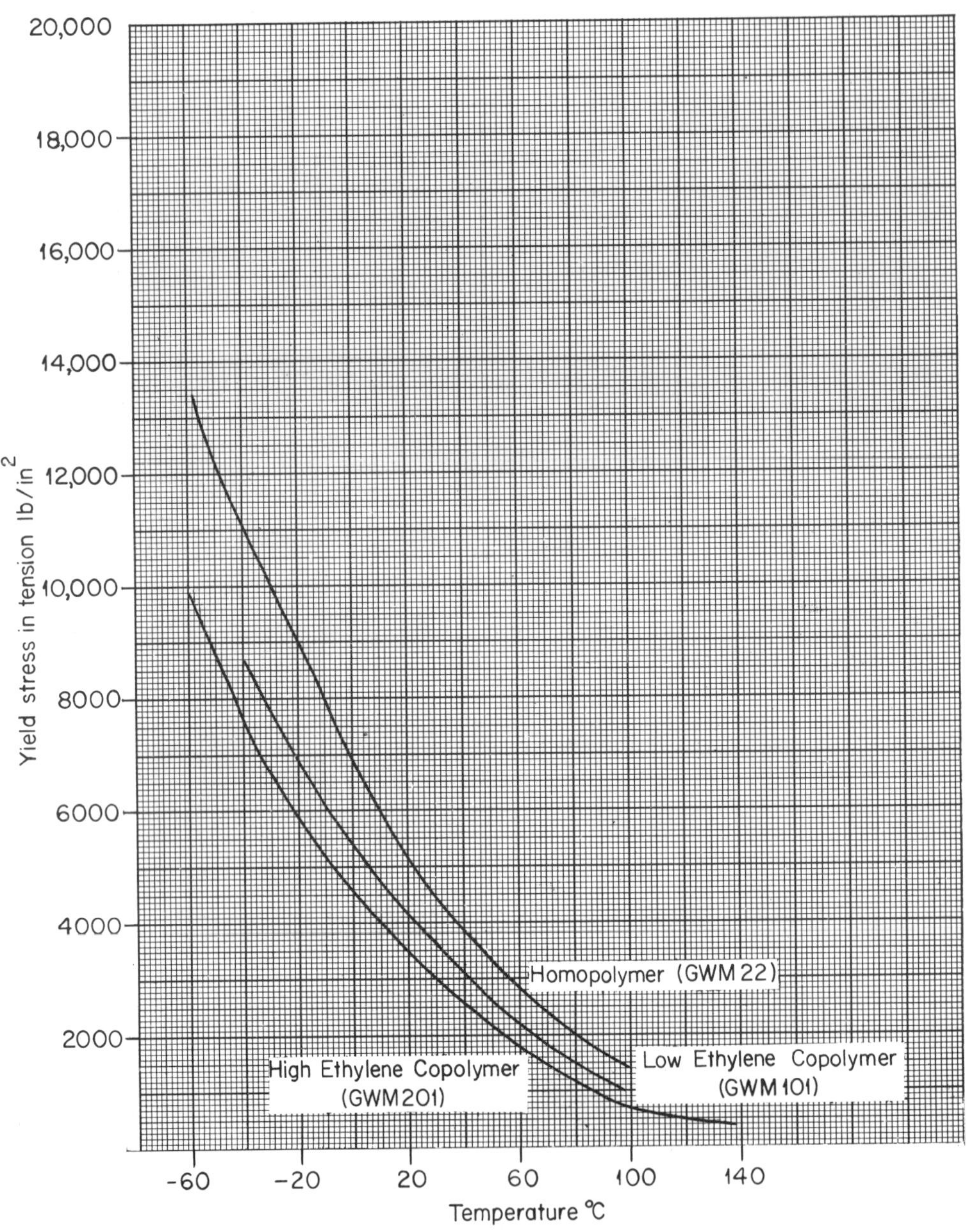


**Figure 7.15.** Yield stress in tension vs temperature: 50% per min straining rate. Propylene homo- and co-polymers ('Propathene' GWM 22, GWM 101, GWM 201)

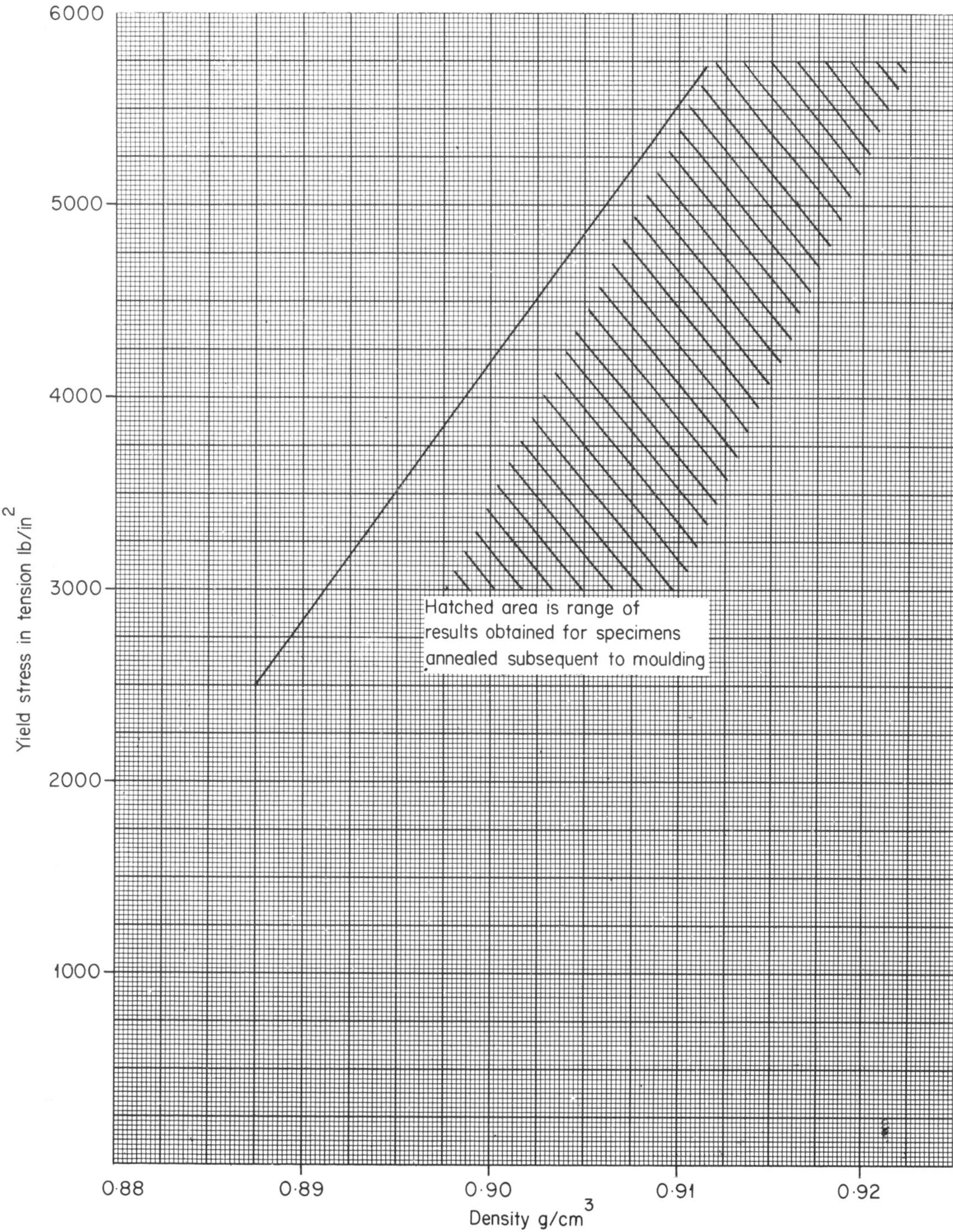


**Figure 7.16.** Yield stress in tension vs density: 50% per min straining rate. Propylene homo- and co-polymers. (Various 'Propathene' grades)

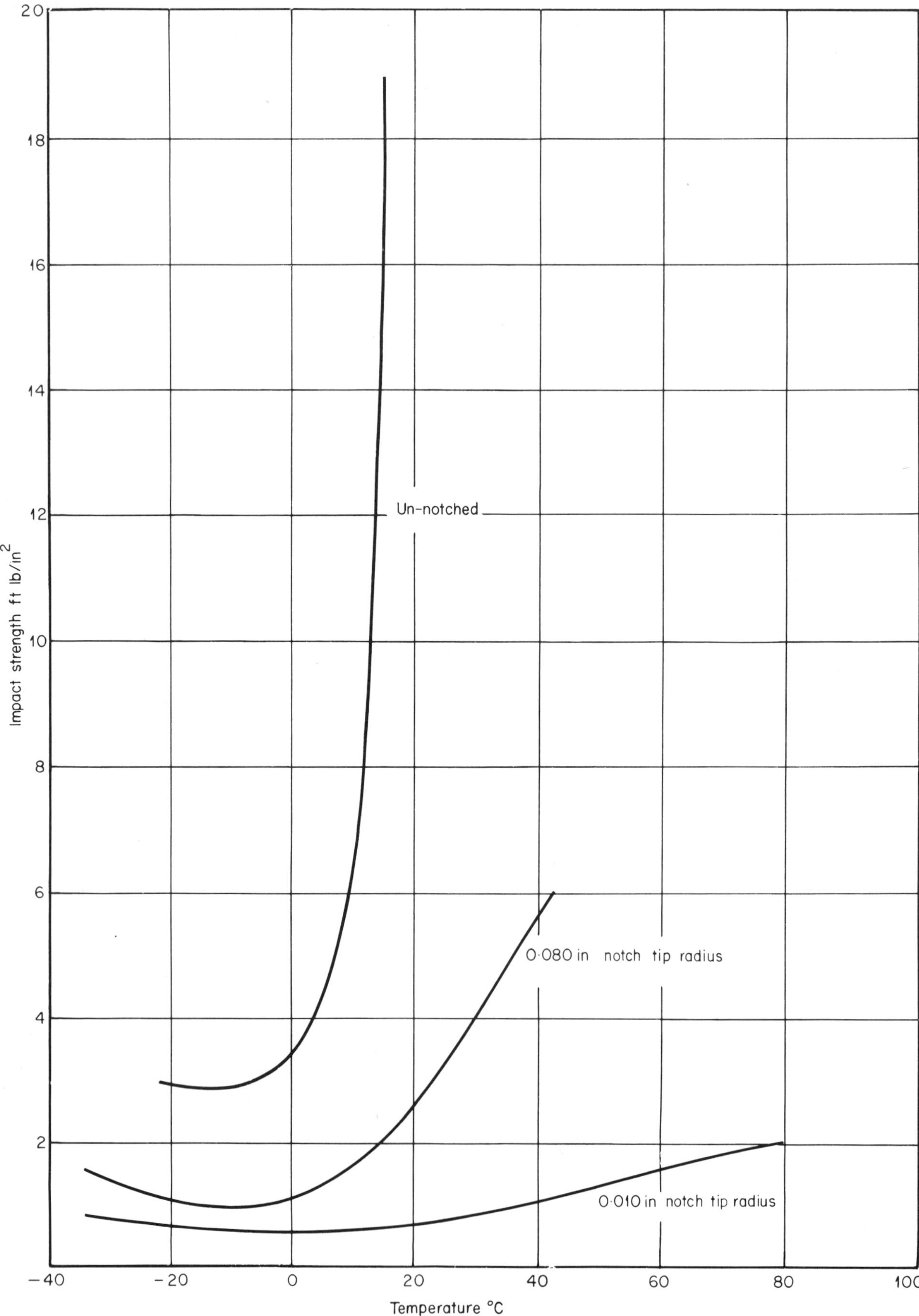


**Figure 7.17.** Impact strength vs temperature. Propylene homopolymer ('Propathene' GWM 22)

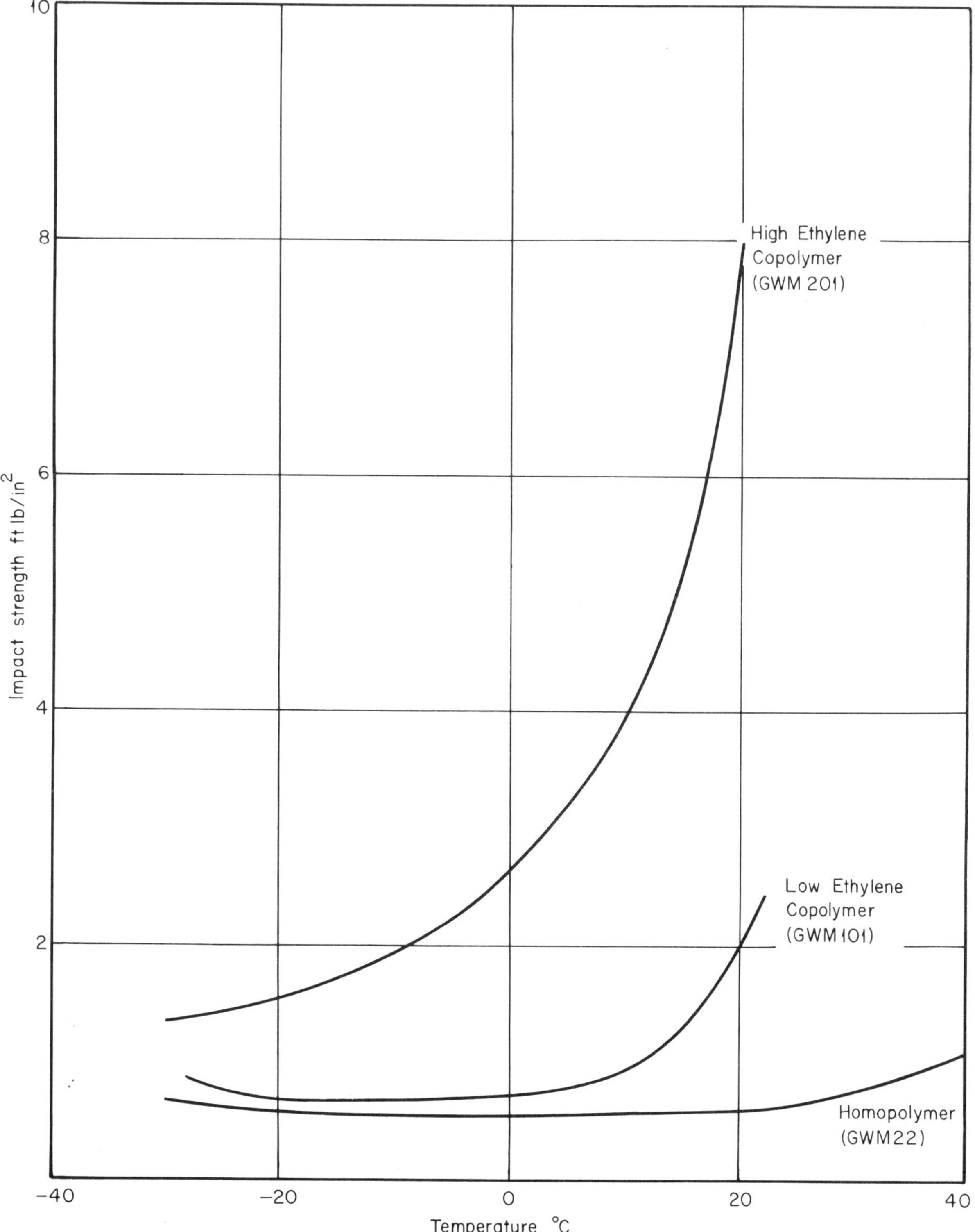


**Figure 7.18.** Impact strength vs temperature: 0·010 in notch tip radius. Comparison of propylene homo- and copolymer mouldings ('Propathene' GWM 22, GWM 101, GWM 201)

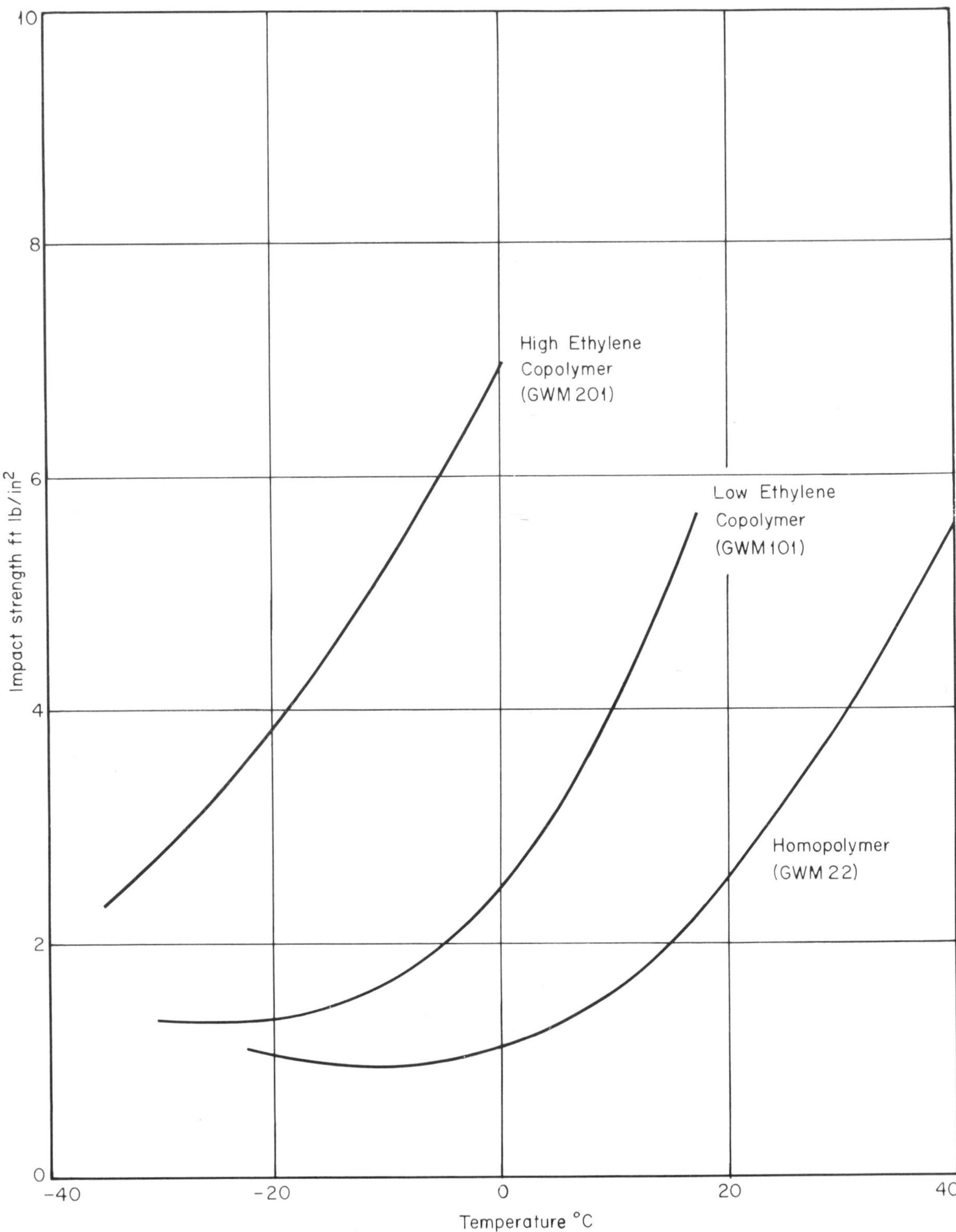


**Figure 7.19.** Impact strength vs temperature: 0·080 in notch tip radius. Comparison of propylene homo- and copolymer mouldings ('Propathene' GWM 22, GWM 101, GWM 201)

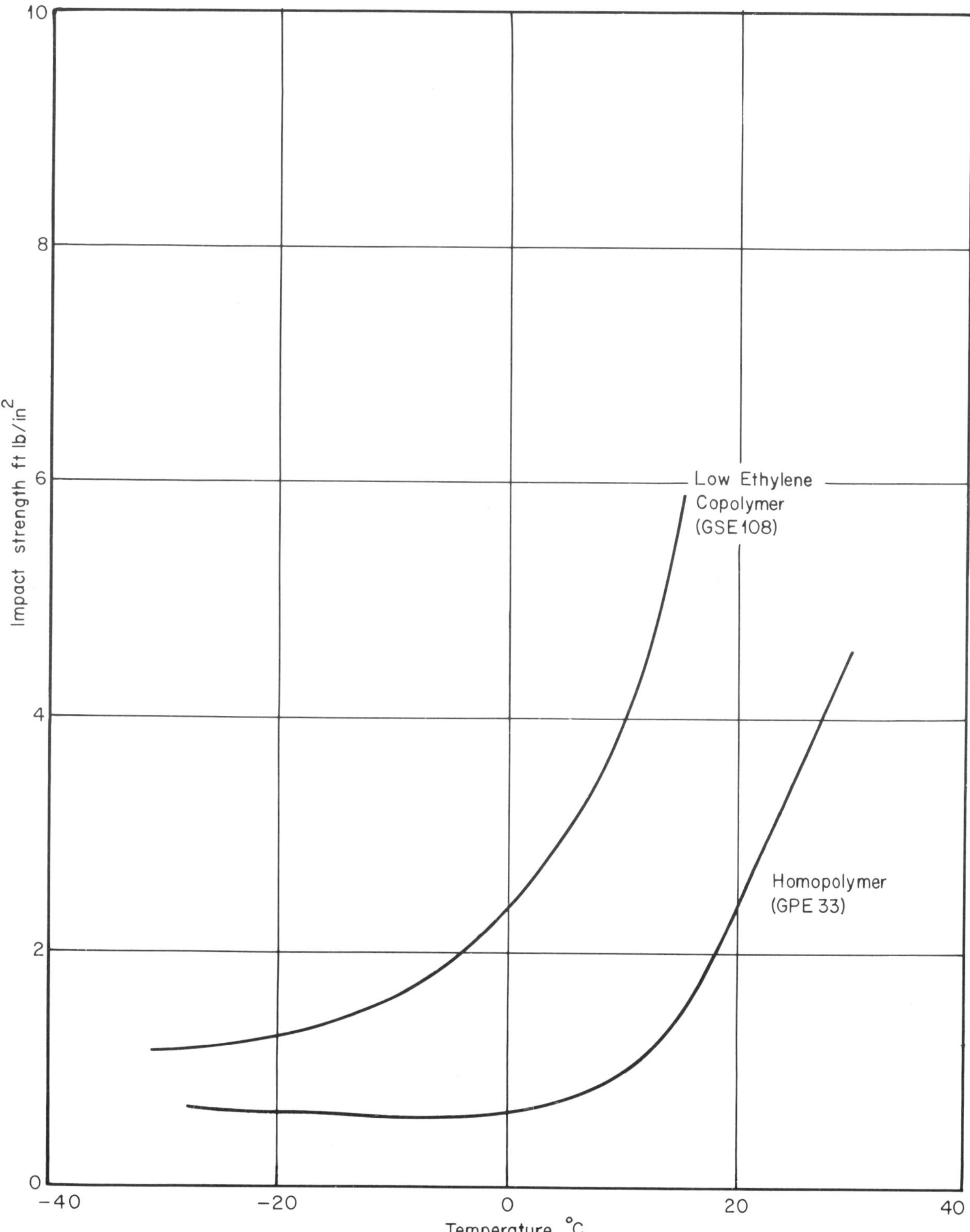


**Figure 7.20.** Impact strength vs temperature: 0·010 in notch tip radius. Comparison of propylene homo- and low ethylene co-polymer extruded sheets ('Propathene' GPE 33, GSE 108)

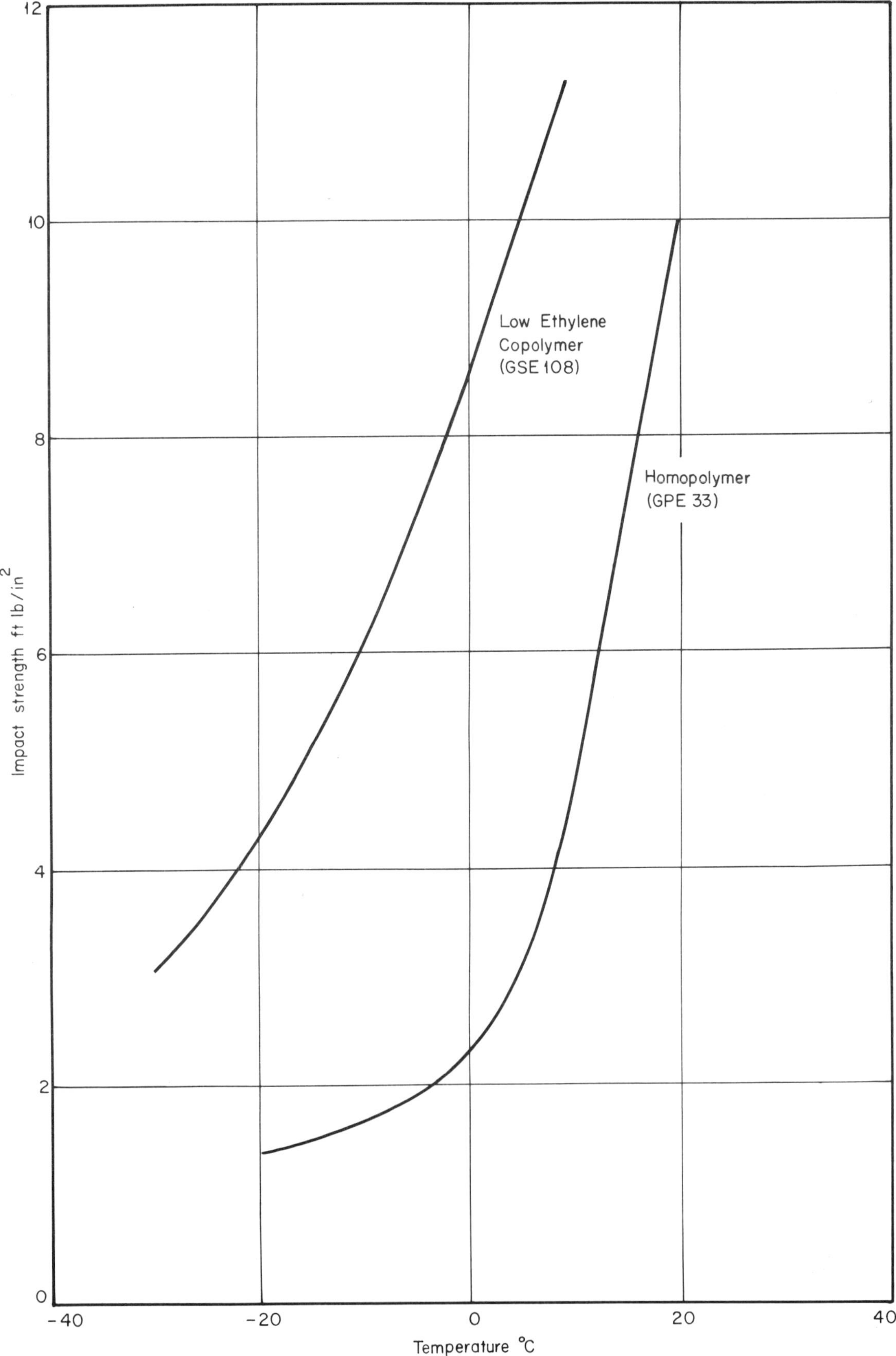


**Figure 7.21.** Impact strength vs temperature: 0·080 in notch tip radius. Comparison of propylene homo- and low ethylene co-polymer extruded sheets ('Propathene' GPE 33, GSE 108)

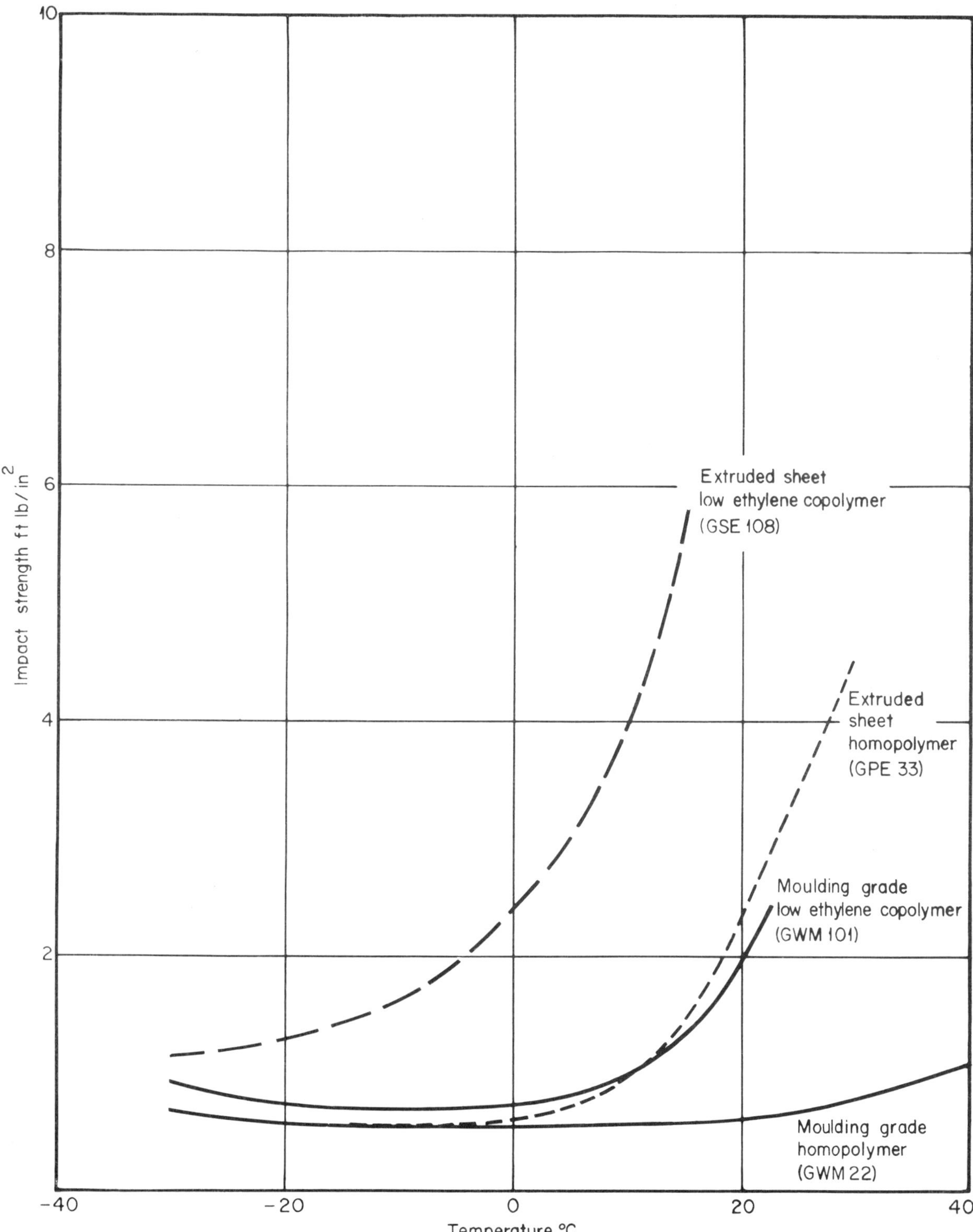


**Figure 7.22.** Impact strength vs temperature: 0·010 in notch tip radius. Comparison of propylene homo- and low ethylene co-polymer mouldings and extruded sheets ('Propathene' GWM 22, GWM 101, GPE 33, GSE 108)

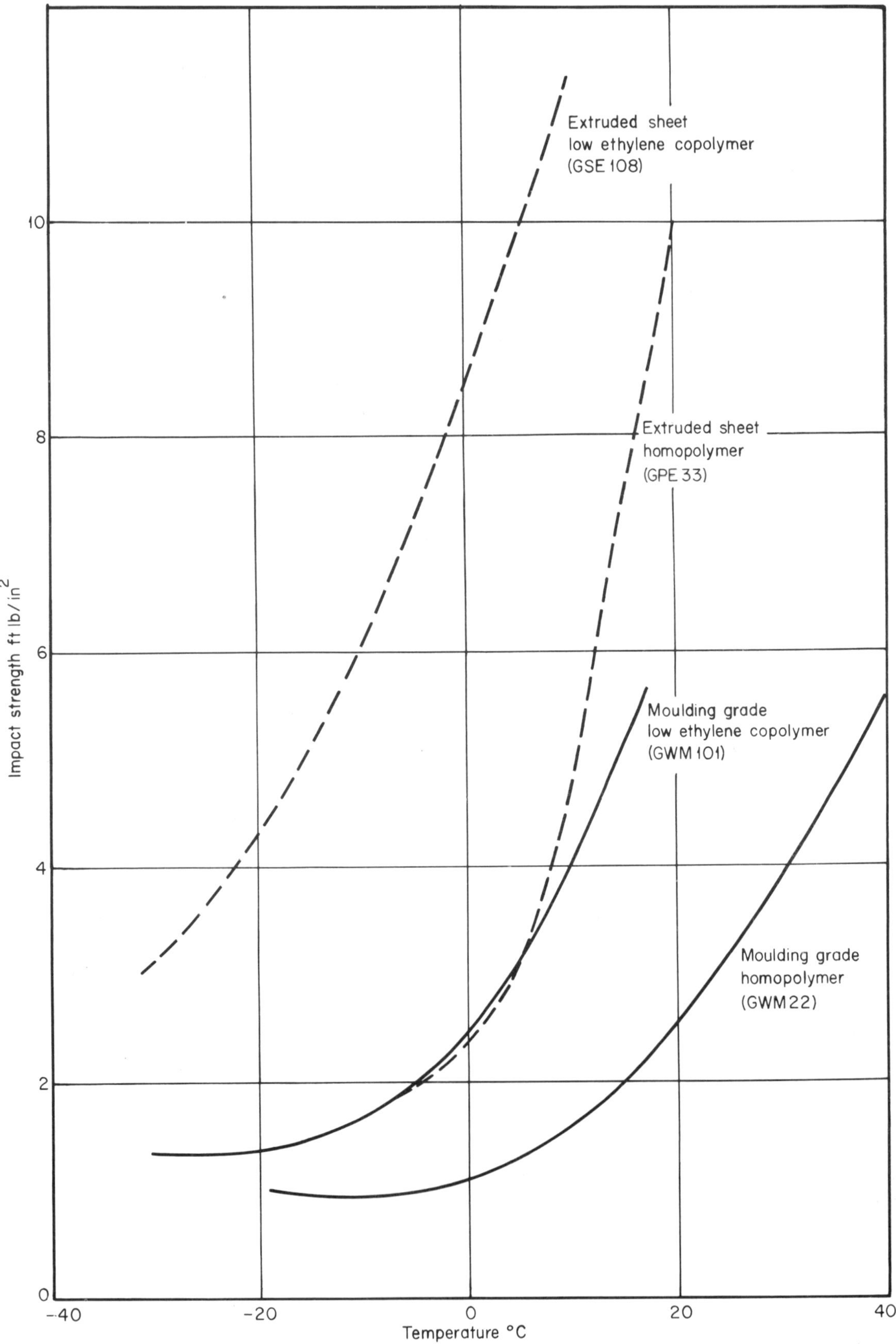


**Figure 7.23.** Impact strength vs temperature: 0·080 in notch tip radius. Comparison of propylene homo- and low ethylene co-polymer mouldings and extruded sheets ('Propathene' GWM 22, GWM 101, GPE 33, GSE 108)

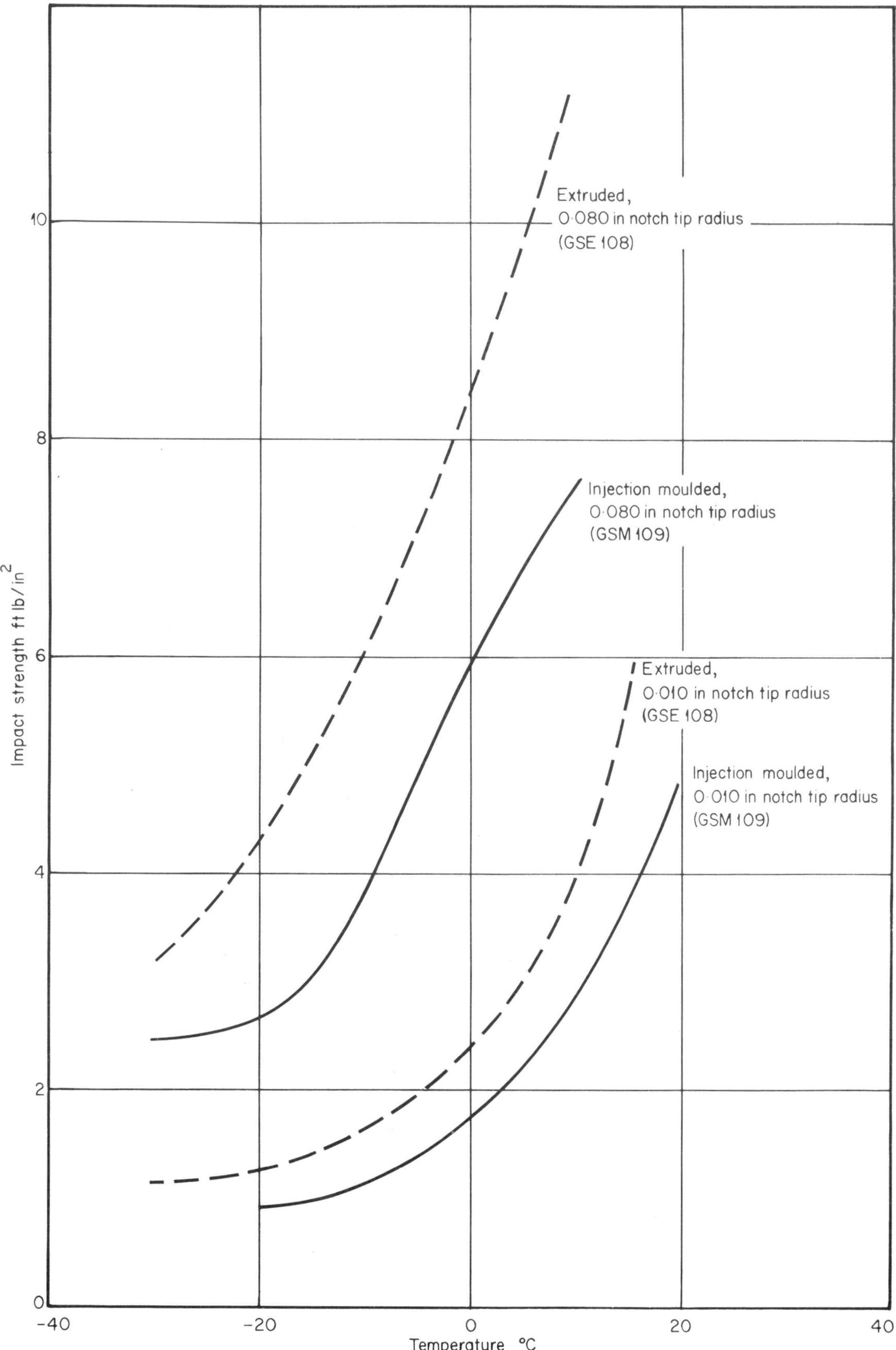


**Figure 7.24.** Impact strength vs temperature. Comparison of injection moulding vs extrusion fabrication. Propylene low ethylene co-polymer ('Propathene' GSM 109, GSE 108)

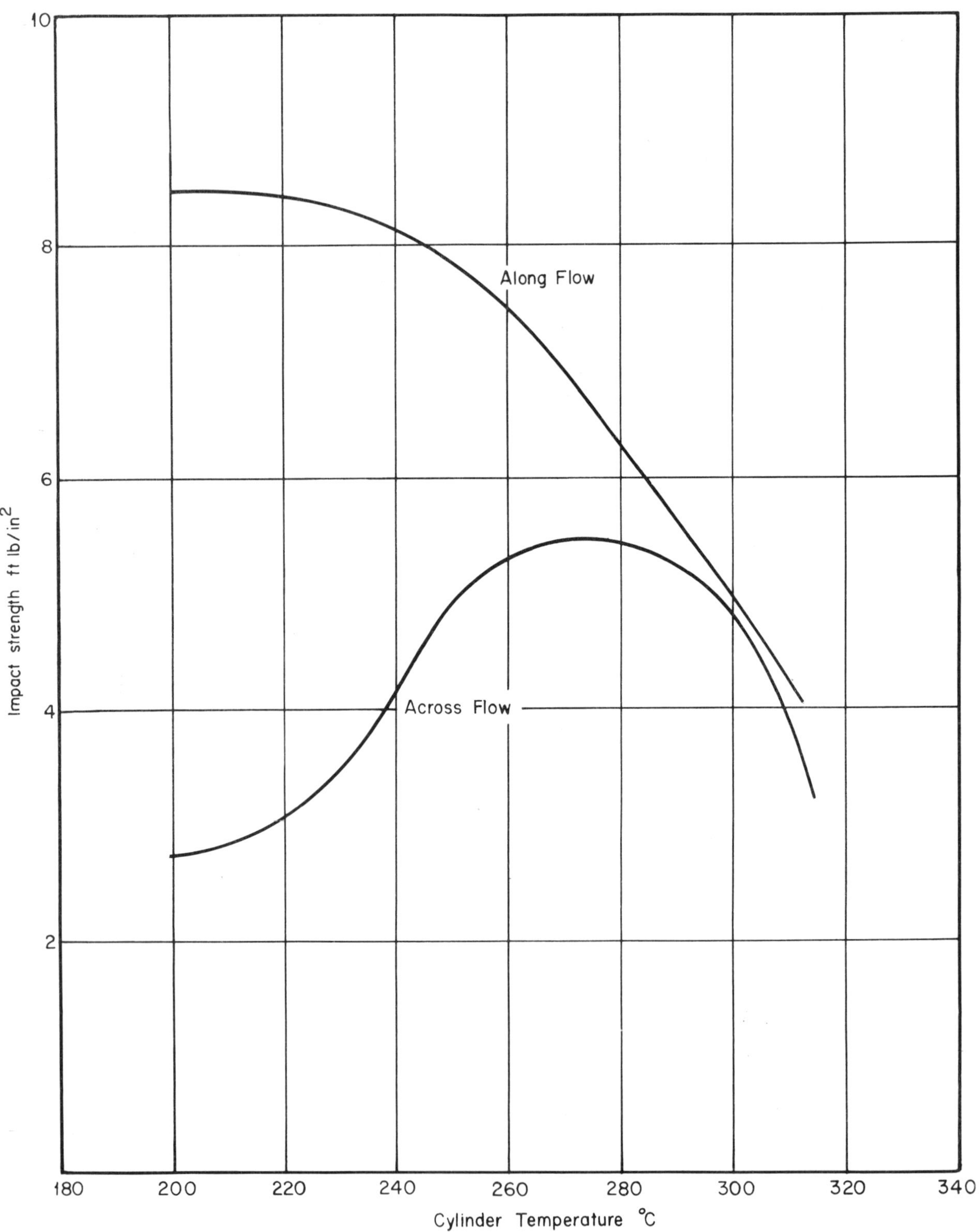


**Figure 7.25.** Impact strength vs injection moulding cylinder temperature: 20°C, 0·080 in notch tip radius. Propylene homopolymer, tested 4 months after moulding ('Propathene' GWM 22)

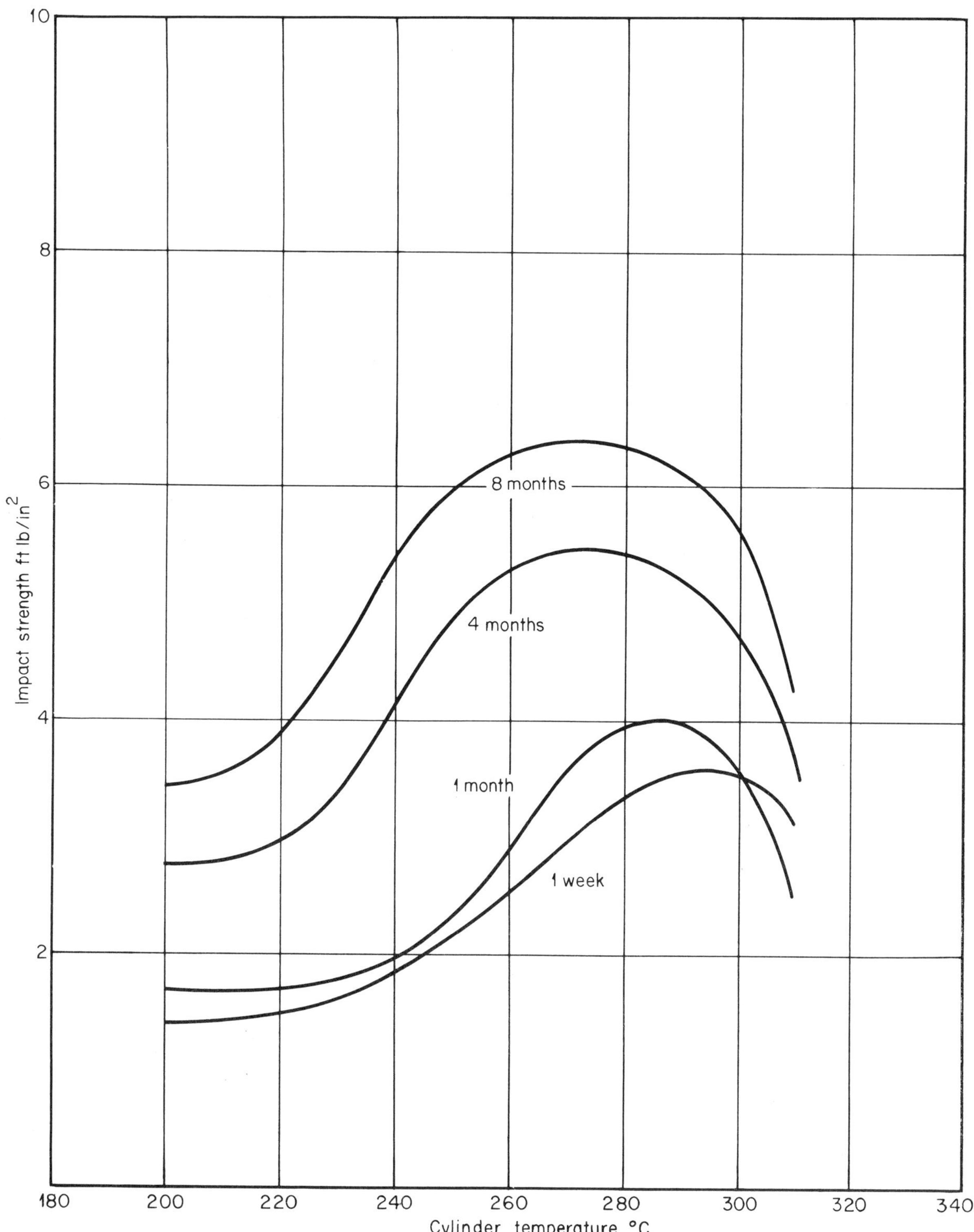


**Figure 7.26.** Impact strength vs injection moulding cylinder temperature: 20°C, 0·080 in notch tip radius. Effect of time of storage at 20°C on impact strength. Propylene homopolymer ('Propathene' GWM 22)

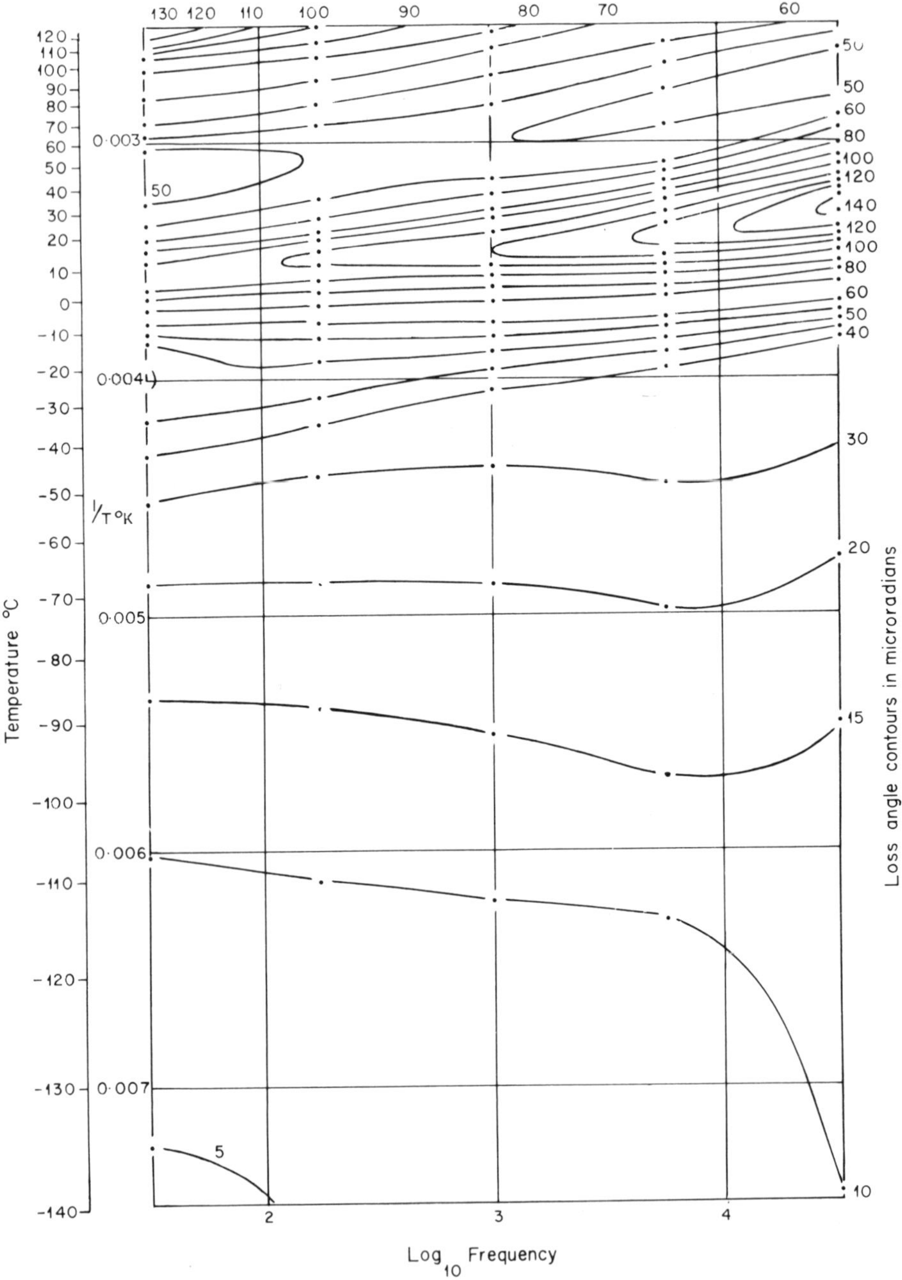


**Figure 7.27.** Loss angle vs frequency and temperature. Propylene homopolymer, dielectric grade

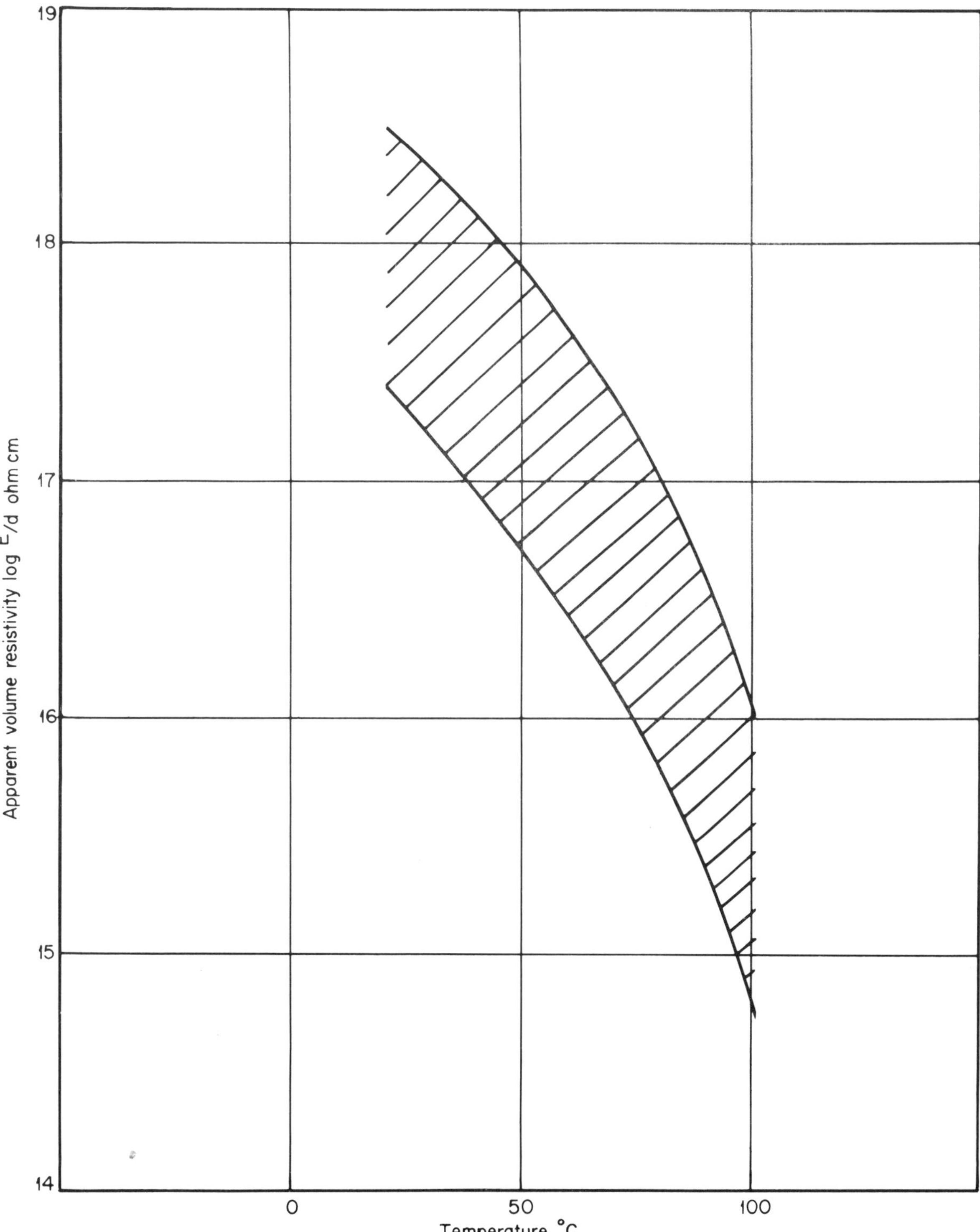


**Figure 7.28.** Apparent volume resistivity vs temperature. Propylene homo- and co-polymers. (Various 'Propathene' grades)

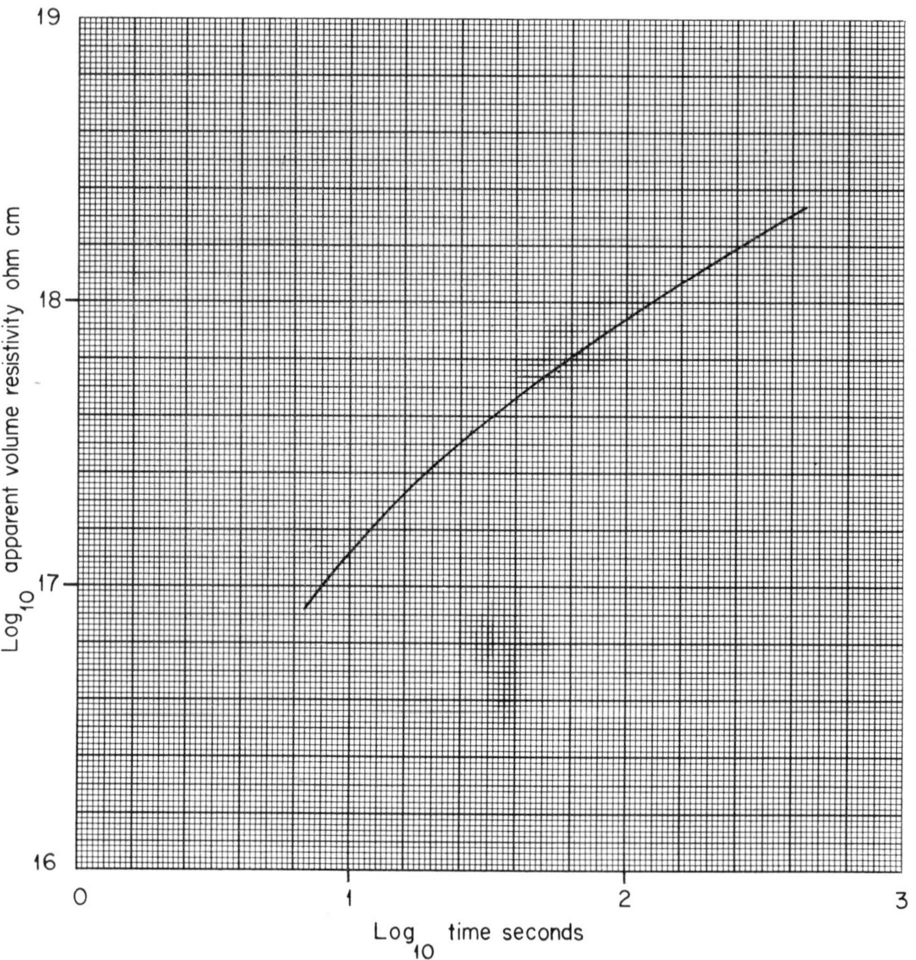


**Figure 7.29.** Apparent volume resistivity vs time of electrification: 20°C. Propylene homopolymer ('Propathene' HWE 24)

# 8

# LOW DENSITY POLYTHENE

## INTRODUCTION

Polythene was discovered accidentally in 1933 during investigations at the Alkali Division of Imperial Chemical Industries Limited on the effects of high pressures on chemical reactions. After reaction of a mixture of ethylene and benzaldehyde, the benzaldehyde was recovered unchanged, and a white, waxy solid was found coating the walls; this solid contained no oxygen and was recognized as a polymer of ethylene. Polythenes produced under these conditions are branched materials of low density. More linear polythenes of higher density can be made at near normal temperatures and pressures by using Ziegler catalysts.

Ethylene can be polymerized to any desired molecular weight, giving, as the molecular weight increases, oils, greases, soft or hard waxes, and finally the tough flexible materials now generally associated with the term polythene. Polythene combines flexibility and toughness with outstanding electrical insulating properties, insensitivity to moisture, and good chemical resistance. Low volume cost and ease of fabrication are added advantages.

Ethylene can be copolymerized with other monomers to give a variety of products. Of these, the copolymers of ethylene and vinyl acetate (EVA copolymers) have, compared with the homopolymers of ethylene, increased flexibility, transparency, gas permeability and low temperature toughness.

## NATURE OF POLYTHENE

The repeating unit of the long-chain molecules of polythene is ($—CH_2—$). The molecules contain also olefinic and other groups, but these amount to less than 0·1 % of the polymer, and there appears to be less than one olefinic bond per molecule. Branch points occur at intervals of every 25–100 chain carbon atoms, the frequency of branching depending on polymerization conditions. Most of the branches are 2 or 4 carbon atoms long, though some may be much longer.

The similarity in structure of the individual molecules makes possible the close packing of parts of the chains, thus giving ordered or crystalline regions (crystallites) which can associate to form spherulites. Because of chain entanglement and the presence of branches, low density polythene contains unordered amorphous regions. The greater the number of branches the smaller the degree of crystallinity. In general, the crystallinity of low density polythene lies in the range 55–70 %.

## GRADES AND FORMS

Polythene is available in many grades, each one specially designed for a particular process or application. These grades differ basically in crystallinity and molecular weight (see section on Properties). Polythene compounds, hot-homogenized mixtures of polymers with additives such as pigments and antioxidants, are produced to fulfil specific requirements, and additive masterbatches can be tumble-blended with the natural polymers before processing to produce effects similar to those obtained from compounds. Polymers and compounds are supplied in granular form, but some grades are available as powders.

Polythene is available also in semifabricated forms such as film, sheet, block, rod and tube.

## PROCESSING

Polythene is easily processed by all the usual techniques, but injection moulding, extrusion and blow moulding are the most important. It is very easy to mould on modern injection moulding equipment, because it has a wide moulding range, excellent flow characteristics and good heat stability and is insensitive to moisture. It can readily be extruded into flat and tubular film, rod, pipe, wire and cable coverings, sheet and tubes for subsequent blowing into bottles and other hollow articles.

By the process of extrusion coating, polythene film can be extruded directly on to one or both sides of other films, paper, hessian or jute to produce laminates which combine the properties of the two materials, e.g. polythene film can be heat sealed, and this facilitates subsequent forming when it is laminated to a film which cannot be heat sealed.

Polythene powders can be converted by rotational moulding into items of quite large size, e.g. cold water cisterns, and can be used also for coating, either by dipping (the fluidized bed technique) or by spraying.

Polythene can be machined easily on conventional woodworking machinery provided that high speeds and sharp tools with acute angles are used. The temperature at the point of working should not exceed 40–50°C.

Polythene sheet and tube can be shaped easily by heating them to within about 10°C of the melting point, and forming the shape by hand. Polythene sheet can be welded using a filler rod and a torch supplying hot nitrogen. Polythene tube can be joined by using a simple spigot and socket joint, by butt jointing or by using polythene couplings. No satisfactory adhesives for polythene are known.

## PROPERTIES

The properties of a particular grade of polythene depend primarily on two factors—molecular weight and degree of crystallinity—and both are controlled during the polymerization process. Because the packing of the molecules in the crystallites is close compared with that in the amorphous regions, the less branched, more highly crystalline polythenes have higher densities. Because density is related to crystallinity, and is easier and more convenient to measure, it is usual to quote the density of a polythene rather than its crystallinity. As with polypropylene, however, density is not a complete index of the structural state of the material.

Similarly, as with polypropylene, the molecular weight is not normally determined and in practice the melt viscosity is obtained and expressed as the melt flow index at 190°C and 2·16 kg. Thus, polythenes are classified in terms of density and MFI.

Commercially available polythenes have densities in the range 0·91 to 0·97 g/cm$^3$ and are conveniently divided into two categories as follows:

| | Density range |
|---|---|
| Low density polythene | 0·91 to 0·94 g/cm$^3$ |
| High density polythene | 0·94 to 0·97 g/cm$^3$ |

The data presented in this book refer essentially to low density polythene, some physical properties of which are given in Table 8.1. By correct choice of density a polythene may be made with properties which lie anywhere within the limits shown in Table 8.1.

In common with many other crystalline plastics polythene may be ‘cold drawn’. In the process the spherulites are broken up and the crystallites oriented along the direction of drawing. As ‘cold drawing’ proceeds, there is a sharp demarcation between the highly oriented material, which has increased transparency, and the remaining undrawn material. This demarcation is visible as a sharp change in cross-section, or ‘neck’.

At any temperature, there is in any sample of polythene an equilibrium ratio of crystalline to amorphous material. At temperatures up to about 70°C the ratio changes little from its room temperature value, but it then begins to decrease rapidly, this decrease becoming very sharp as the final melting point is approached. Above the final melting point the polymer is an amorphous mass which exhibits both viscous and elastic behaviour.

Because polythene consists of very long molecules it is difficult for it to achieve an ordered (crystalline) state on being cooled from an amorphous melt to a solid. It is possible, however, especially for the less branched polythenes, to achieve fairly high crystallinities by cooling slowly from melt to solid. Slow cooling also allows the crystallites to organize themselves together into exceptionally large spherulites, in contrast to rapidly cooled polythene which contains very much smaller spherulites and is consequently more transparent.

For ethylene/vinyl acetate copolymers, the properties are determined by the proportion of vinyl acetate randomly copolymerized as well as by the factors mentioned above. An increase in vinyl acetate content increases the flexibility of the resultant product, but decreases the softening temperature. This limits, therefore, the degree of flexibility which can usefully be achieved. The vinyl acetate content affects the density too, but density is not used to classify these polymers because it is a complex quantity which depends on both vinyl acetate content and crystallinity.

### *Deformation*

The deformational behaviour at 20°C of polythene of density 0·9220 g/cm$^3$ is given by the tensile creep curves in Figure 8.1. The corresponding isochronous stress vs. strain, isometric stress vs. time and tensile creep modulus vs. time curves are given in Figures 8.2, 8.3 and 8.4. These data are appropriate to a polythene which has been annealed by cooling at 5°C/h from 145°C[1]. Because of the very large range of polythenes available, and the major influence of thermal and storage history, the set of curves in Figures 8.1 to 8.4 constitute only a rough guide to likely performance and should be corrected for polymer type and specimen history.

Fortunately the creep behaviour at long times is surprisingly straightforward, particularly if the creep results are plotted as log strain vs. log time, instead of the usual strain vs. log time. Such curves assume the form of two nearly straight sections, the latter one having the lower slope. The time at which the discontinuity occurs depends on the type of polythene and the thermal history. Some examples of this effect are shown in Figures 8.5 and 8.6. It has been found that provided the molecular weight is not very low, the experimental points will lie on a straight line that can be extrapolated with considerable confidence. The observation that the discontinuity in slope occurs at constant strain values for thermal conditioning and constant time values for conditioning at 20°C is, so far, unexplained.

Figure 8.7 shows the very large effect of different thermal histories on the 100 second isochronous stress vs. strain curve. An increase in the stress–strain ratio is accompanied by an increase in density usually. Important though this is from the user's viewpoint, Figure 8.8 in which the same data are plotted against logarithmic axes, demonstrates a more important principle, and the similarity of shape of the two curves enables the curve for any intermediate state to be drawn provided one datum point is available.

[1] This annealing treatment is the standard method (ASTM D1928-65T) which produces almost the maximum achievable crystallinity. The resulting crystallinity (and density) is higher than that normally encountered in service components because the usual fabrication processes involve rapid cooling. Such test samples are, therefore, not typical of the particular type of polymer but more rapid cooling gives samples that continue to change in properties during subsequent storage or testing. A full discussion of the ramifications of thermal history and test techniques is inappropriate in this book but it will be found in, for example, *Polythene* (Eds. Renfrew and Morgan), 2nd ed., Iliffe, London, 1960.

TABLE 8.1

| Property | Units | Typical Value |
|---|---|---|
| Density | g/cm$^3$ | 0·913–0·940 |
| Melt flow index (190°C and 216 kg) | | 0·5–100 |
| Crystalline melting point | °C | 109 to 125 |
| Coefficient of linear thermal expansion | | |
| −20 to 20°C | per °C | $1{\cdot}4\times10^{-4}$ |
| | per °F | $0{\cdot}8\times10^{-4}$ |
| 20 to 80°C | per °C | $2{\cdot}4\times10^{-4}$ |
| | per °F | $1{\cdot}3\times10^{-4}$ |
| Thermal conductivity, 20 to 80°C | cal/cm s °C | $7$ to $10\times10^{-4}$ |
| | B ThU in/ft$^2$ h °F | 2 to 2·9 |
| Specific heat | | |
| 20 to 80°C | cal/g °C | 0·52–0·65[a] |
| | B ThU/lb °F | 0·52–0·65[a] |
| 20 to 100°C | cal/g °C | 0·60–0·75[a] |
| | B ThU/lb °F | 0·60–0·75[a] |
| Flammability | | burns slowly with drips |

[a] Lower values correspond to higher density materials.

A similar change occurs in the stress vs. strain curve if an initially quenched specimen is stored at room temperature, Figure 8.9. These changes are accompanied by density increases, similar to those arising from thermal annealing, but the magnitudes of the stress–strain ratios are not uniquely determined by the specimen density. This is a disappointing complication, because density is one of the quantities used in the specification and quality control of polythenes.

The same pattern of behaviour persists over the whole range of polythenes, but it is noticeable, as shown in Figure 8.10, that the 100 second isochronous stress–strain relationships gradually become more curved as the reference density[1] increases, that is to say as the non-linearity becomes more marked. If the stress–strain ratio at some arbitrary small strain, say 0·2%, is plotted as a function of density for specimens annealed in the standard manner the relationship of Figure 8.11 emerges, but because of the differing degrees of non-linearity the relationship is not unique and changes if the ratio is calculated at a different strain. Within certain obvious limits, Figure 8.11 can be used to adjust the deformation data on Figures 8.1 to 8.4 so that they are applicable to other polythenes of different reference density, but such corrections will be valid only for highly annealed samples (e.g. they are inapplicable for density differences arising from different thermal histories) and for relatively small strains.

Creep strains up to at least 10% are fully recoverable in a manner similar to that reported for polypropylene. The experimental data available are not, however, suitable for inclusion here.

The deformational behaviour at 60°C of polythene of reference density 0·921 g/cm$^3$ is given by the tensile creep curves in Figure 8.12. The corresponding isochronous stress vs. strain, isometric stress vs. time and tensile creep modulus vs. time curves are given in Figures 8.13, 8.14 and 8.4 respectively. The deformational behaviour at temperatures other than 20°C and 60°C can be obtained by interpolation between the 20°C and 60°C data or estimated from the tensile creep modulus (at 0·2% strain) vs. temperature curves given in Figure 8.15.

Figure 8.16 shows that the recovery at 60°C is high, even after long periods of creep.

[1] Reference density is the density of a specimen annealed in the manner described in the previous footnote.

*Environmental Effects*

When certain low molecular weight grades of polythene are under stress and in contact with certain environments, brittle cracking may occur. The stress may be either a direct result of the application of external forces or be latent in the polythene through fabrication strains, or a combination of these separate effects. The term used to describe this phenomenon is 'Environmental Stress Cracking' and as such, in the broadest sense, can be considered to embrace the behaviour of polythene exposed to a wide range of environments from air, which is comparatively inactive and in the presence of which cracks develop slowly, to much more active environments, e.g. silicone fluids and surfactants in the presence of which cracks develop much more quickly. Generally speaking, the phenomenon has been limited to describe situations where no obvious chemical attack such as swelling of the polythene occurs and as such, therefore, it has been considered basically physical in nature.

The cracks initiated at stress raisers in the polythene propagate in an apparently brittle fashion. If, however, the morphology of the fracture surface is resolved under high power magnification, the description 'brittle' is seen to be very much an oversimplification.

The effect of environmental stress cracking is shown in Figure 8.17 by the behaviour at 20°C of a highly annealed sample of an injection moulding grade of polythene (density 0·9255 g/cm$^3$) stressed under conditions of constant load in the presence of both an active environment, silicone oil, viscosity 200 cs, and an inactive one, air. As can be seen in Figure 8.17 the specimen used for this test was parallel sided with deliberate notches in the sides.

In the presence of silicone oil, cracks begin to propagate quickly, leading to a marked reduction in strength, whereas the corresponding effect is delayed in the presence of air alone, leading to premature failures only at much longer times. At the same applied stress, the corresponding crack growth rates are of the order of $10^{-2}$ cm/s and $10^{-7}$ cm/s for silicone oil and air respectively.

By the use of creep data, it is possible to measure the macroscopic strain—away from the stress raisers—developed in the specimen at the onset of crack propagation. The complementary strain vs. log time plot suggests that the phenomenon of environmental stress cracking is a strain-dominated one leading to a limiting strain for a particular situation below which cracking is unlikely to occur in a finite time. For the example quoted this strain was of the order of 1 %.

The effect of increasing molecular weight is generally to make polythene more resistant to this type of failure, leading to higher limiting strains for the same conditions of test. The effects of temperature, density and thermal history, however, are more complex and it is inappropriate to consider these, in the detail that would be necessary, in this book.

*Impact Behaviour*

Impact resistance is not normally a serious problem with low density polythene because it yields before breaking, except under very severe conditions. Figure 8.18 shows the Charpy impact strengths of two injection moulding grades as functions of temperature. Even with these materials it is necessary to use sharp notches (tip radius 0·010 in) and low temperatures (−15°C and below) in order to obtain brittle fracture.

The impact strength of polythene depends on the melt flow index and density of the material. Rough guides to the impact behaviour, based on these two parameters, have been derived from measurements on many polythenes essentially in the density range 0·915 g/cm$^3$ to 0·945 g/cm$^3$, and all made by the high pressure process, or by blending low density polythene with high density material. If an arbitrary brittleness temperature is defined as the temperature at which the impact strength in the standard Charpy test with a 0·010 in notch tip radius reaches 5 ft-lb/in$^2$, results for 18 different polythenes indicate that

(a) within the density range 0·915 to 0·946 g/cm$^3$ approximately, an increase in specimen density of 0·005 g/cm$^3$ causes an increase in brittleness temperature of 22°C, and
(b) within the melt flow index range 2·5 to 70, an increase of melt flow index by a factor of 10 causes an increase in brittleness temperature of 27°C.

The validity and accuracy of these generalizations are examined in Figure 8.19 where all the available data have been corrected to a melt flow index of 20. The solid line represents the least squares fit of the available data: it may be argued that a straight line is not the best fit to the points but at least it provides easy working rules for design.

*Electrical Properties*

Polythene as an insulating material is one of the best available, and it is found in many insulating applications ranging from submarine telecommunication cables to less exacting ones such as low power distribution cables and 'hook up' wiring. In some cases it is possible, by a careful choice of fillers, to pigment polythene without any marked deterioration in its dielectric properties. In addition to its insulating properties, or rather as another consequence of the structure and character of polythenes, these plastics have low power factors, amongst the lowest of any plastics material, and the permittivity follows the Clausius–Mosotti Law relating it to density: with density ranging from 0·915 g/cm$^3$ to $\sim$ 0·96 g/cm$^3$ the permittivity varies from 2·27 to 2·36 and the effect of temperature on permittivity again follows the density change with temperature.

A loss tangent contour map for low density polythene is given in Figure 8.20. A graph showing the apparent volume resistivity as a function of time at room temperature is given in Figure 8.21.

*Optical Properties*

Polythenes can show a wide range of optical properties: some, like refractive index, depend on the density, whilst the transparency may range from high, in some thin, highly quenched packaging films, to translucency in mouldings of low density materials, to virtual opacity in high density materials. With this range of behaviour it is possible only to exemplify the optical properties of a particular sample of low density polythene of density 0·9148 g/cm$^3$, produced by rapid cooling of a film sample.

The refractive index of the material was 1·5162 at 5461 Å and the temperature coefficient of refractive index −0·00024 per °C.

The transparency characteristics are of more importance and are given below, for the same low density sample. Values are quoted for transparency behaviour corrected for reflection and surface defects. Thickness 1·02 mm:

| | |
|---|---|
| Direct transmission factor | 42% |
| Direct transmission factor corrected to 1 mm thickness | 45% |
| Scattering coefficient | 8·3 cm$^{-1}$ |
| Forward scattered fraction (ASTM D1003) | 27% |

The data given define a moderately transparent material: light transfer of a reasonable level is also possible.

| | |
|---|---|
| Total transmission factor | 85% |
| Total reflection factor | 11% |

*Chemical Properties*

Polythene exhibits the very high resistance to chemical attack typical of polyolefines. Its chemical resistance is complementary to that of metals, i.e. polythene is resistant to alkalis, aqueous solutions, non-oxidizing inorganic acids, and to a lesser extent to concentrated oxidizing acids, all of which cause corrosion in metals, but it may be affected by organic solvents which usually have little, if any, effect on metals. Where acid attack is likely to occur it may well be accelerated by the presence of ultraviolet light.

At room temperature, polythene is almost completely insoluble in all organic solvents, although some absorption, softening or embrittlement may occur. At temperatures above 70°C, polythene dissolves in common solvents such as xylene, toluene, amyl acetate, trichloroethylene, paraffin and turpentine, but is not dissolved by glycerine, ether, carbon disulphide, or acetone.

Ethylene/vinyl acetate (EVA) copolymers are, in general, more susceptible to solvent and chemical attack, but considerably more resistant to environmental stress cracking than homopolymers of the same MFI.

TABLE 8.2

| | Low density polythene (0·92 g/cm³) | EVA Copolymers | |
|---|---|---|---|
| Equilibrium water content (immersed) | 0·015% | 0·015% | |
| Gas permeabilities (cm³ cm/cm² s cm Hg) × 10⁹ at 20°C and 0% rh | | | |
| Carbon dioxide | 1·7 | 3·2 | |
| Oxygen | 0·3 | 0·4 | |
| Nitrogen | 0·1 | 0·2 | |
| Water vapour permeability (g/m² 24 h) at 38°C and 90% rh for 0·001 in thickness | 14 | 50 | |
| Resistance to: | | | |
| Mineral acids (dilute) | Excellent | Good | |
| Mineral acids (conc.) | Good (except with oxidizing acids) | Poor | |
| Alkalis | Excellent | Excellent | |
| Solvents: | | | |
| alcohols | Good | Good | Some may present a cracking hazard |
| ketones | Fair | Poor | |
| aromatic hydrocarbons | Poor | Poor | |
| chlorinated hydrocarbons | Poor | Poor | |
| Detergent solutions | Fair | Excellent | |
| Greases and oils | Good | Fair | |

## APPLICATIONS

Immediate recognition of the outstanding dielectric properties of polythene led to its early use as an insulator, particularly in high frequency A.C. and underwater cables. Electrical applications still account for a significant amount of polythene, but the largest single outlet is now film, for which polythene's combination of toughness, good barrier properties, ease of fabrication and heat sealing, and freedom from toxic hazards makes it ideally suited. Mouldings, extrusions (such as pipe and sheet) and blow mouldings (bottles and containers) all exploit the properties of toughness, flexibility and chemical resistance to varying degrees.

Ethylene/vinyl acetate copolymers are relatively new and few applications are established. Their growing use would seem to depend on their combination of low temperature flexibility and toughness, high gas permeability and absence of plasticizer (compared with soft PVC).

*Typical examples* (low density polythene)

Injection mouldings:
Domestic ware, bobbins, grommets, washers.
Film:
Packaging, waterproof membranes, weatherproof cladding, reservoir lining, temporary glazing.
Cables:
Insulation, wire covering, cable sheathing.
Pipe:
Cold water plumbing, agricultural irrigation, emergency water lines, beer hose, chemical plant and effluent piping.
Blow mouldings:
Squeeze bottles, carboys, large containers.
Rotational mouldings:
Large containers, particularly those required in small quantities.
Suggested applications (EVA):
Tubing for food and chemicals.
Puffer packs.
Disposable gloves.
Meat wrap (stretch film).
Hot melt adhesives.
Cable sheathing.

## LIST OF FIGURES

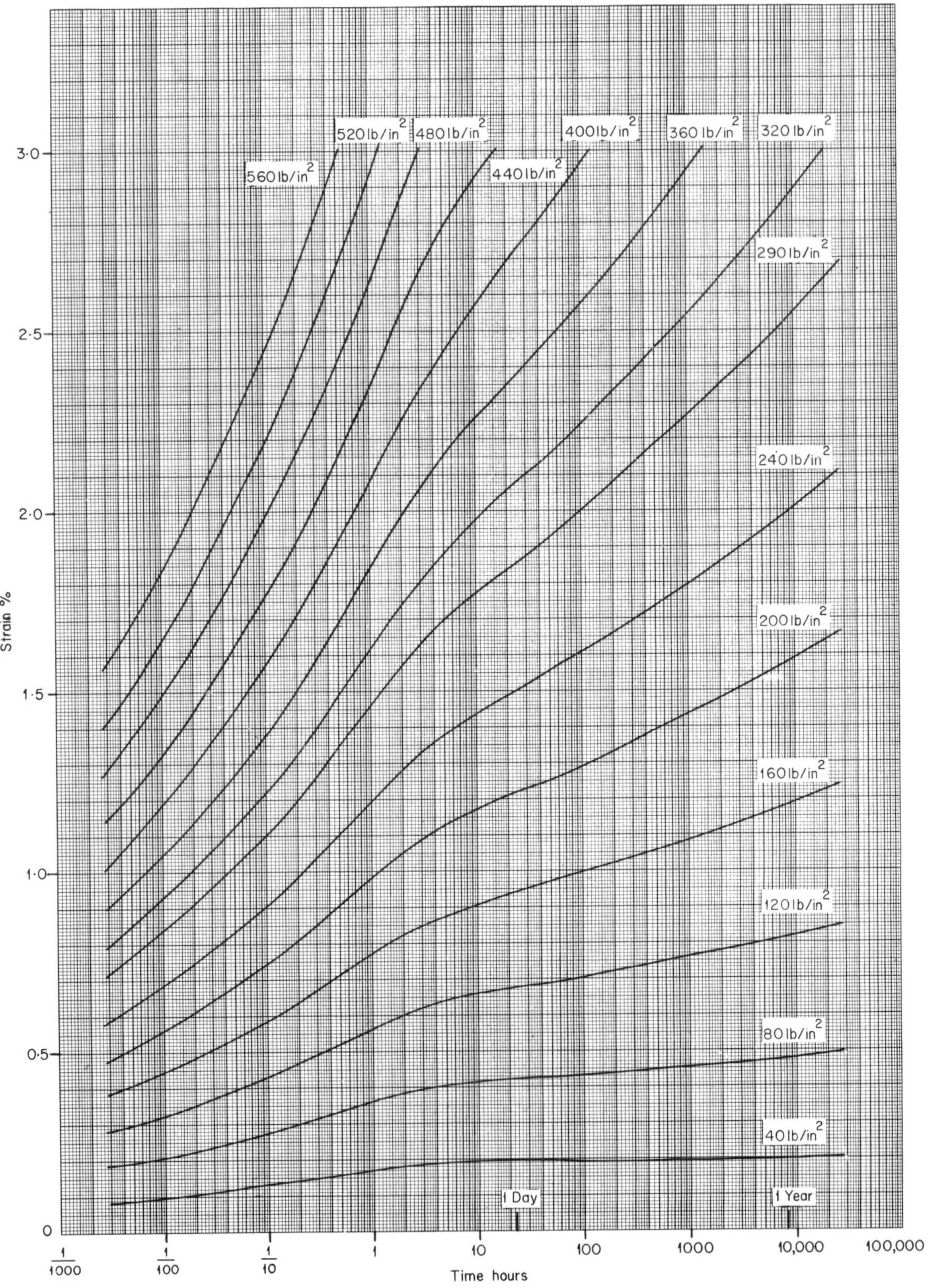


**Figure 8.1.** Creep curves in tension: 20°C. Polythene, density 0·9220 g/cm³ ('Alkathene' WJG 11)

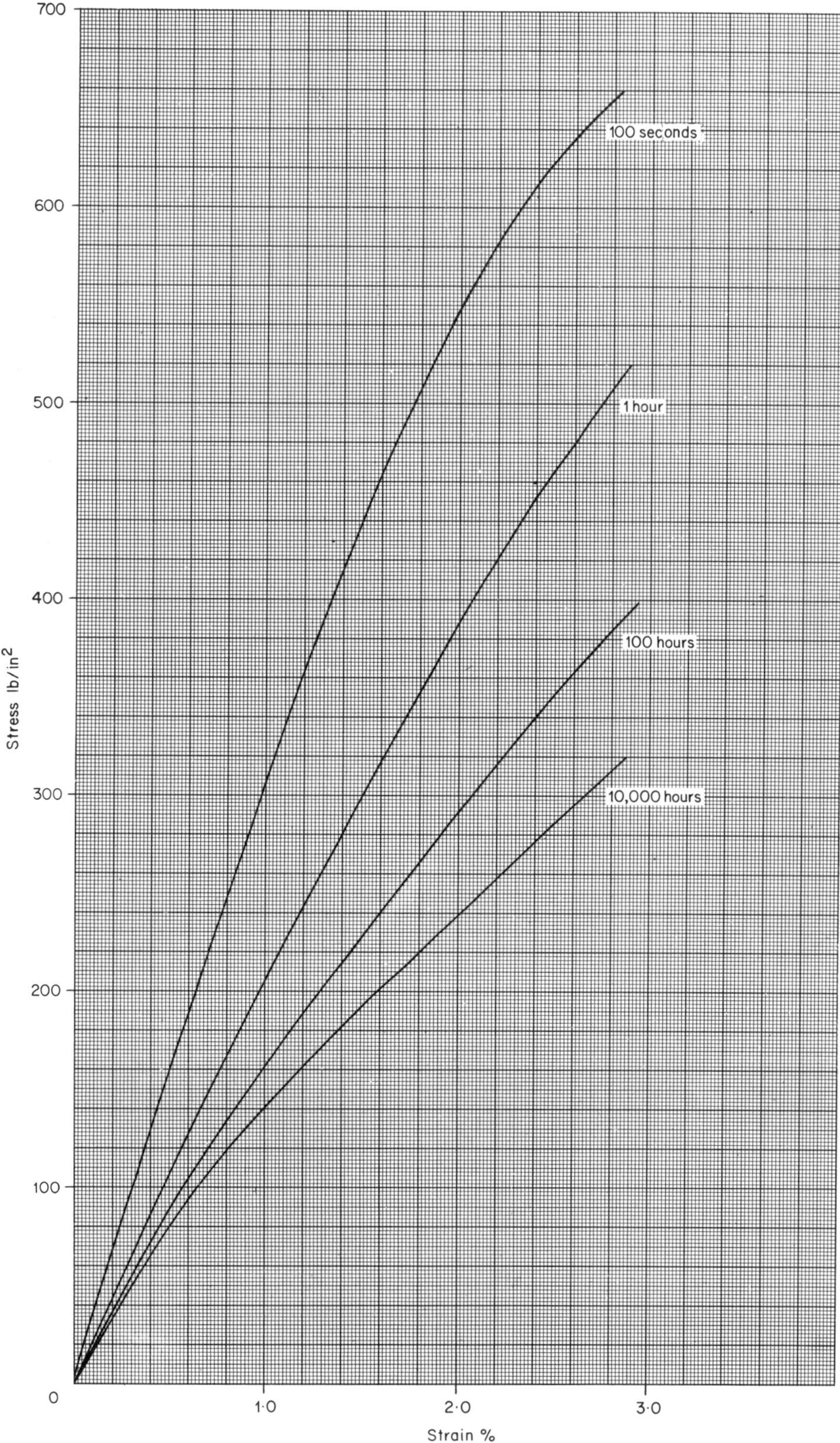


**Figure 8.2.** Isochronous stress vs strain curves: 20°C. Polythene, density 0·9220 g/cm$^3$ ('Alkathene' WJG 11)

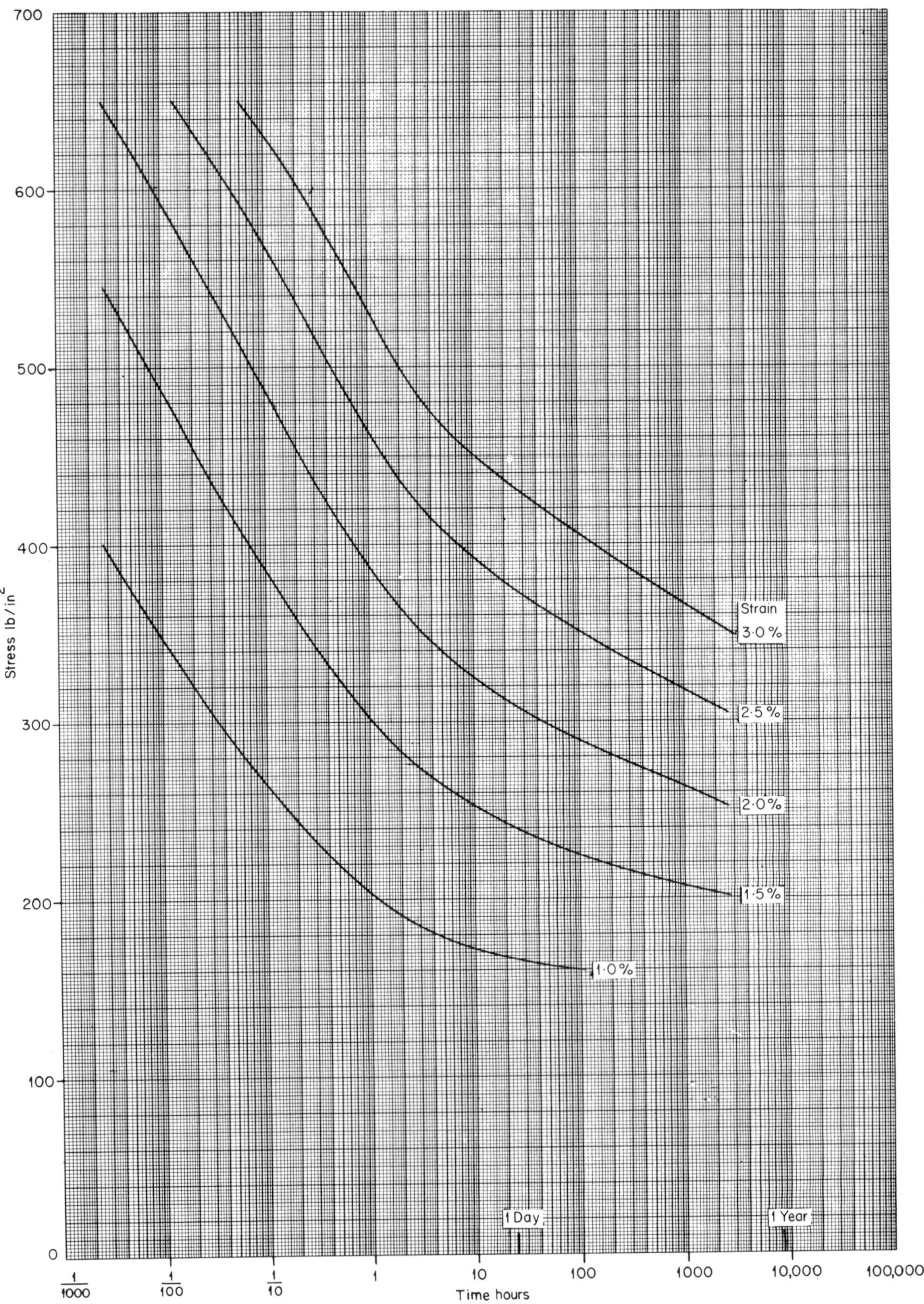


**Figure 8.3.** Isometric stress vs time curves: 20°C. Polythene, density 0·9220 g/cm$^3$ ('Alkathene' WJG 11)

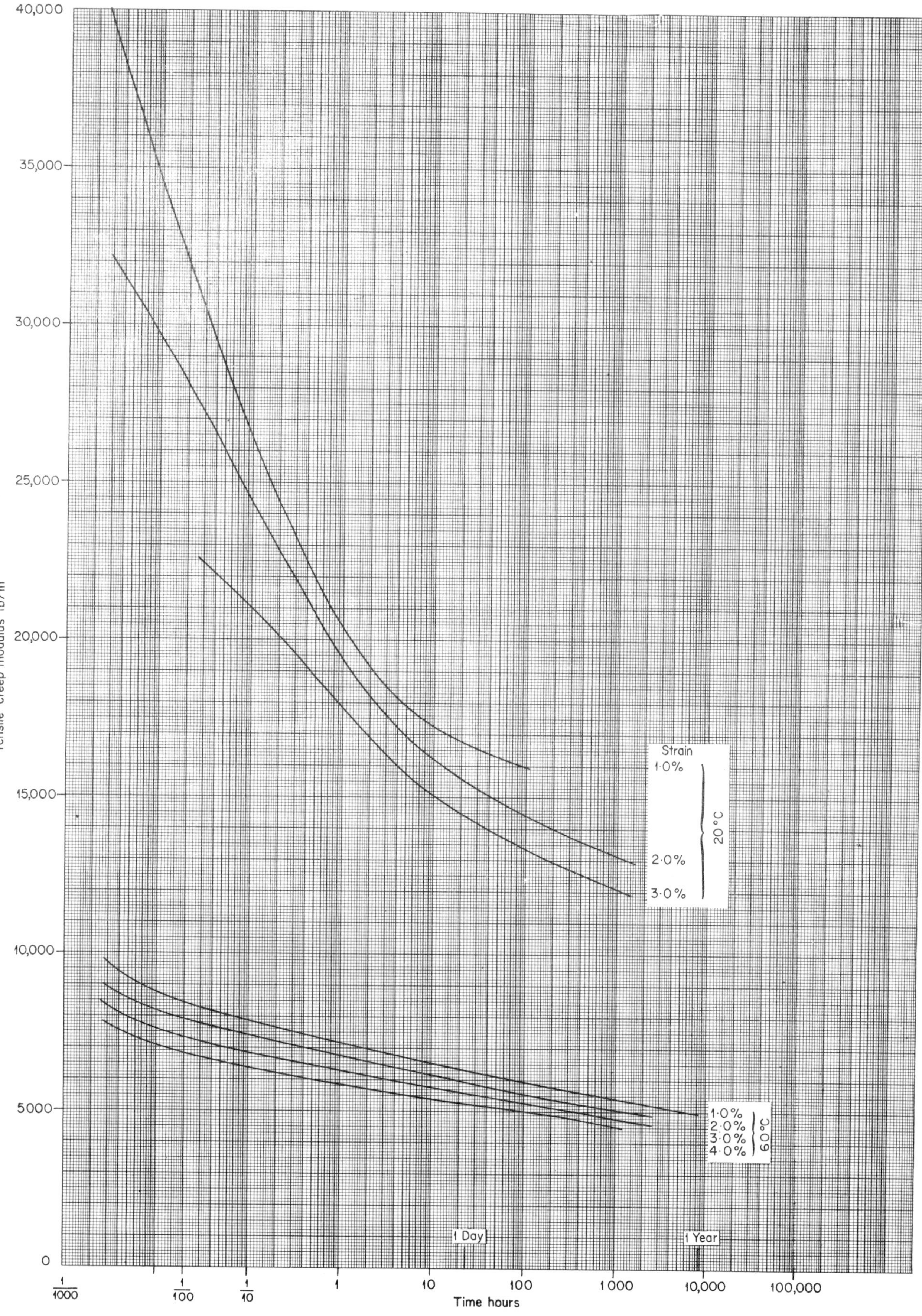


**Figure 8.4.** Tensile creep modulus vs time curves: 20°C and 60°C. Polythene, density 0·9220 g/cm³ at 20°C ('Alkathene' WJG 11)

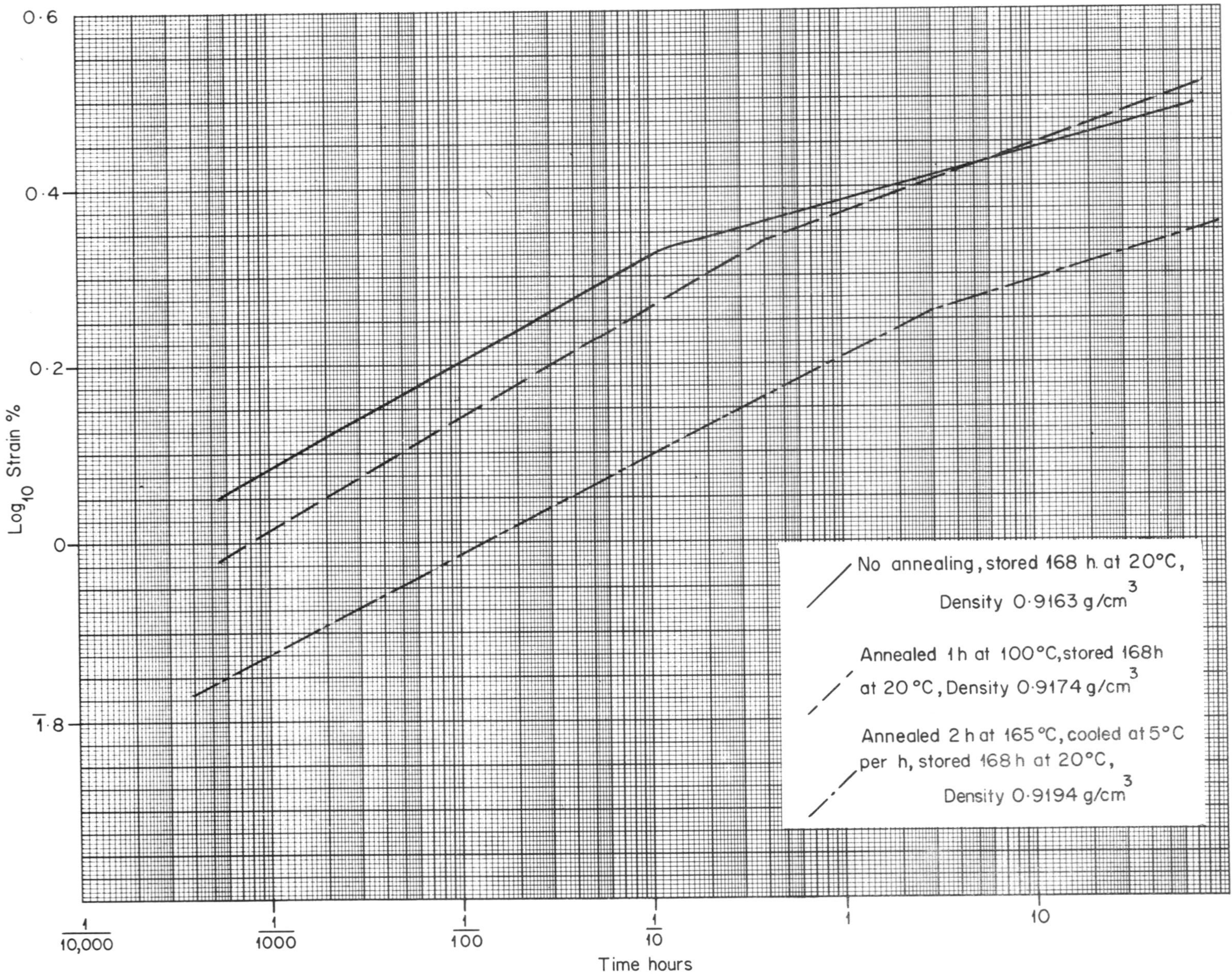


**Figure 8.5.** Creep curves in tension: 20°C, 290 lb/in$^2$, logarithmic axes. Effect of high temperature annealing. Polythene ('Alkathene' WRM 19)

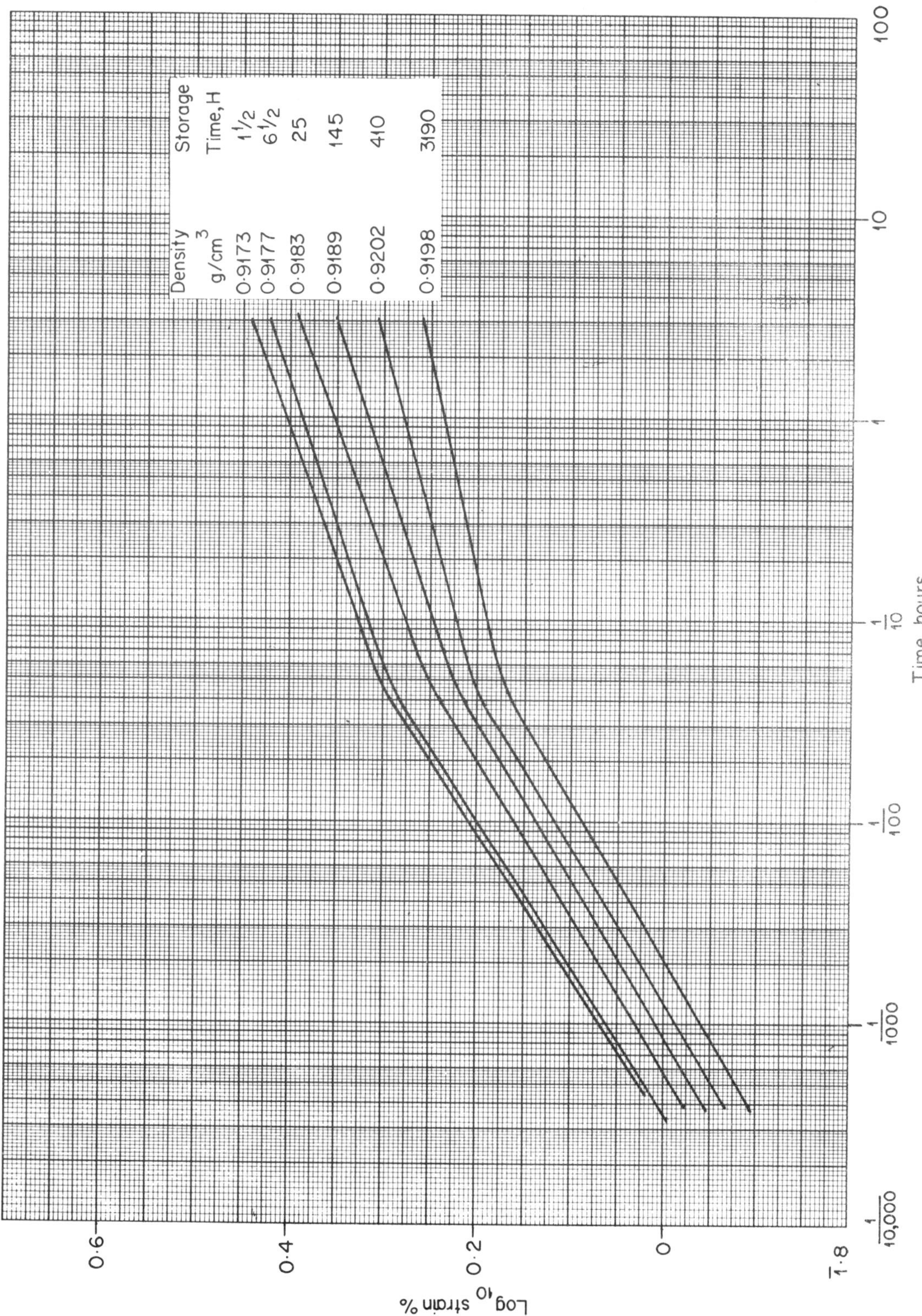


**Figure 8.6.** Creep curves in tension: 20°C, 290 lb/in$^2$, logarithmic axes. Effect of conditioning at 20°C. Polythene ('Alkathene' WJG 11)

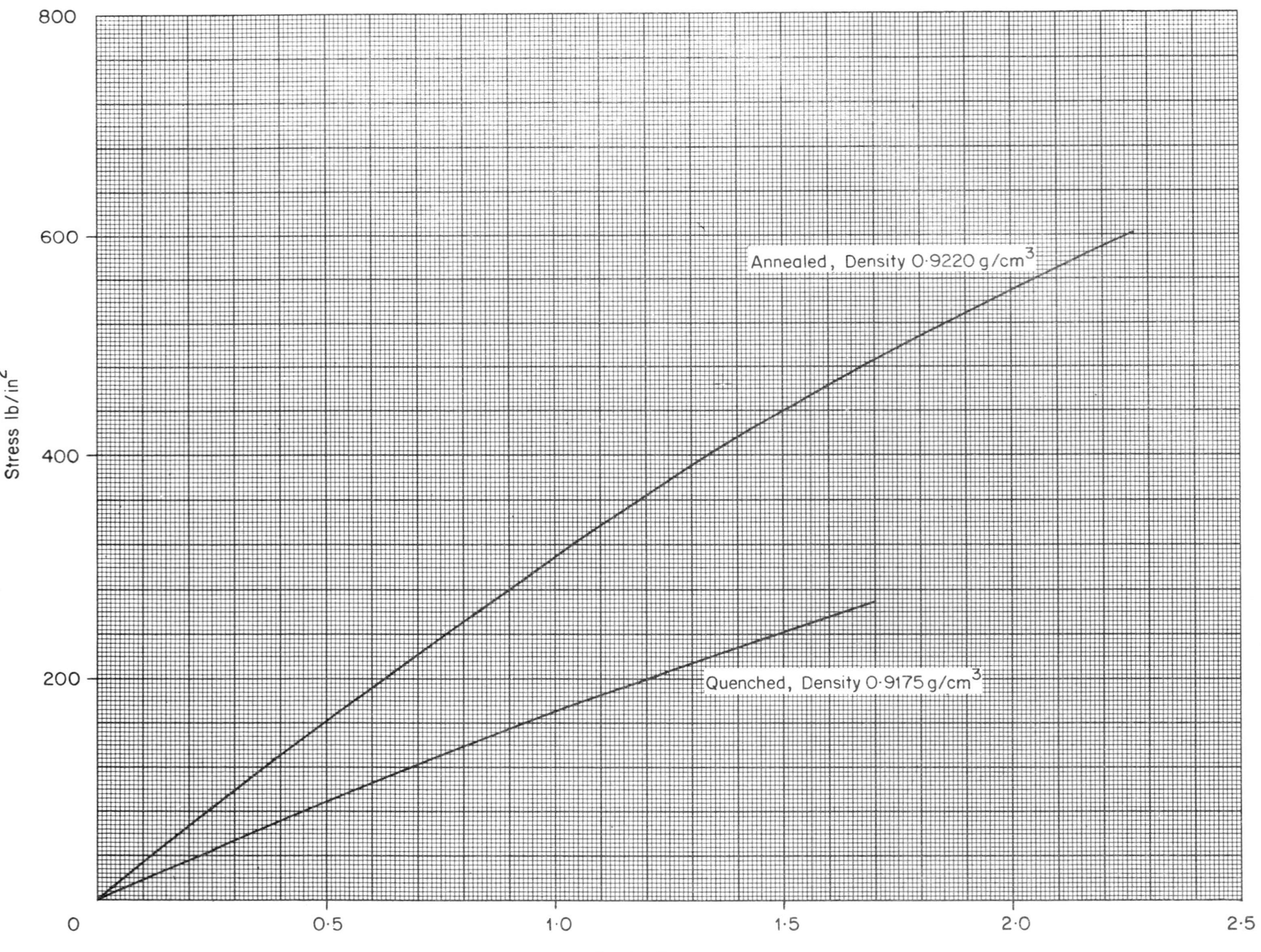


**Figure 8.7.** Isochronous stress vs strain curves: 20°C, 100 sec. Effect of different thermal treatments. Polythene ('Alkathene' WJG 11)

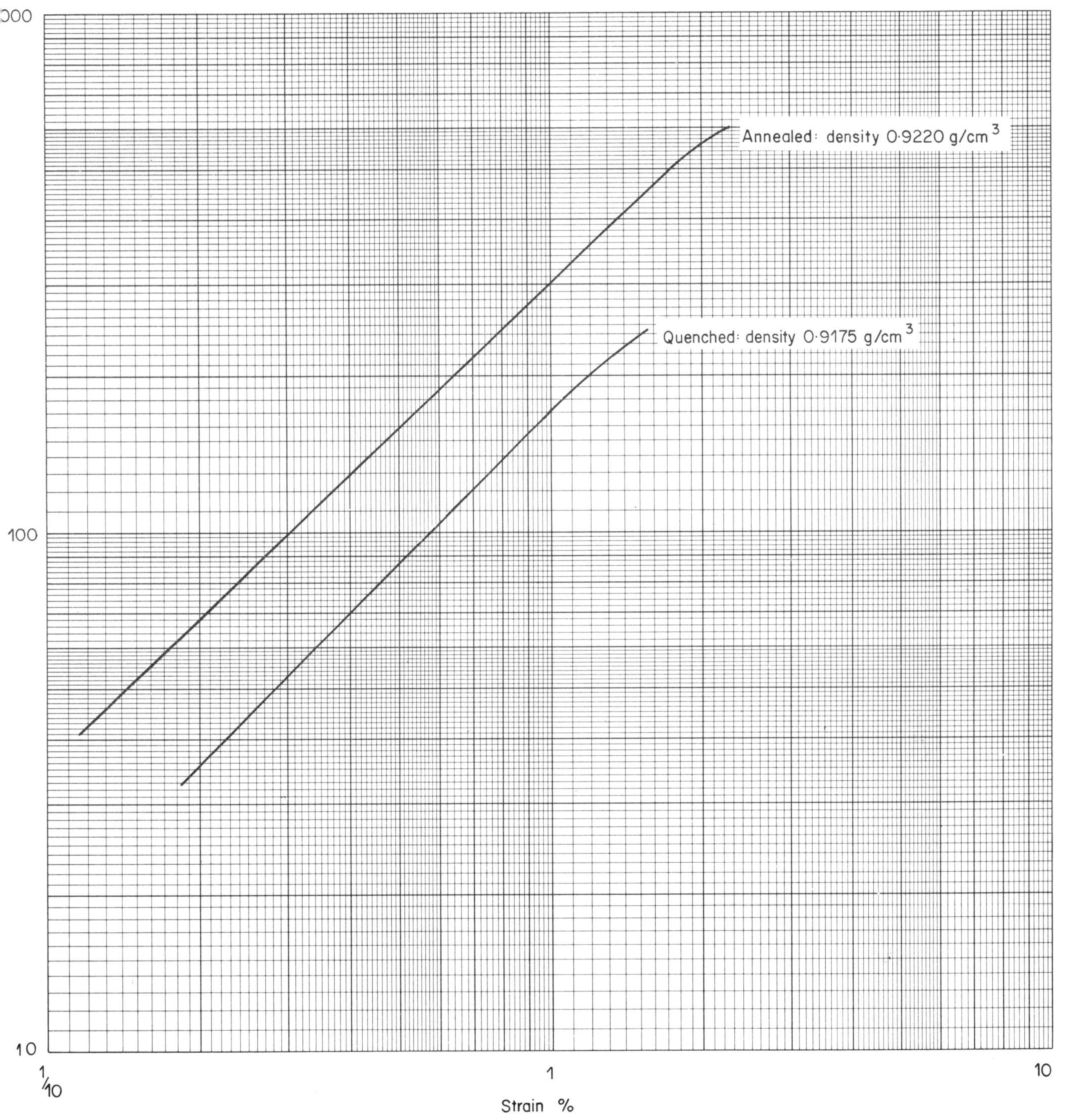


**Figure 8.8.** Isochronous stress vs strain curves: 20°C, 100 sec, logarithmic axes. Effect of different thermal treatments. Polythene ('Alkathene' WJG 11)

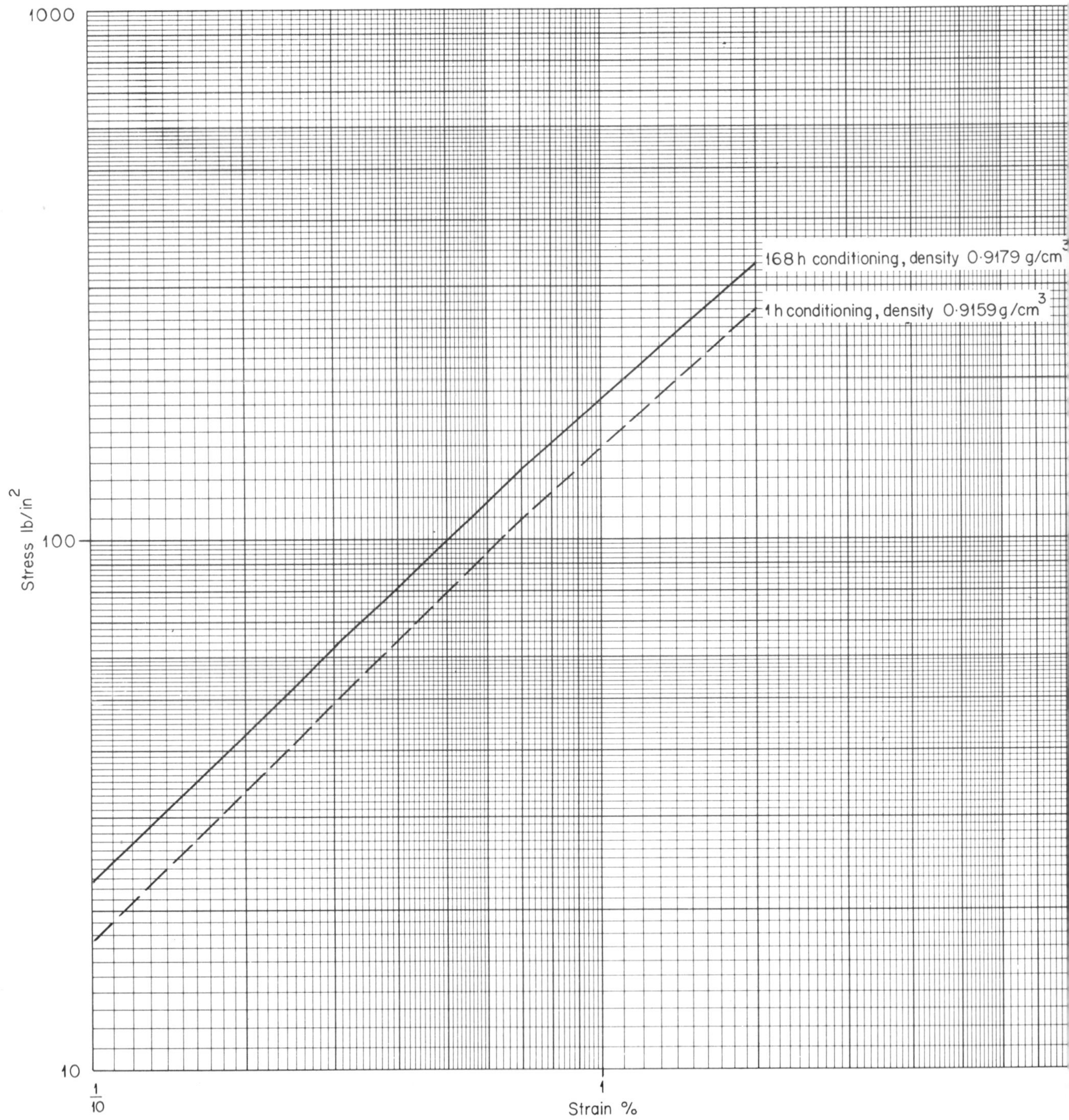


**Figure 8.9.** Isochronous stress vs strain curves: 20°C, 100 sec, logarithmic axes. Effect of conditioning at 20°C on specimens quenched from 190°C. Polythene ('Alkathene' XDG 33)

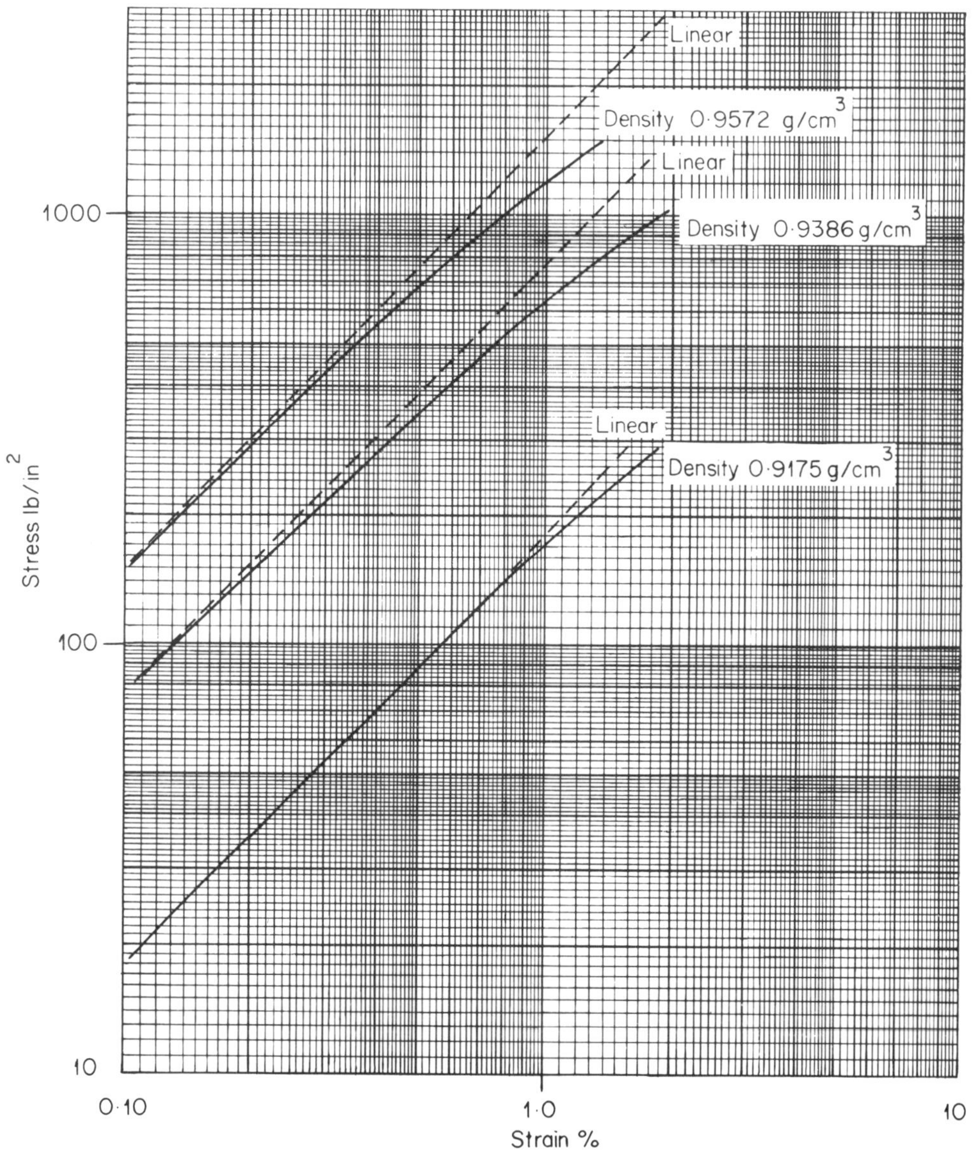


**Figure 8.10.** Isochronous stress vs strain curves: 20°C, 100 sec, logarithmic axes. Effect of density on non-linearity, specimens quenched from 190°C. Polythene, low density and high density grades

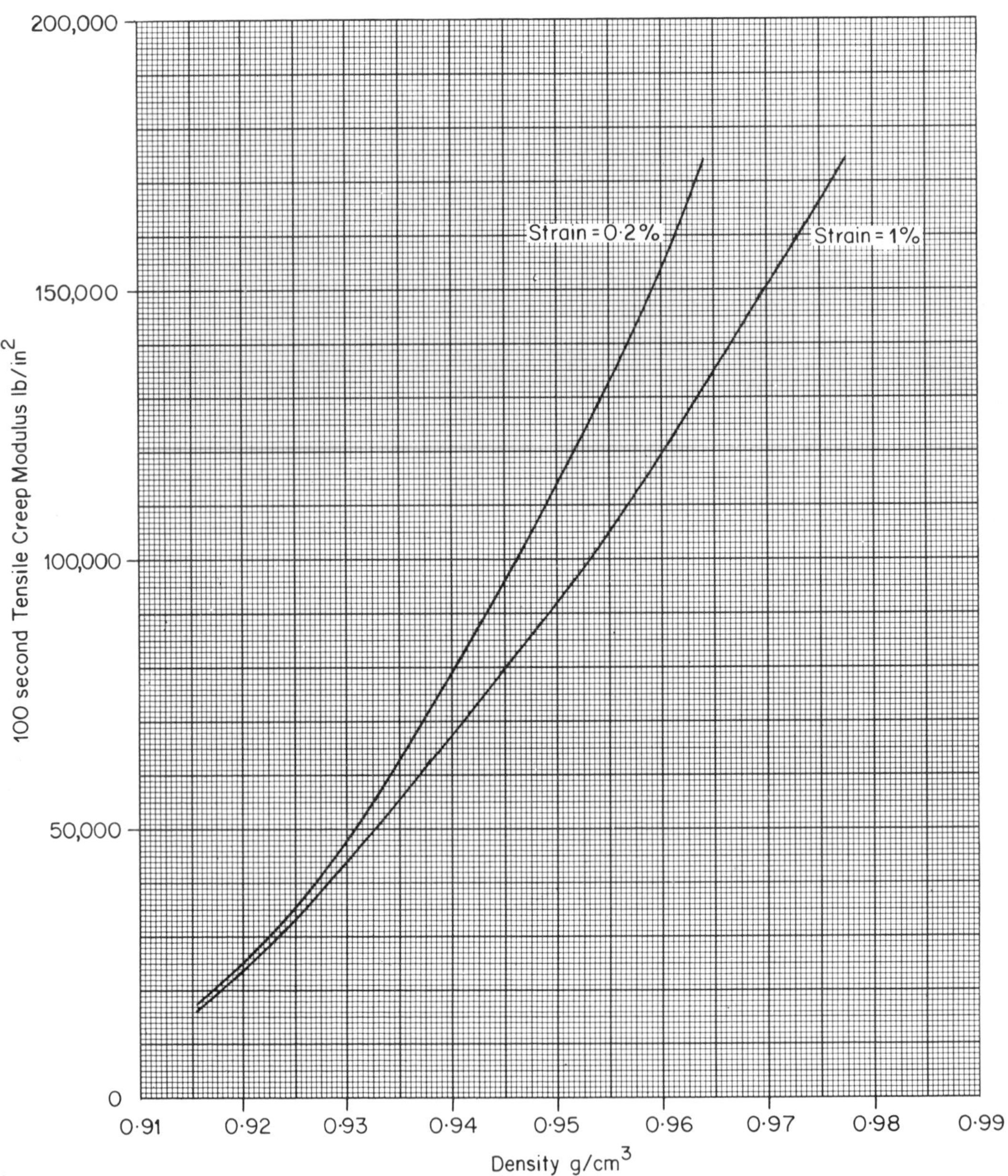


**Figure 8.11.** Tensile creep modulus vs density: 20°C, 0·2 % and 1 % strain. Polythene, low density and high density grades

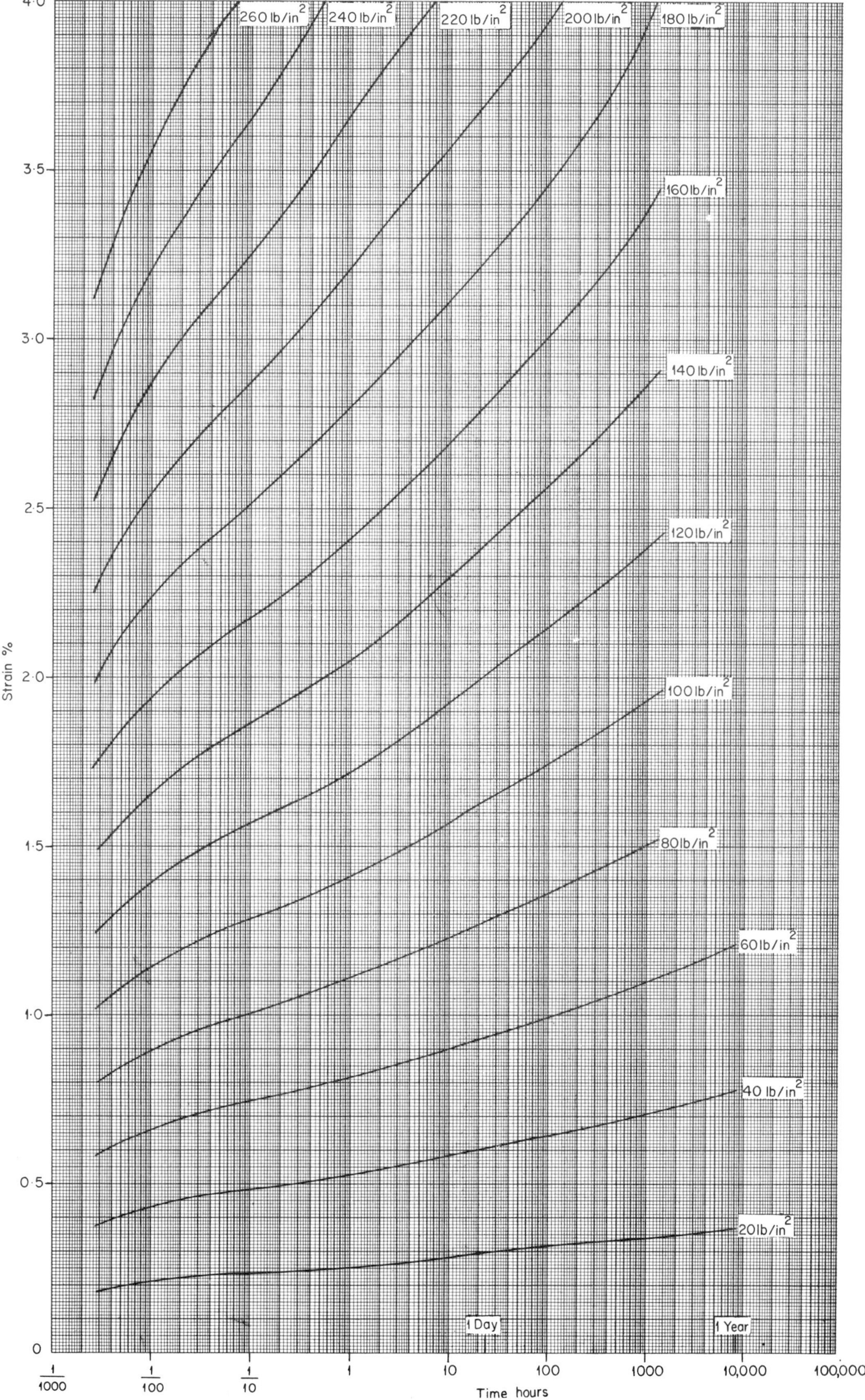


**Figure 8.12.** Creep curves in tension: 60°C. Polythene, density 0·921 g/cm³ at 20°C ('Alkathene' WJG 11)

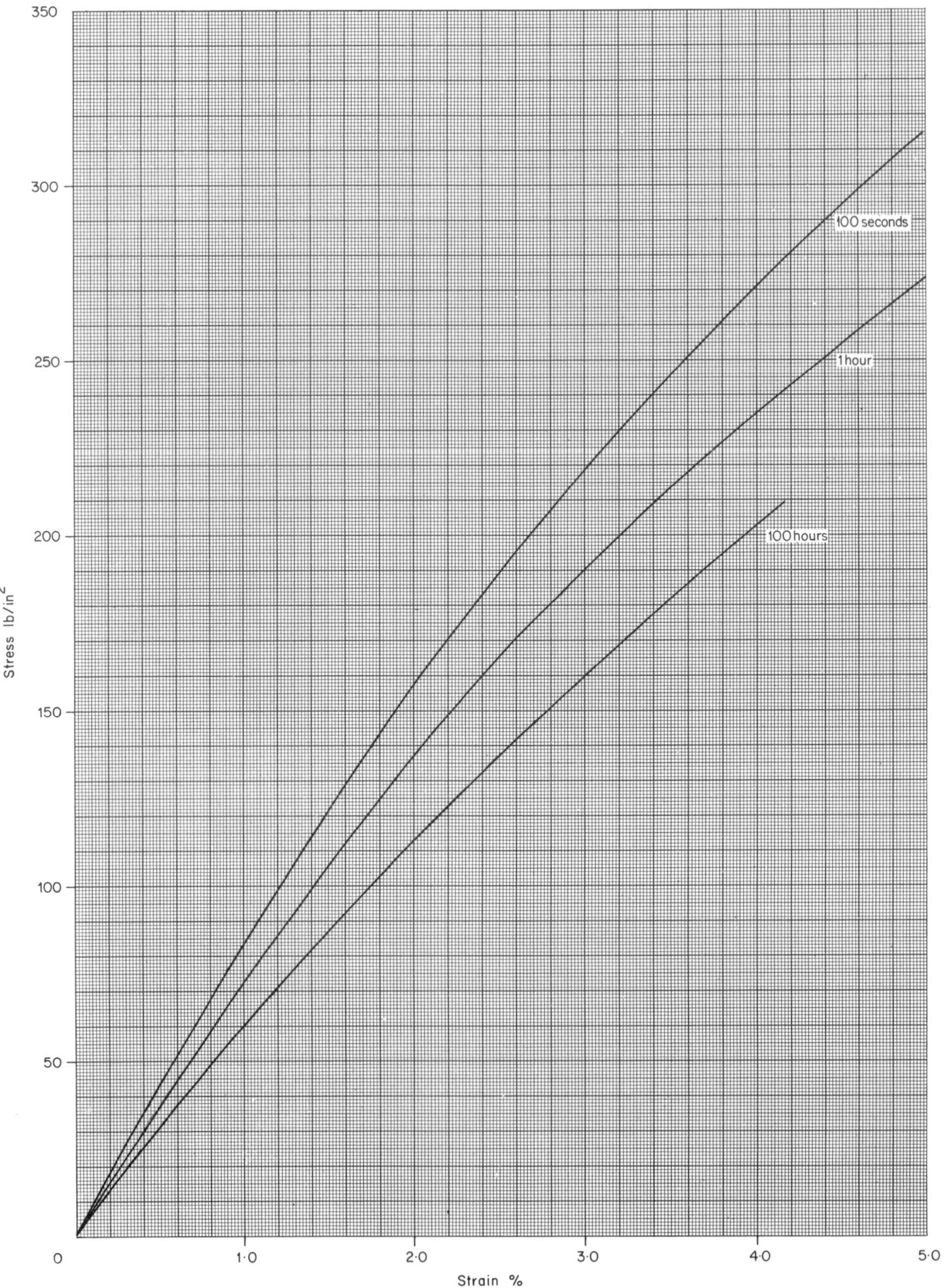


**Figure 8.13.** Isochronous stress vs strain curves: 60°C. Polythene, density 0·921 g/cm$^3$ at 20°C ('Alkathene' WJG 11)

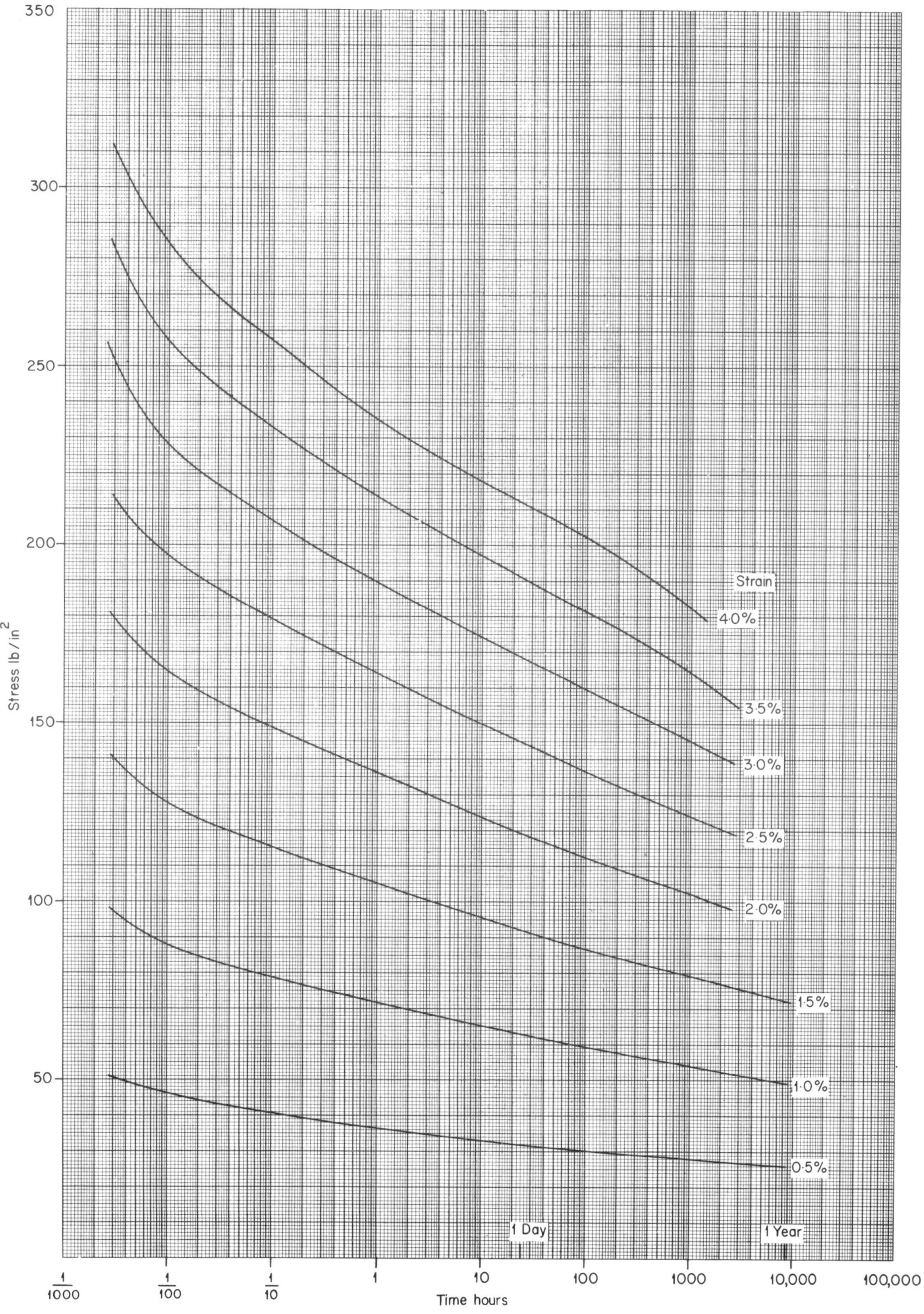


**Figure 8.14.** Isometric stress vs time curves: 60°C. Polythene, density 0·921 g/cm³ at 20°C ('Alkathene' WJG 11)

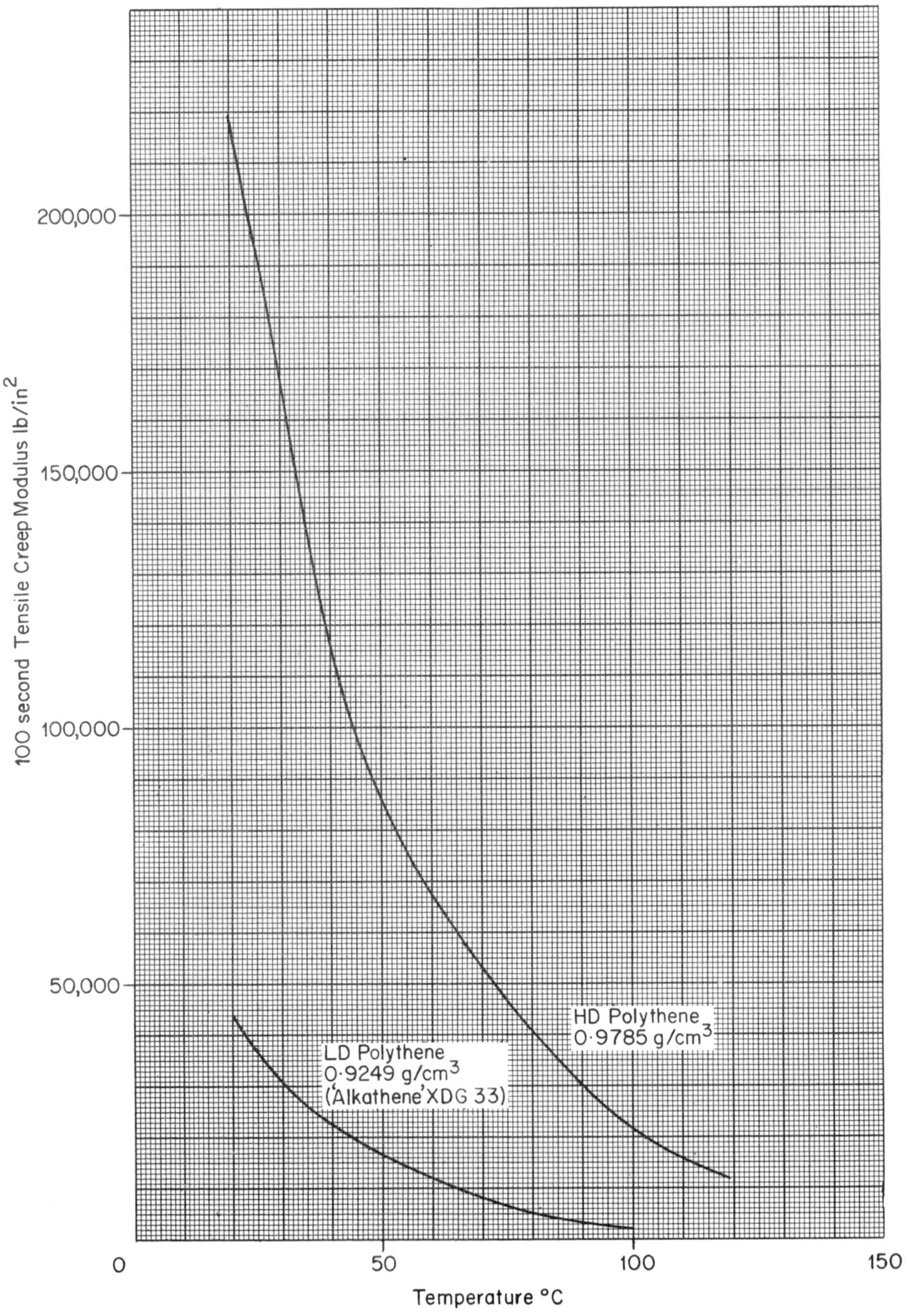


**Figure 8.15.** Tensile creep modulus (100 sec, 0·2% strain) vs temperature. Polythenes, density 0·9249 g/cm$^3$, 0·9785 g/cm$^3$ at 20°C ('Alkathene' XDG 33 and HD Polythene)

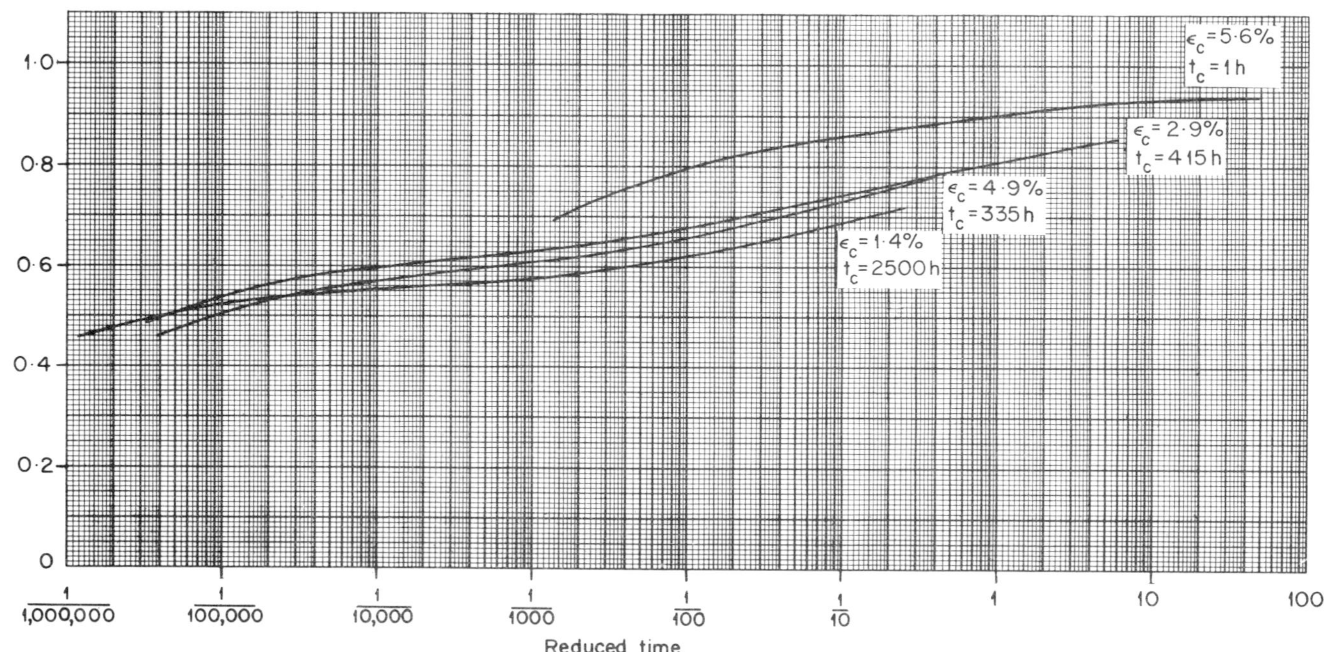


**Figure 8.16.** Recovery from creep in tension: 60°C. Polythene, density 0·921 g/cm$^3$ at 20°C ('Alkathene' WJG 11)

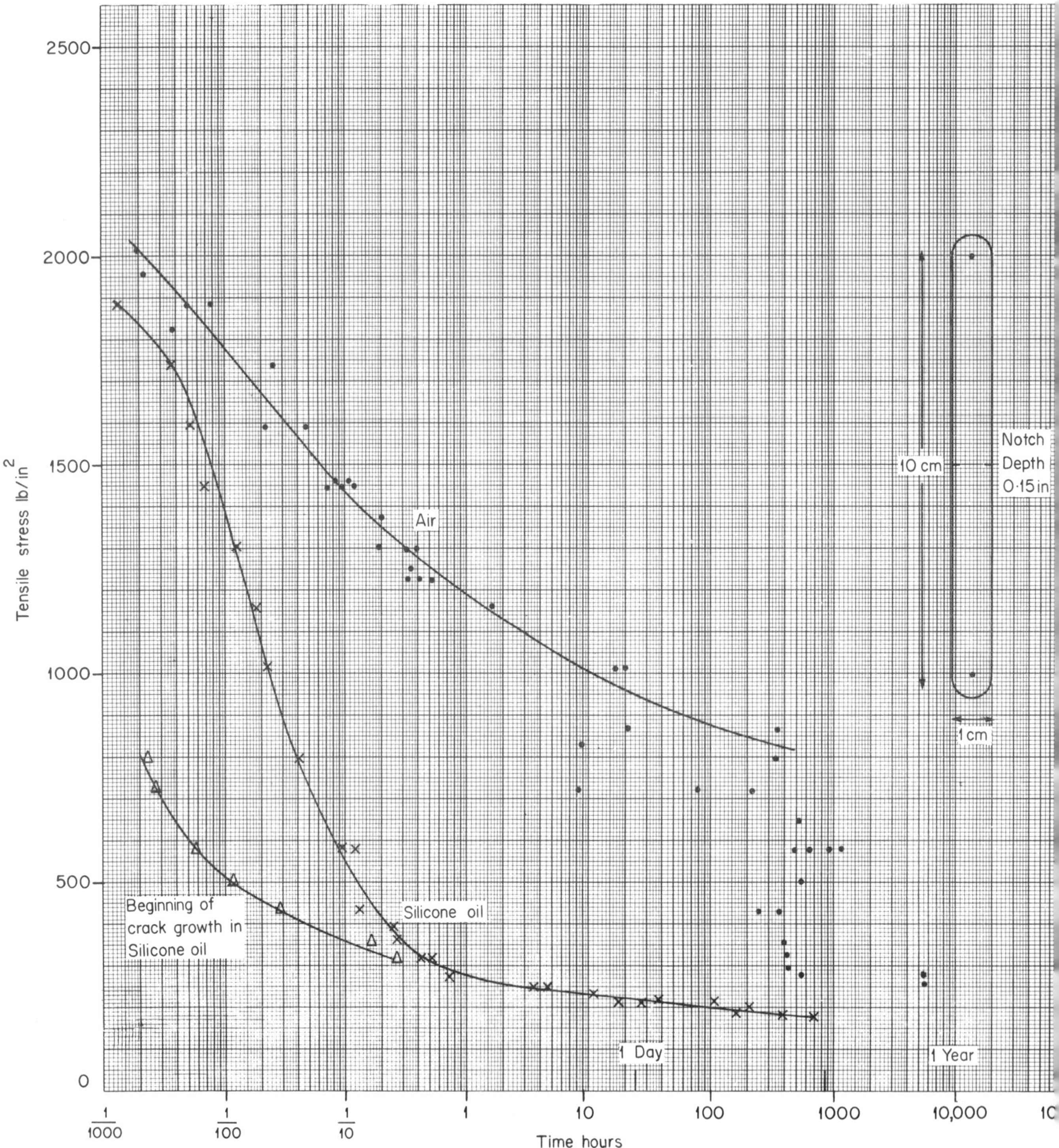


**Figure 8.17.** Creep rupture stress in tension vs time to failure: 20°C. Effect of environmental stress cracking: agent silicone oil 200 CS. Polythene, density 0·9255 g/cm³, MFI 20 ('Alkathene' XRM 40)

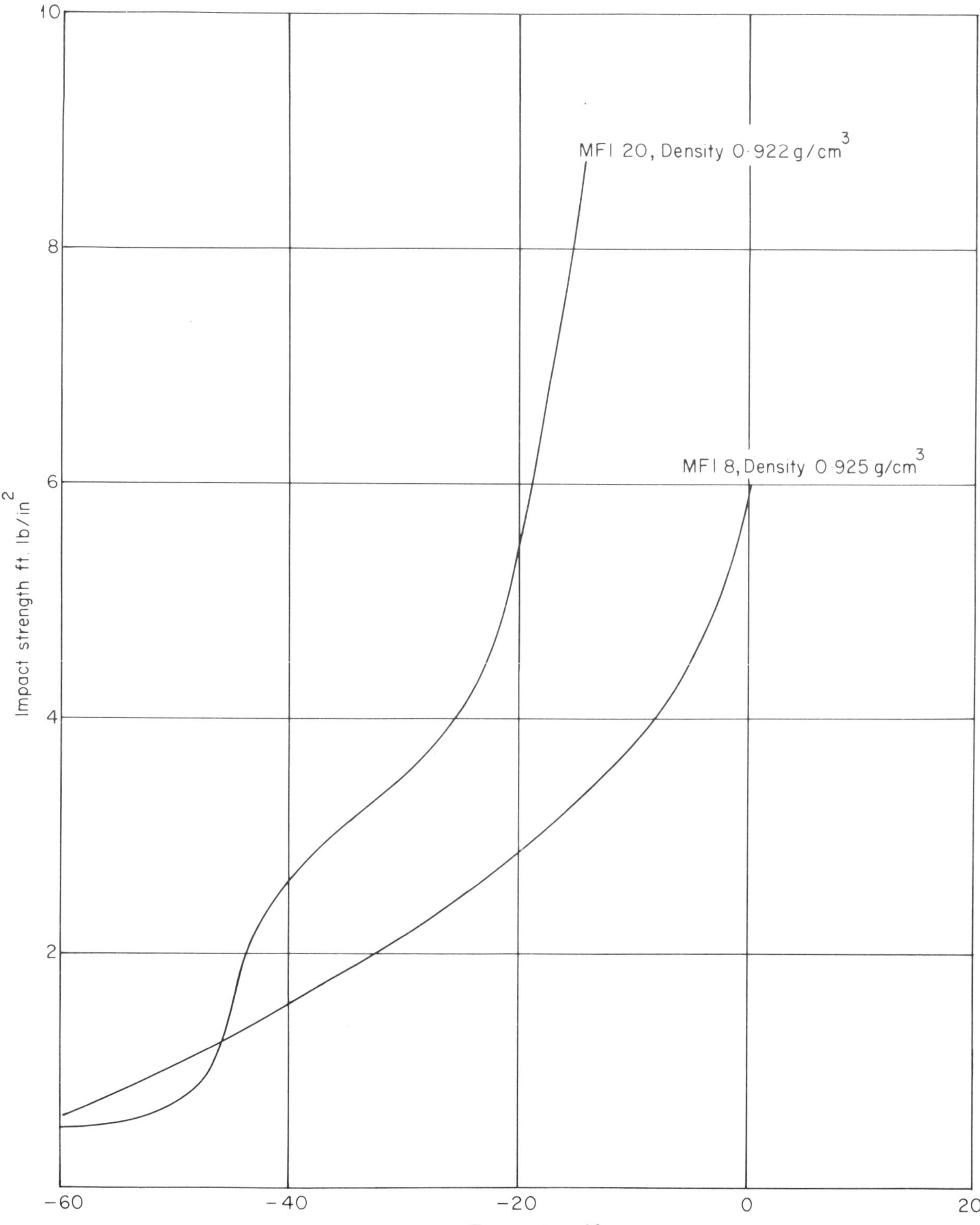


**Figure 8.18.** Impact strength vs temperature: 0·010 in notch tip radius. Polythenes of densities and MFI stated ('Alkathene' XRM 21, Q 916)

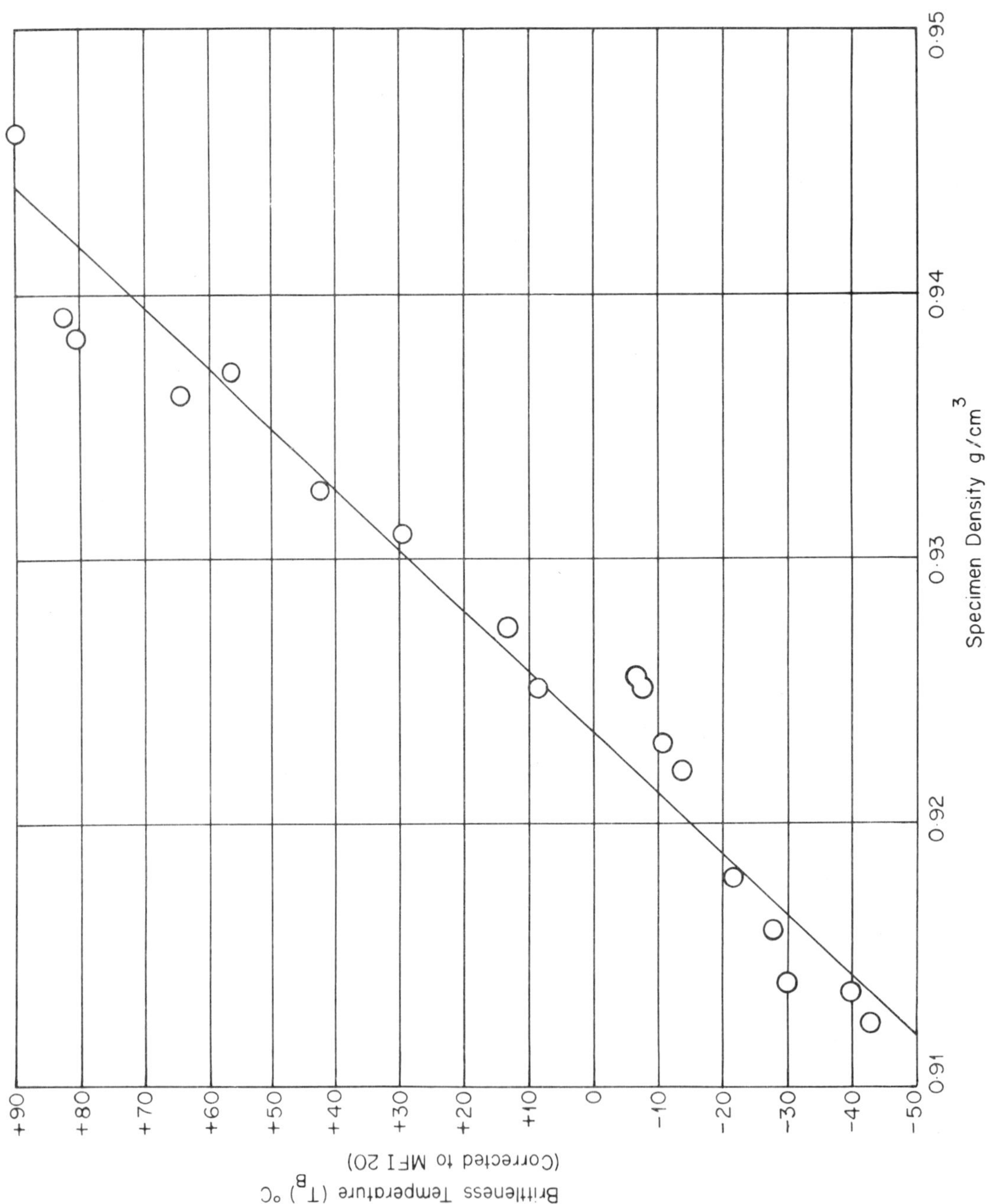


**Figure 8.19**. 'Brittleness' temperature (0·010 notch tip radius, MFI 20, strength 5 ft. lb/in$^2$) vs specimen density. Polythenes made by high pressure process and some LD/HD blends (various 'Alkathene' grades)

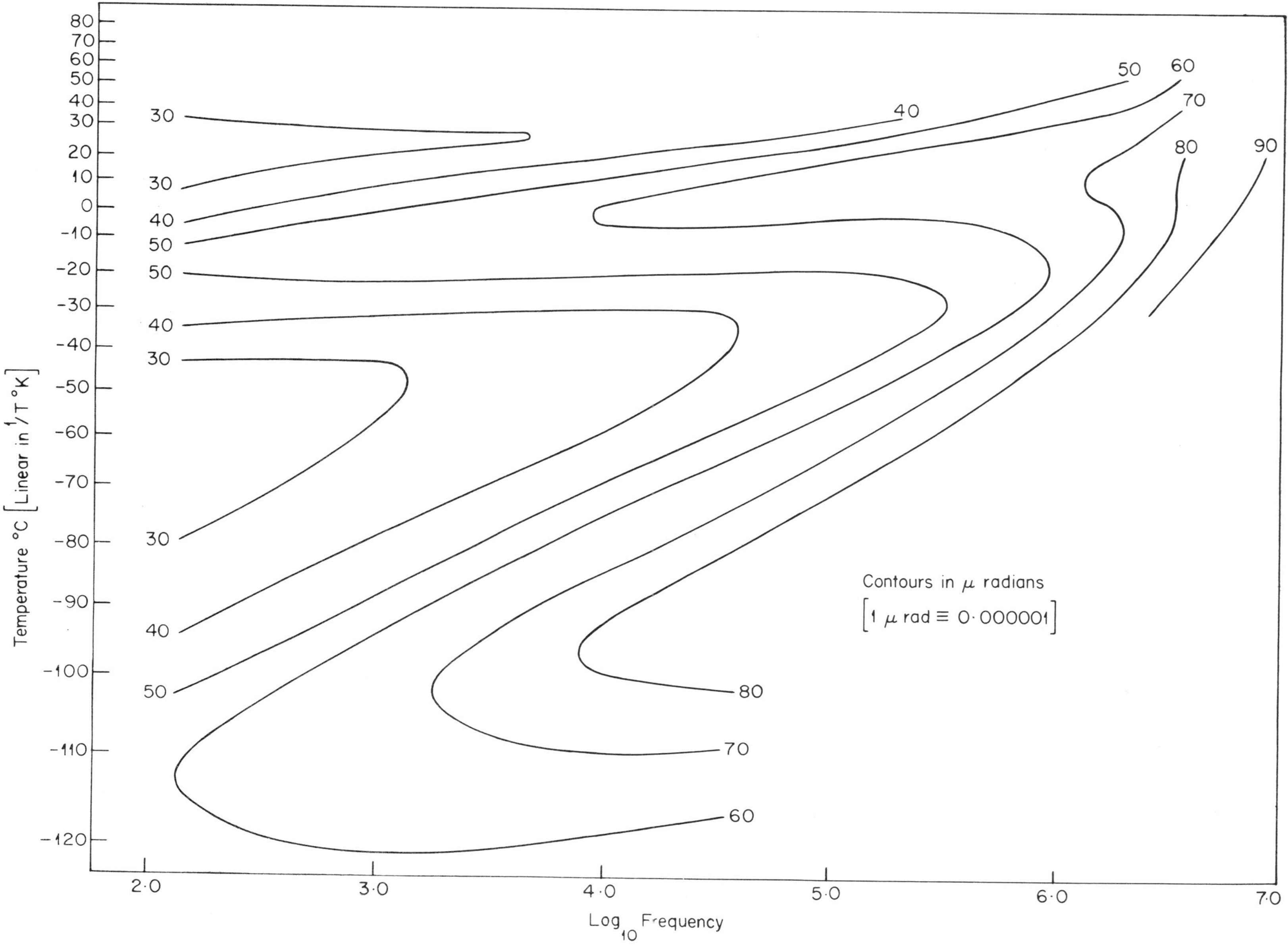


**Figure 8.20.** Loss angle vs frequency and temperature. Polythene, dielectric grade, density 0·928 g/cm$^3$ at 20°C ('Alkathene' Q 3145)

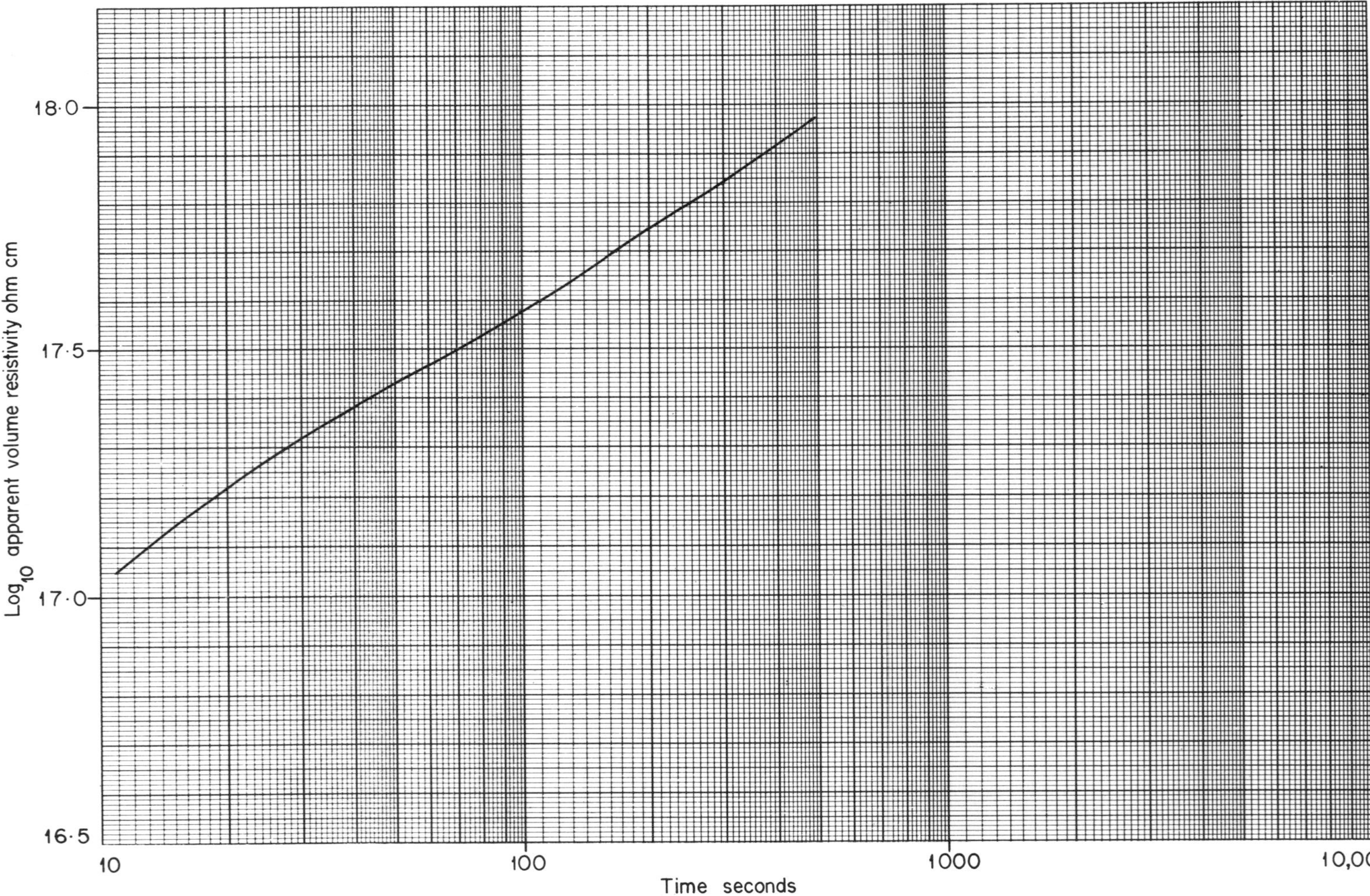


**Figure 8.21.** Apparent volume resistivity vs time of electrification: 20°C. Polythene, density 0.921 g/cm$^3$ ('Alkathene' WJG 11)

# 9

# POLY(4-METHYLPENTENE)

## INTRODUCTION

4-methylpentene-1 is derived from propylene and, like propylene, it can be polymerized in catalyst systems of the Ziegler type to give crystalline, stereoregular polymers. Such polymers were first reported in 1955, but commercial development had to wait a further ten years.

## NATURE

The structure of poly(4-methylpentene) is predominantly isotactic. In normal fabricated articles the crystallinity is about 40 %, although in annealed specimens of the raw material it can be as high as 65 %.

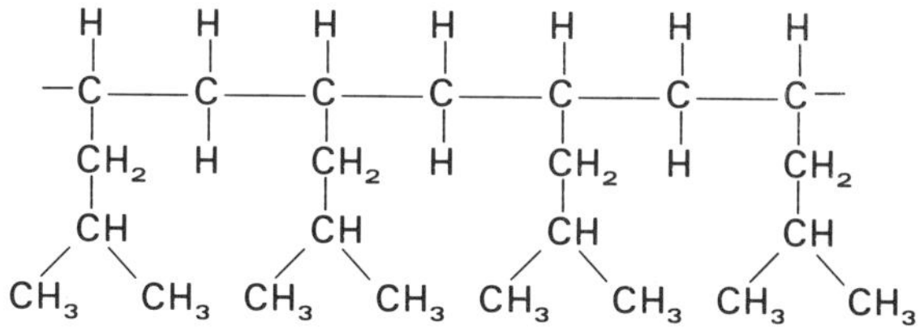

```
   H      H      H      H      H      H      H
   |      |      |      |      |      |      |
 —C——————C——————C——————C——————C——————C——————C—
   |      |      |      |      |      |      |
  CH2     H     CH2     H     CH2     H     CH2
   |             |             |             |
  CH            CH            CH            CH
 /  \          /  \          /  \          /  \
CH3  CH3     CH3  CH3      CH3  CH3      CH3  CH3
```

## GRADES AND FORMS

A series of plastics based on poly(4-methylpentene) can be made. The first to be made available is in two basic forms, transparent and white opaque, either as free-flowing powders or as granules.

In fabricated form, poly(4-methylpentene) is commercially available as rod.

## PROCESSING

Present grades have melt flow properties designed to make them suitable principally for injection moulding, extrusion and blow moulding. They process easily on conventional equipment, the operation being characterized by a high melting temperature, a narrow melting range and a low melt viscosity with significant thixotropy. The melt should not be kept at elevated temperatures for too long, otherwise discolouration and brittleness in the final product can result.

Compression moulding can be used to make block and sheet for machining and welding into prototypes. Sheet can be press formed, but not readily vacuum formed. Extruded film and foil, however, can be vacuum formed, and also oriented provided the temperature is strictly controlled.

Articles made from poly(4-methylpentene) can be finished by printing, painting or metallizing, but only after special preparation of the surface.

## PROPERTIES

Poly(4-methylpentene) exhibits some basic properties typical of a polyolefine, but in addition it possesses some distinctive properties which give it a special role in plastics technology. Some physical properties are given in Table 9.1.

TABLE 9.1

| Property | Units | Typical Value |
|---|---|---|
| Density | $g/cm^3$ | 0·83 |
| Crystalline melting point | °C | 245 |
| Coefficient of linear thermal expansion | | |
| − 50 to 120°C | $°C^{-1}$ | $1{\cdot}2 \times 10^{-4}$ |
| | $°F^{-1}$ | $0{\cdot}7 \times 10^{-4}$ |
| Thermal conductivity, 20 to 70°C | cal/cm s °C | $4 \times 10^{-4}$ |
| | B ThU in/ft² h °F | 1·2 |
| Specific heat, 20 to 150°C | cal/g °C | 0·52 |
| | B ThU/lb °F | 0·52 |
| Flammability | | burns slowly with drips |

### *Deformation*

The deformational behaviour at 20°C of poly(4-methylpentene) is given by the tensile creep curves at three stress levels in Figure 9.1. The corresponding 100 second isochronous stress vs. strain curve is given in Figure 9.2, together with a similar curve for the transparent grade of material. The data in this figure are replotted against logarithmic axes in Figure 9.3. The data of Figures 9.1 to 9.3 are appropriate to an atmosphere of 65% rh, but, as might be expected for a polyolefine with negligible water absorption, the results are not dependent on relative humidity.

Recovery data are given in Figure 9.4, showing as usual the high degree of recovery common with thermoplastics.

With the amorphous phase transition in the temperature range 30°C to 50°C, the deformational properties are very time dependent and sensitive to slight variations in temperature when tested near this temperature range. The 100 second tensile creep modulus (at 0·2% strain) vs. temperature curve given in Figure 9.5 shows this effect. Curves for polypropylene homopolymer and high density polythene are included for comparison, and it can be seen that poly(4-methylpentene) suffers a greater increase in deformation with temperature than these two plastics. However, for higher temperatures above the amorphous phase transition of poly(4-methylpentene) this plastic is superior to polypropylene homopolymer and high density polythene. Such information is relevant for deformational behaviour only and does not guarantee that the plastic will tolerate high temperature conditions for long times, for instance with regard to oxidation resistance. However, it is shown in Figure 9.6 that the recovery after short times of loading, i.e. 100 seconds, and at temperatures in the range 170°C to 210°C, is excellent.

*Limiting Stress*

For long times under load poly(4-methylpentene) fails in essentially a ductile manner, the failure being preceded by crazing or surface cracking, and microvoiding or whitening. As with other plastics, less is known of the factors affecting failure than for deformation. Some data available of tensile creep rupture stress vs. time to failure are given in Figure 9.7.

The effect of temperature on the stress at failure has been determined only in a conventional load elongation test at $\sim 50\%$ extension per minute. Data obtained by this method for the dependence on temperature of the yield stress are given in Figure 9.8, and these show again the marked dependence of mechanical properties on temperatures around ambient conditions, and then the comparatively little further effect up to high temperatures.

*Dynamic Fatigue and Environmental Effects*

Limited dynamic fatigue results show no breakaway from the static fatigue curve up to times of about two days.

Unlike polythene, cracks initiated in stressed poly(4-methylpentene) in an active environment do not propagate in an apparently brittle fashion. Although cracks might be initiated subsequent growth can be impeded, and experience of practical applications indicates that poly(4-methylpentene) is less susceptible to detrimental environments than, say, polythene. However, some agents such as detergent solutions do appear to increase the chance of brittle failure.

*Impact Behaviour*

Figure 9.9 shows the resistance to impact, determined by the Charpy-type impact test, as a function of temperature. With either sharply or bluntly notched specimens, there is little dependence of impact strength on the temperature in the region $-20°C$ to $+40°C$ whilst unnotched specimens improve with temperature very markedly in the region of room temperature. The essential conclusion to be drawn from this work is that, under most practical conditions, the resistance to impact of currently available grades of poly(4-methylpentene) is not likely to be high.

*Electrical Properties*

Like the previous plastics considered, polythene and polypropylene, poly(4-methylpentene) is hydrocarbon in character and has a permittivity of 2·12 at 20°C calculable from the density and the Clausius–Mosotti relationship. The dielectric losses, however, do depend on frequency and temperature and a contour map versus these two variables is given in Figure 9.10 where the loss is expressed in microradians. Figure 9.11 is a 20°C section through the contour map of Figure 9.10 but the data have been obtained over a wider frequency range. With low losses, the values obtained are critically dependent on the formulation of the plastics under consideration, with respect to antioxidant, processing aids etc.

The apparent volume resistivity is given as a function of the time of electrification in Figure 9.12.

Other information obtained indicates that these plastics are not likely to be limited by their electrical properties in any but the most critical applications.

*Optical Properties*

Poly(4-methylpentene) is a material of some clarity: some modified materials are highly transparent, and can be employed in a wide variety of optical applications, because they combine polyolefine properties and high softening point with transparency. Because the modified materials will always be preferred to the unmodified homopolymer for optical uses, only the properties of these modified materials are presented here.

*Refraction*

Refractive index data at 20°C for a variety of spectral wavelengths are presented in Table 9.2. The dispersion curve of refractive index vs. spectral wavelength is given in Figure 9.13.

The derived value of reciprocal dispersive power sometimes known as constringence $V_d$ is 44·9 and the critical angle 43° 25′ for sodium D lines.

TABLE 9.2

| Wavelength (Å) | Refractive index |
|---|---|
| 4358 | 1·47557 |
| 4861 | 1·47274 |
| 5461 | 1·47168 |
| 5893 | 1·46542 |
| 6563 | 1·46539 |

The temperature coefficient of refractive index is shown in Figure 9.14 for the sodium D lines: the derived temperature coefficient of refractive index is −0·00024/°C.

*Transparency*

Modified 4-methylpentene plastics can achieve a very high transparency as shown by the following data for material approximately $\frac{1}{8}$ in in thickness (actually 2·896 mm) and for mercury light of wavelength 5461 Å. Values corrected for surface defects for the various transparency properties have been obtained by 'oiling out' the surface with olive oil.

| | |
|---|---|
| Direct transmission factor | 96–97% |
| Direct transmission factor corrected to 1 mm thickness | 99% |
| Scattering coefficient | 0·11 $cm^{-1}$ |
| Forward scattered fraction (ASTM D1003) | 2·5 to 3·0% |

Reflection from a poly(4-methylpentene) surface at normal incidence should be $[(n-1)/(n+1)]^2$, that is, of the order of 3·5%. However, measurement of this quantity is very sensitive to surface irregularities, even when extreme care is taken to prepare a smooth surface.

*Light Transfer*

With its inherently low absorption demonstrated in the transparency characteristics, poly(4-methylpentene) is likely to have good light transfer properties.

*Chemical Properties*

Poly(4-methylpentene) is highly resistant to inorganic environments, being appreciably affected only by strong oxidizing agents. It shows good resistance to many organic chemicals, although some are absorbed with consequent loss in strength, stiffness and clarity and some present a cracking hazard. Its permeability to gases and water vapour is considerably higher than that of other polyolefines.

TABLE 9.3

| | |
|---|---|
| Equilibrium water content (immersed) | |
| 20°C | <0·05% |
| 60°C | <0·05% |
| Gas permeabilities (cm³ (N.T.P.) cm/cm² s cm Hg) × $10^9$ | |
| Oxygen | 2·7 |
| Nitrogen | 0·7 |
| Water vapour permeability (g/m² 24 h) at 38°C and 90% rh for 0·001 in thickness | 95–110 |
| Resistance to: | |
| Mineral acids (dil.) | Excellent |
| Mineral acids (conc.) | Excellent (slightly attacked by conc. oxidizing acids) |
| Alkalis | Excellent |
| Solvents: | |
| alcohols | Excellent |
| ketones | Fair–good |
| aromatic hydrocarbons | Poor |
| chlorinated hydrocarbons | Poor |
| Detergent solutions | May show stress cracking |
| Greases and oils | Excellent |

## APPLICATIONS

Methylpentene polymers are being used and evaluated in five main areas: lighting, electrical, medical, packaging and applications demanding a combination of clarity and chemical resistance. They are not recommended for continuous outdoor exposure.

*Typical Examples*

High intensity internal lighting fittings, metallized lamp reflectors.
High frequency coaxial connectors and electronic components.
Hospital ware, syringes, tubing and connectors.
Containers for prepacked foods.
Laboratory ware, sight glasses, waste traps, catch pots.
Textile bobbins.

## LIST OF FIGURES

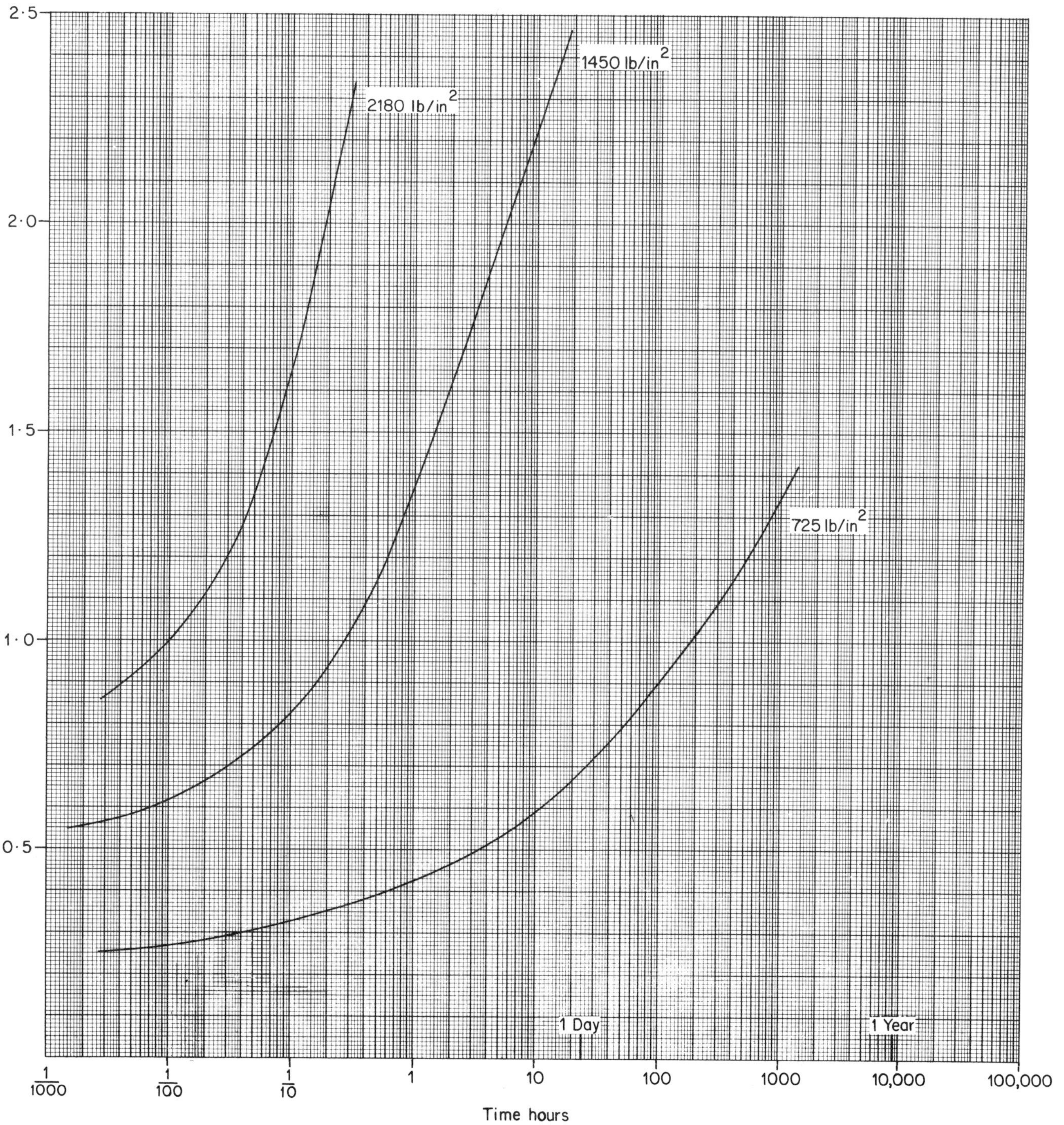


**Figure 9.1.** Creep curves in tension: 20°C. Poly(4-methylpentene) ('TPX'-RO)

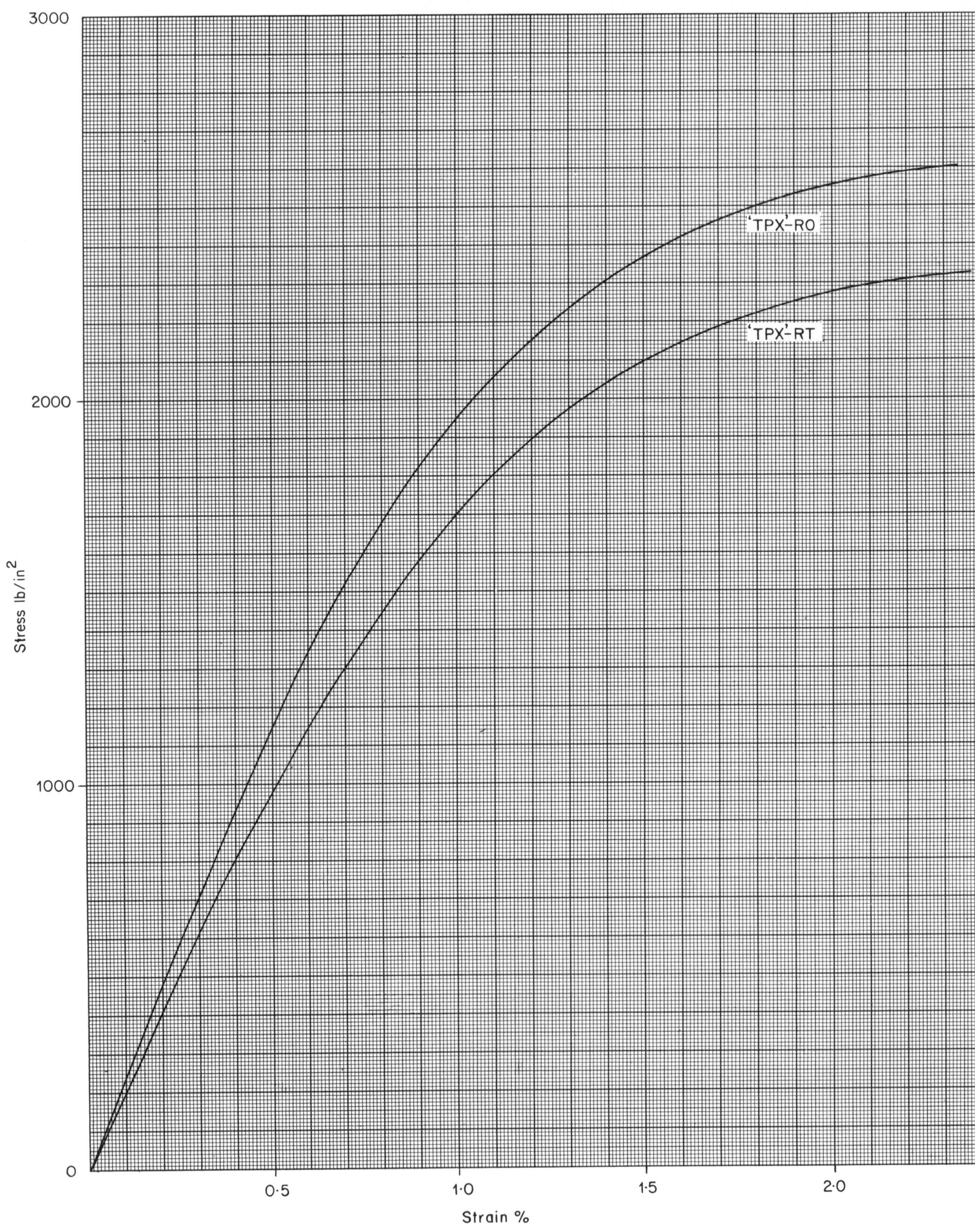


**Figure 9.2.** Isochronous stress vs strain curves: 20°C, 100 sec. Poly(4-methylpentene), ('TPX'-RO, 'TPX'-RT)

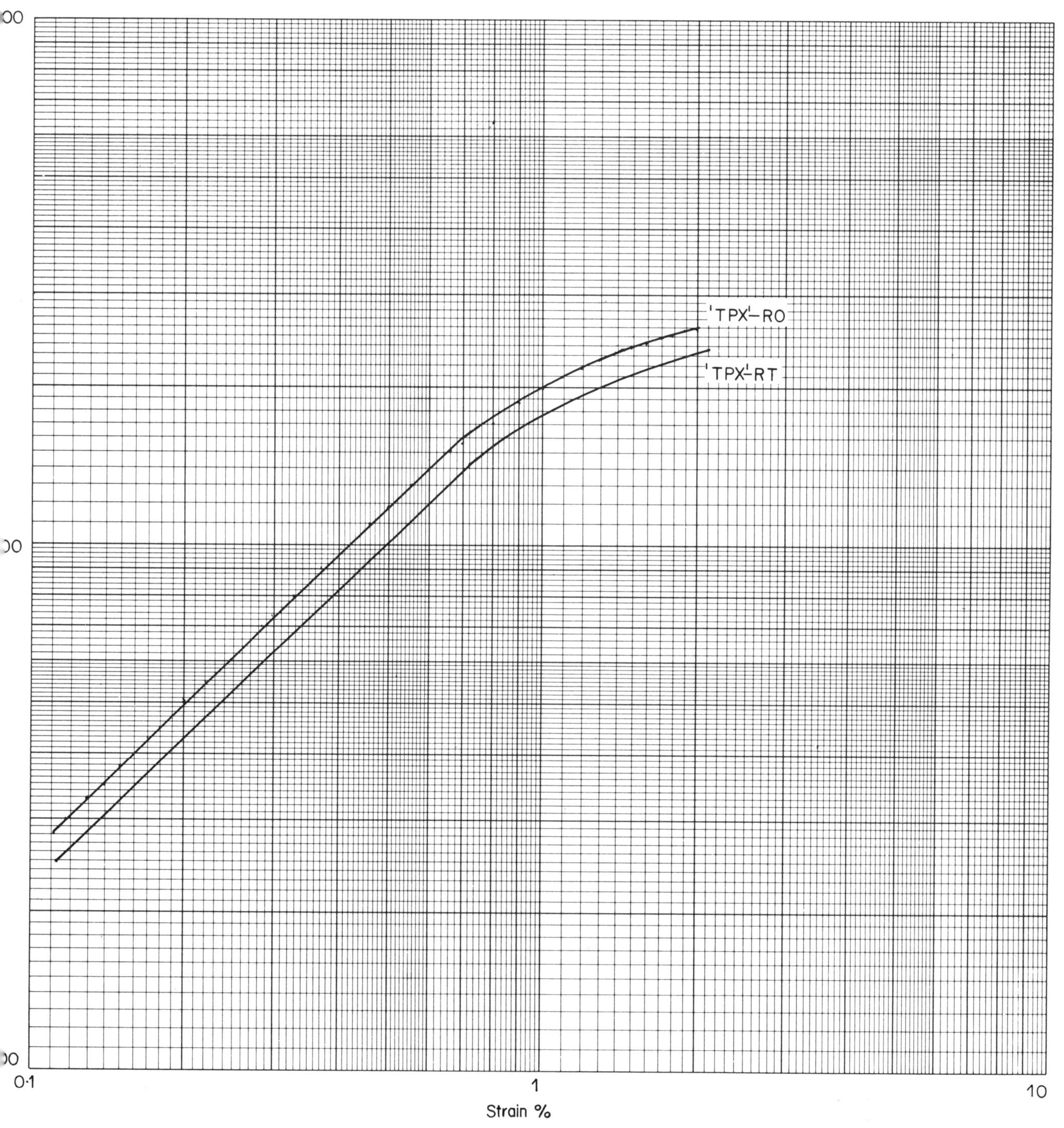


**Figure 9.3.** Isochronous stress vs strain curves: 20°C, 100 sec, logarithmic axes. Poly(4-methylpentene) ('TPX'-RO, 'TPX'-RT)

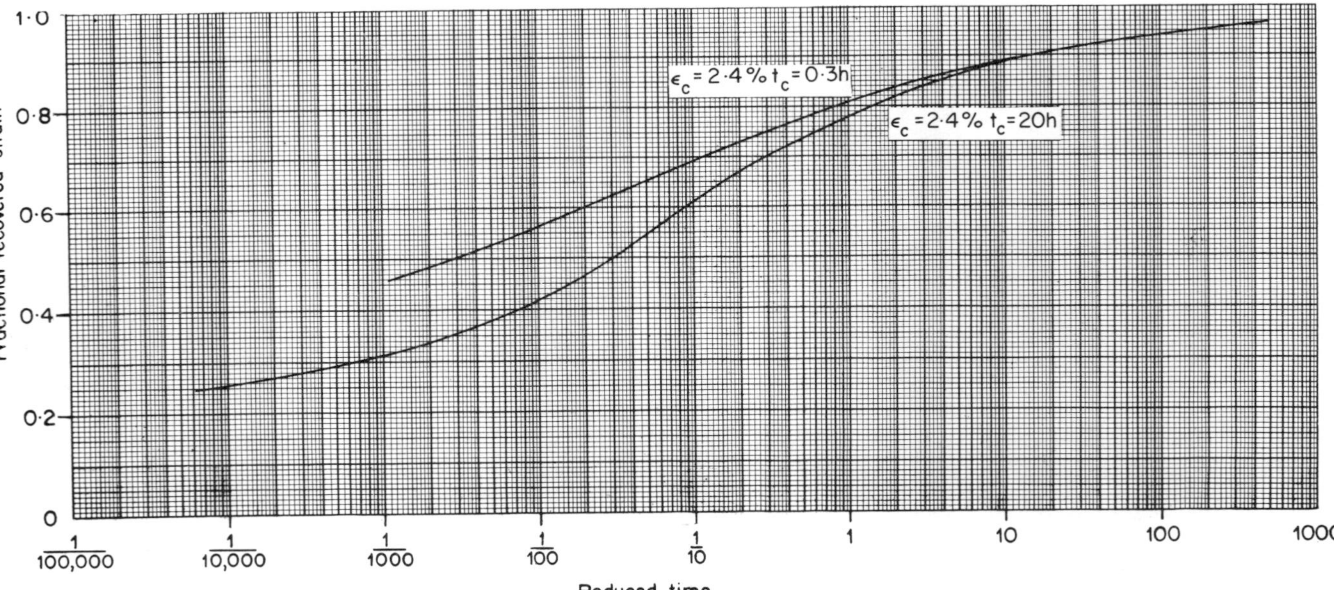


**Figure 9.4** Recovery from creep in tension: 20°C. Poly(4-methylpentene) ('TPX'-RO)

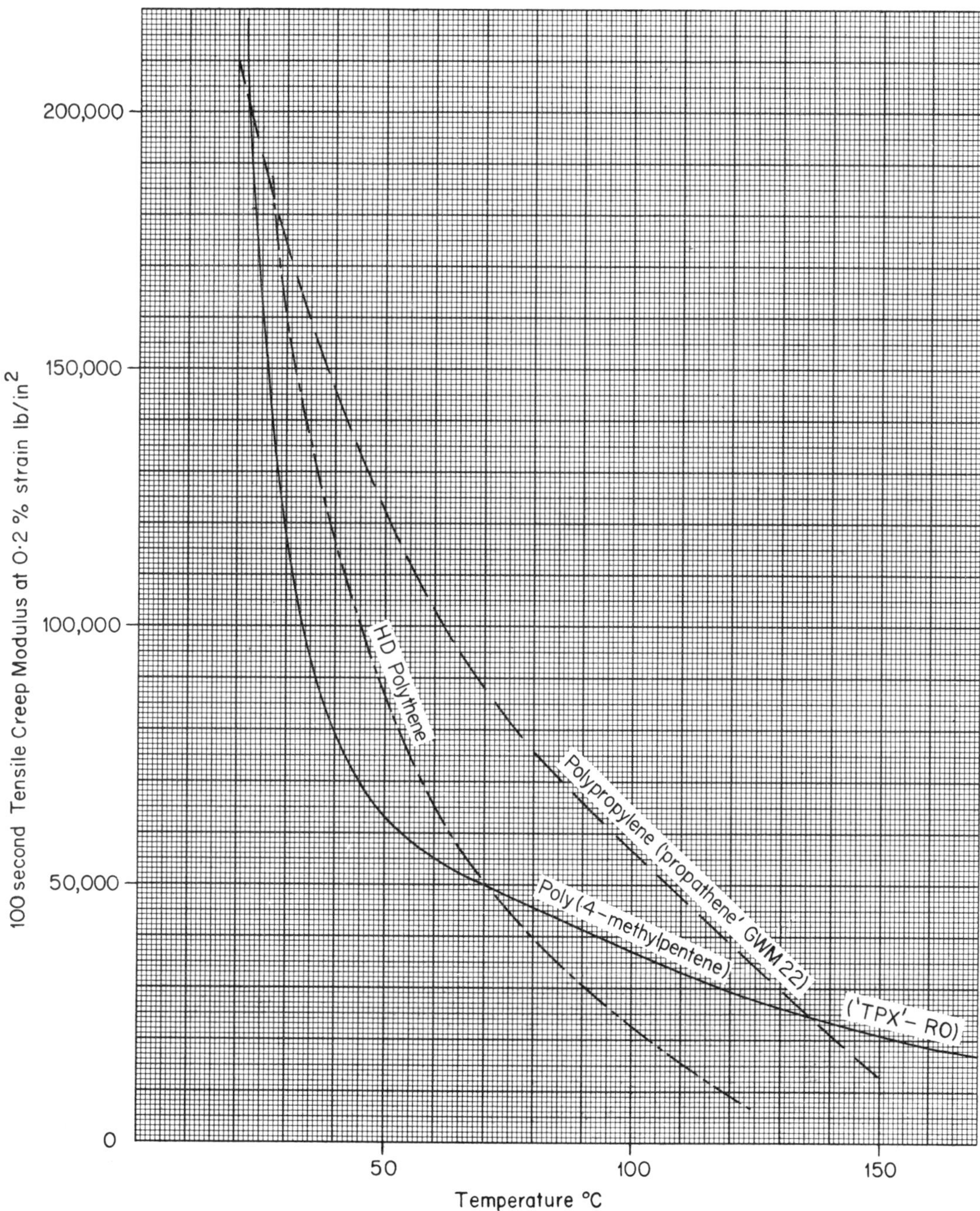


**Figure 9.5.** Tensile creep modulus (100 sec, 0·2% strain) vs temperature. Comparison of poly(4-methylpentene) with polypropylene and HD polythene ('TPX'-RO)

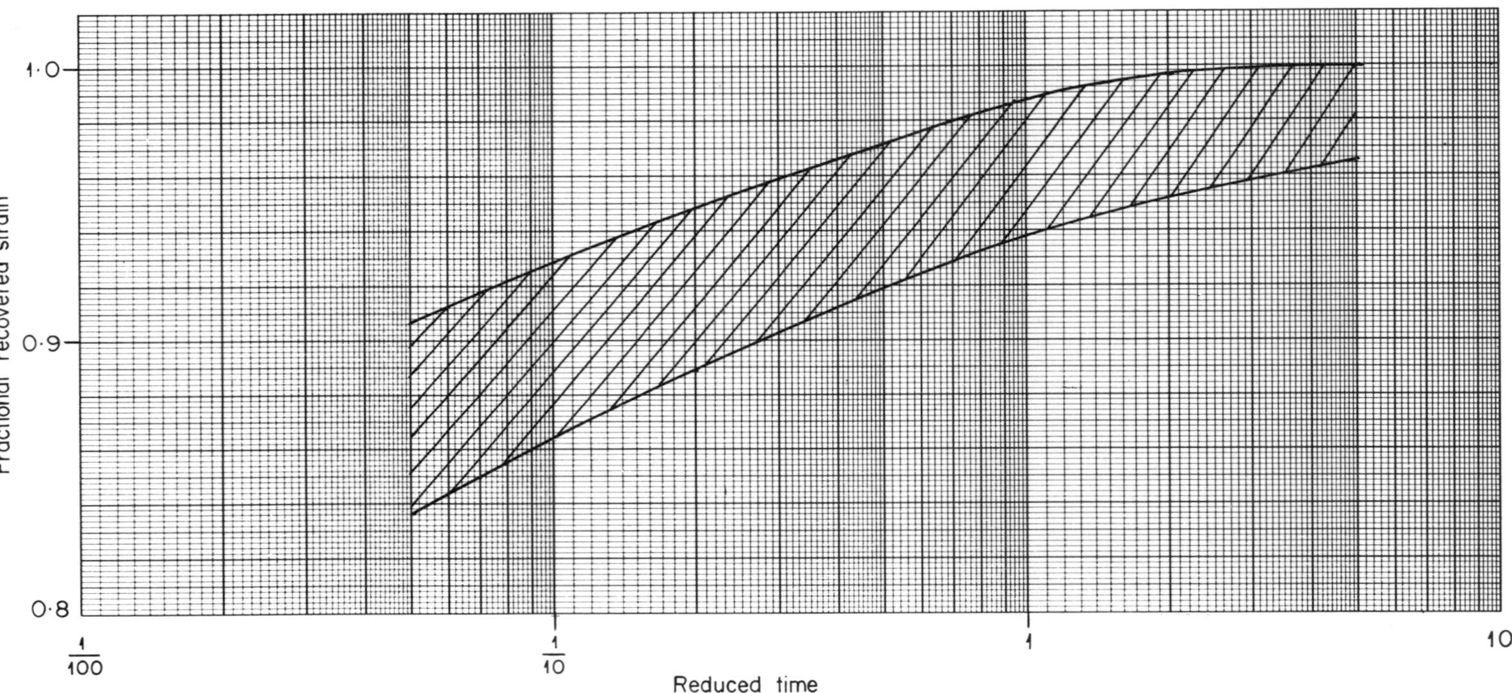


**Figure 9.6.** Recovery from creep in tension: temperatures 170°C–210°C. Poly(4-methylpentene) ('TPX'-RO)

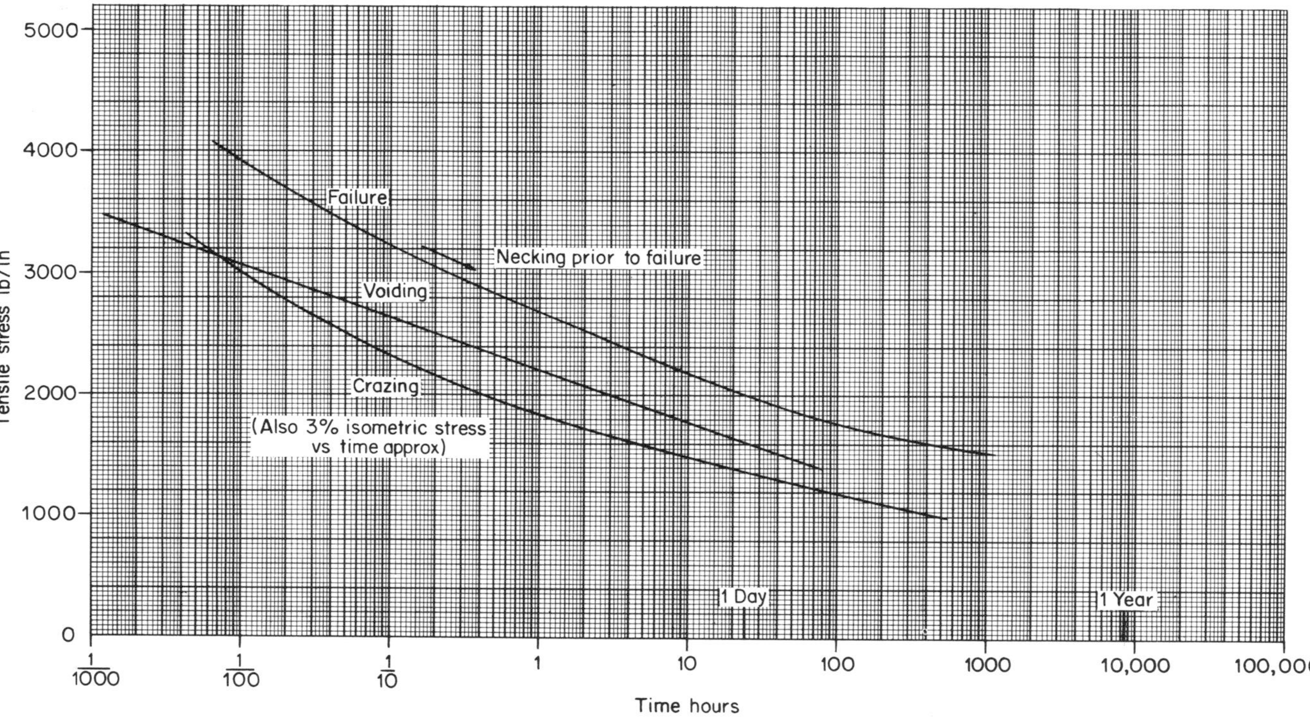


**Figure 9.7.** Creep rupture stress in tension vs time to failure: 20°C. Poly(4-methylpentene) ('TPX'-RT)

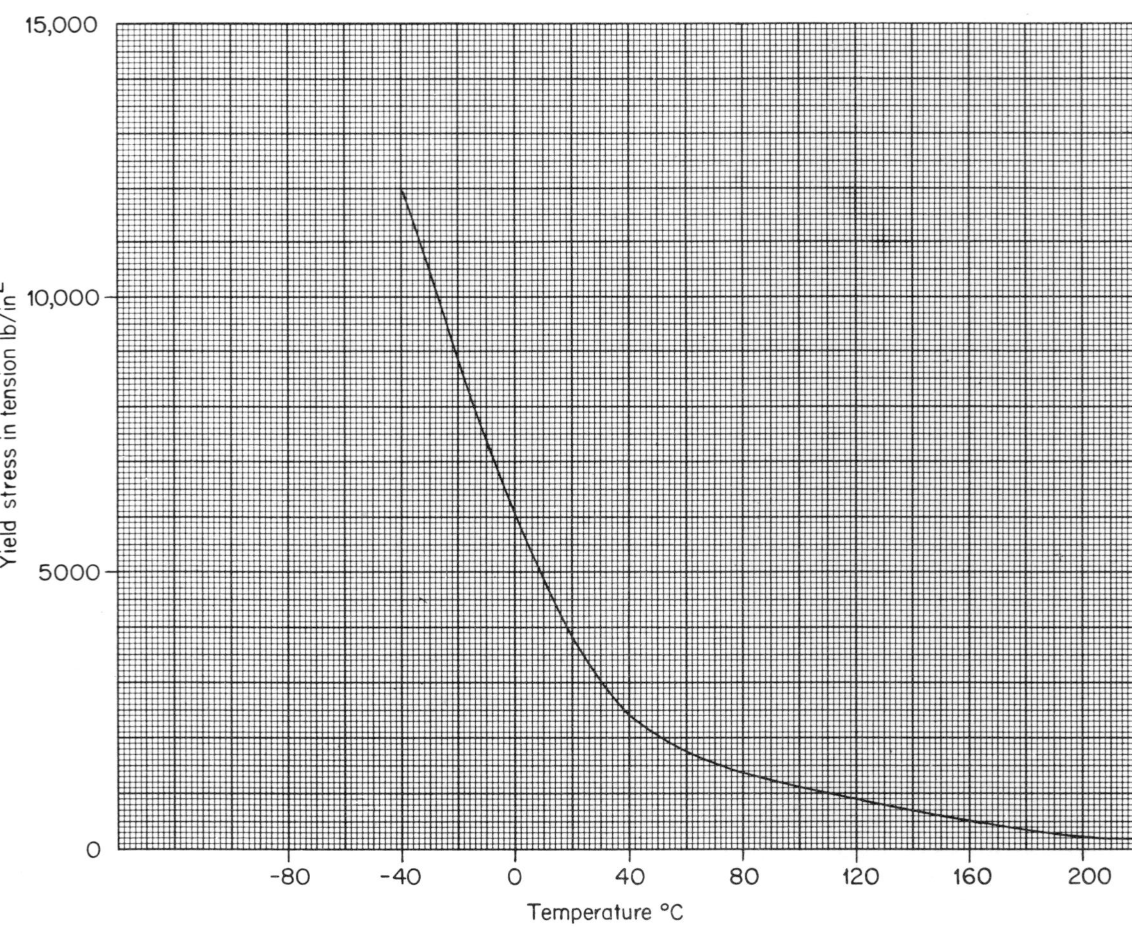


**Figure 9.8.** Yield stress in tension vs temperature: 50% per min straining rate. Poly(4-methylpentene) ('TPX'-RO)

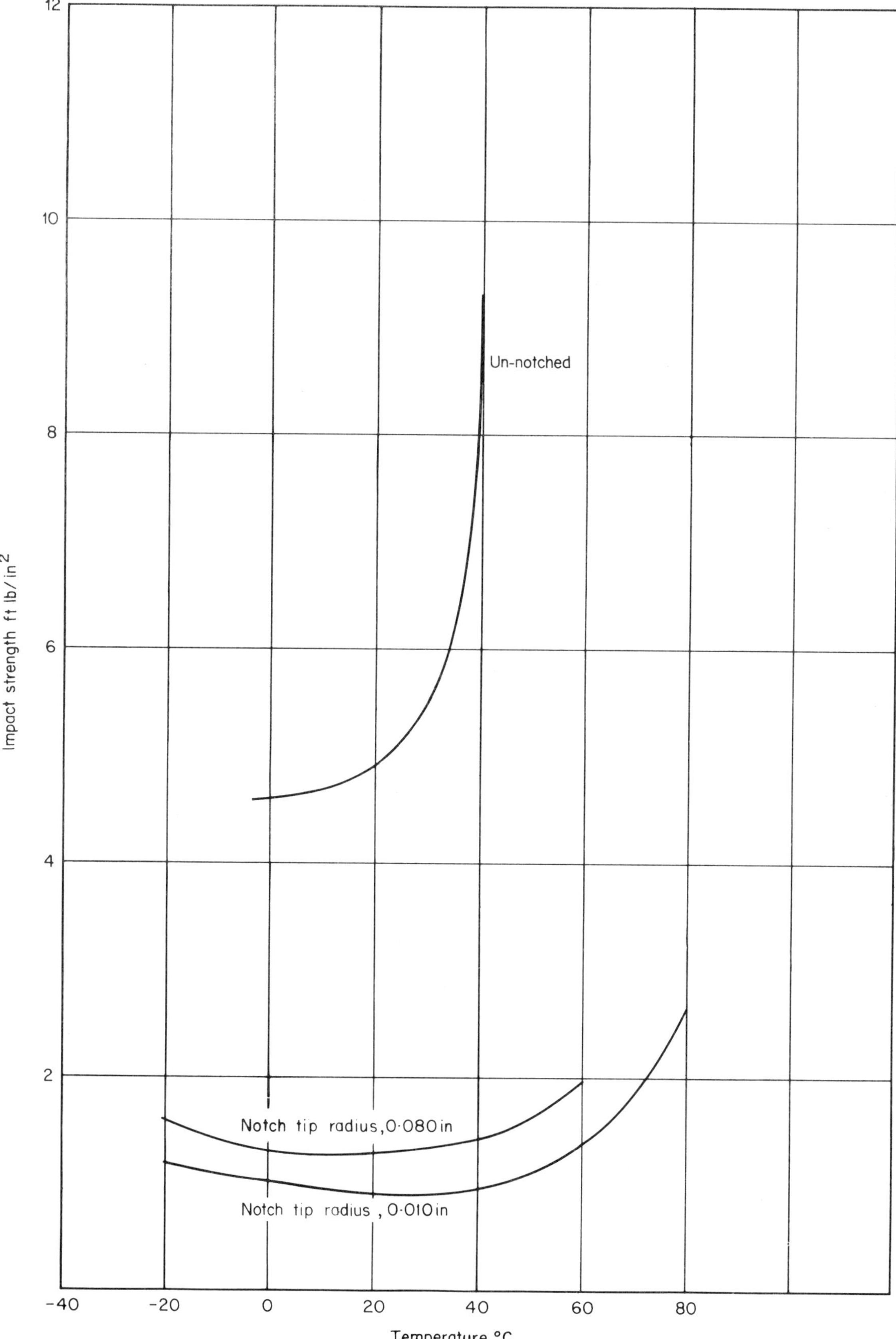


**Figure 9.9.** Impact strength vs temperature. Poly(4-methylpentene) ('TPX'-RT)

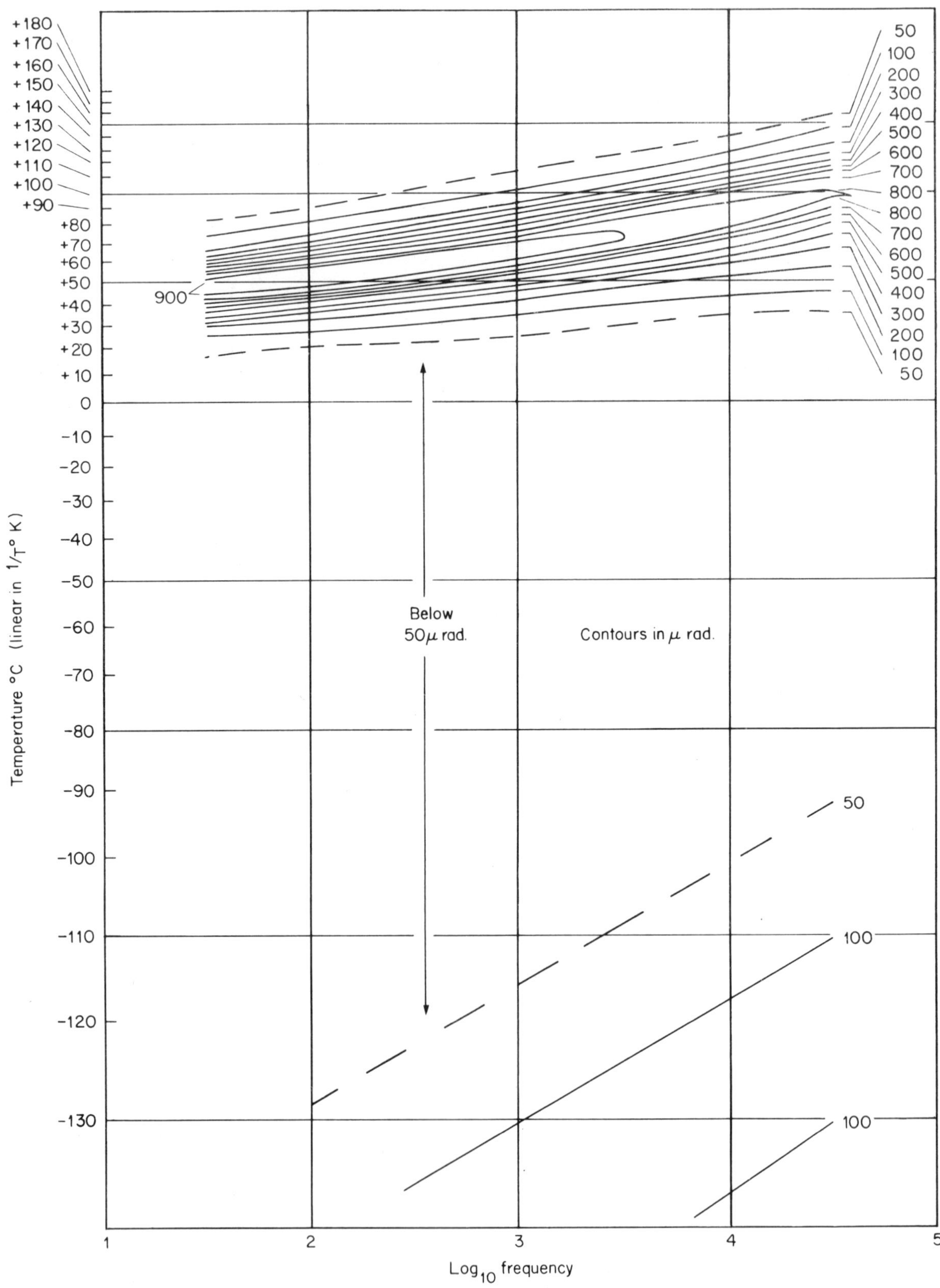


**Figure 9.10.** Loss angle vs frequency and temperature. Poly(4-methylpentene) ('TPX'-RT)

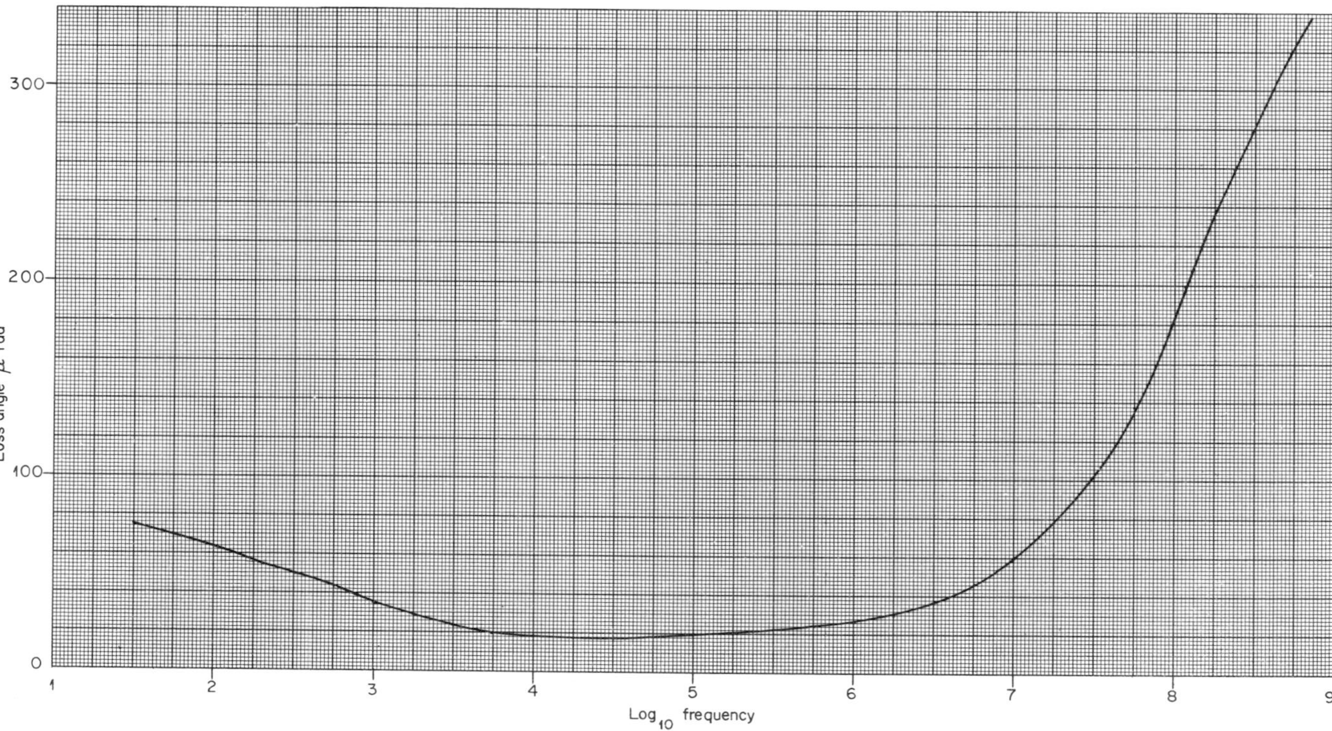


**Figure 9.11.** Loss angle vs frequency: 20°C. Poly(4-methylpentene) ('TPX'-RT)

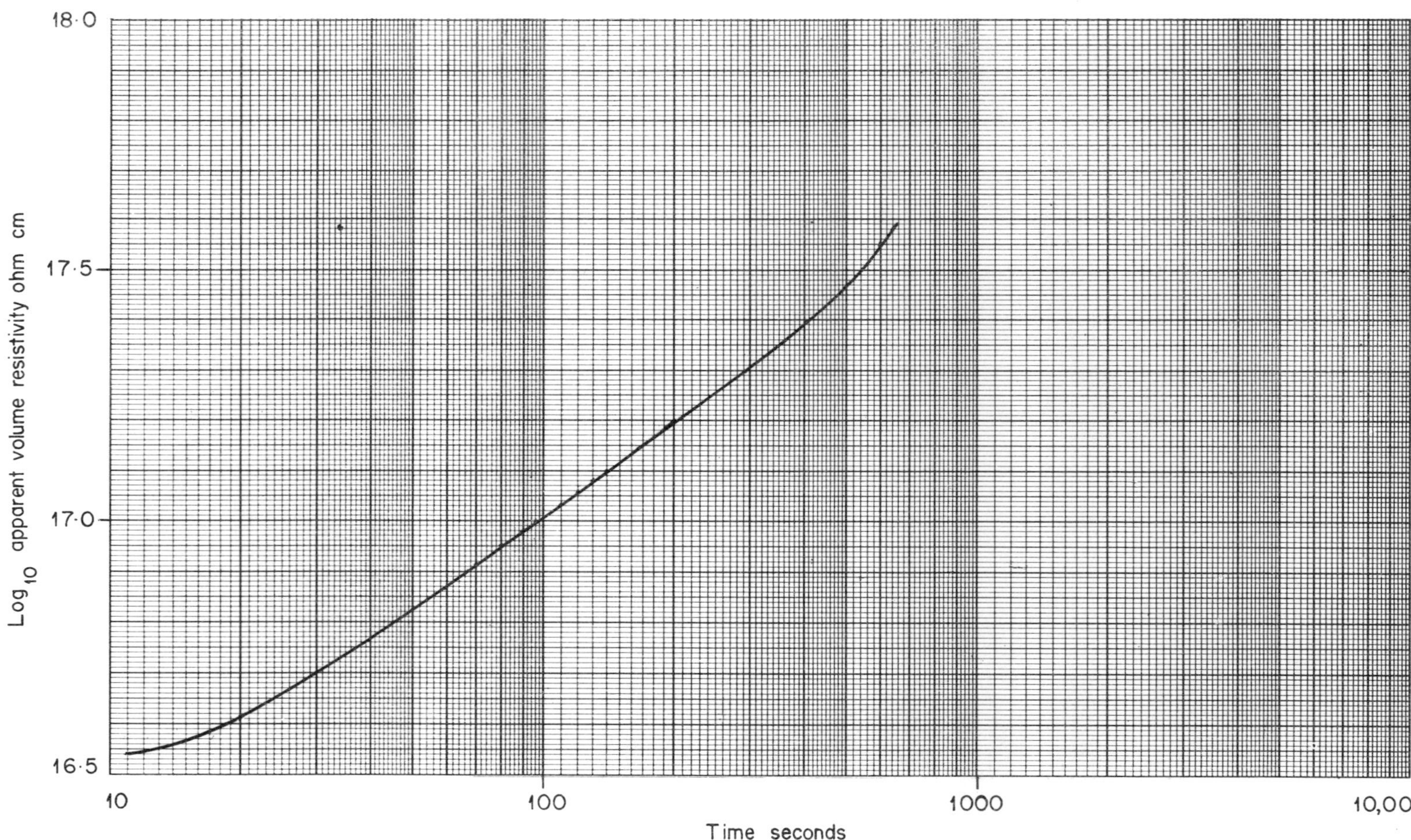


**Figure 9.12.** Apparent volume resistivity vs time of electrification: 20°C. Poly(4-methylpentene) ('TPX'-RT)

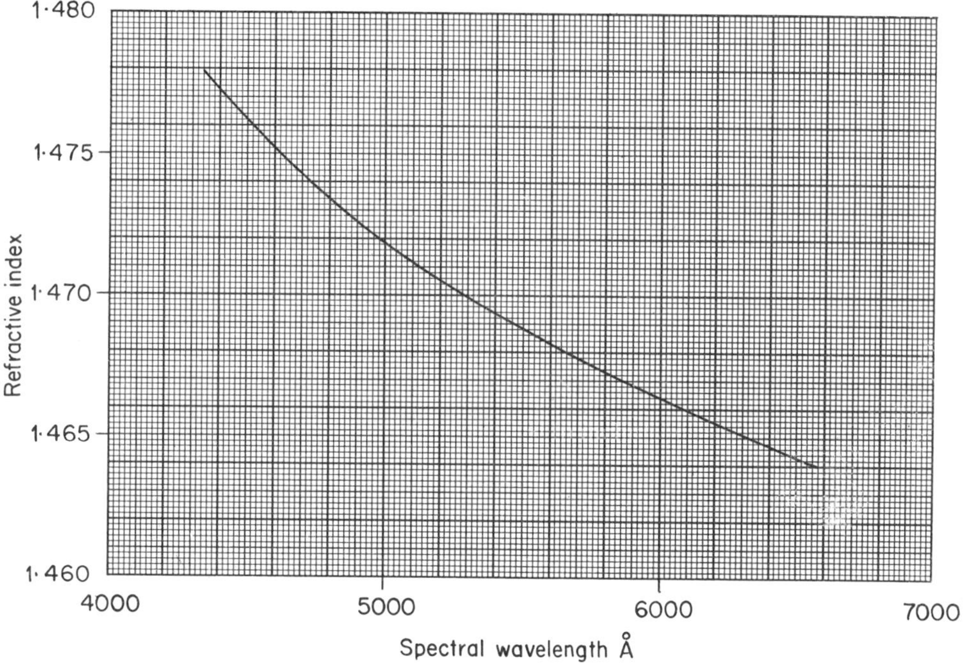


**Figure 9.13.** Optical dispersion curve: 20°C. Poly(4-methylpentene) ('TPX'-RT)

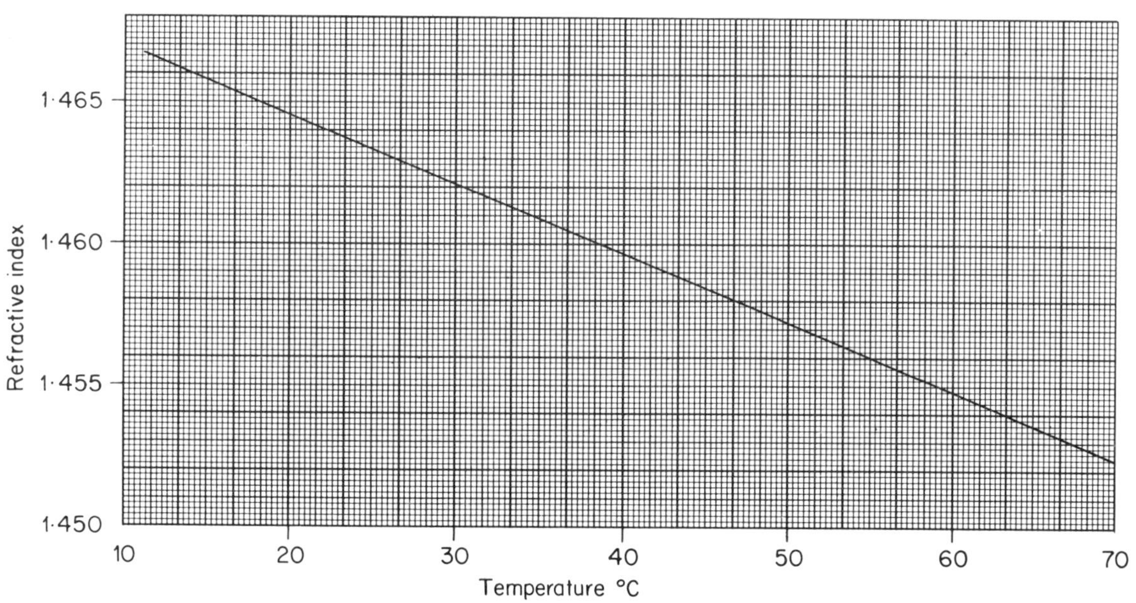


**Figure 9.14.** Refractive index vs temperature curve: wavelength 5893 Å. Poly(4-methylpentene) ('TPX'-RT)

# 10

# ACETAL COPOLYMER

## INTRODUCTION

Acetal copolymer is one of a family of thermoplastics which are similar in structure to the polyolefines, the principal difference being that the backbone of the acetal polymer chain consists of alternate carbon and oxygen atoms. Several members of the acetal family have been made, but undoubtedly the most useful, and therefore the most important, are the homopolymers and copolymers of formaldehyde, commonly referred to as acetals or acetal resins, less frequently as polyacetals, and occasionally as polyformaldehydes or polyoxymethylenes.

High polymers of formaldehyde were first observed in 1859 by the Russian, Butlerov, but it was not until almost exactly a hundred years later that a useful polymer was developed which did not revert to formaldehyde on heating. The copolymers of formaldehyde, known as acetal copolymers, were developed soon afterwards; the monomer used is, in fact, trioxane.

The acetals are outstanding amongst thermoplastics for their high rigidity, which is combined with high tensile yield strength, good fatigue strength, and high temperature resistance. They have also excellent dimensional stability in a wide range of environments.

## NATURE OF ACETALS

Both acetal copolymer and homopolymer are highly crystalline materials. The main difference in their structure is that whereas the homopolymer consists of simple polyformaldehyde chains, the copolymer chains contain occasional ethylene units.

$$\begin{array}{ccccccc} & H & & H & & H & \\ & | & & | & & | & \\ - & C & -O- & C & -O- & C & -O- \\ & | & & | & & | & \\ & H & & H & & H & \end{array}$$

Homopolymer

$$\begin{array}{ccccccccccc} & H & & H & & H & & H & & H & \\ & | & & | & & | & & | & & | & \\ - & C & -O- & C & - & C & -O- & C & -O- & C & -O- \\ & | & & | & & | & & | & & | & \\ & H & & H & & H & & H & & H & \end{array}$$

Copolymer

The ends of the homopolymer molecule have to be reacted with an organic group to hinder reversion to formaldehyde; the copolymer, however, has a built-in stabilizing system in that the chains terminate in a stable ethylenic residue, and even if the chain should become ruptured depolymerization will proceed only as far as the next ethylene unit. The copolymer therefore has the advantage of being more stable, particularly at high temperatures and in the presence of hot water or alkaline solutions.

Further discussion here is restricted to the copolymer, and the data given have been obtained from copolymer samples.

## GRADES AND FORMS

Acetal copolymer is available in grades suitable for injection moulding, extrusion and extrusion blow moulding. Ultraviolet stabilized grades for outdoor use are also available: these stabilized grades are, however, not often required because black, unstabilized grades have excellent weathering resistance, and white and pastel coloured compounds also are resistant to outdoor exposure. Glass-filled acetal copolymer can be used when maximum stiffness is required.

Acetal copolymer is supplied as free-flowing, cylindrical pellets. In its natural form it is a translucent white colour, but coloured articles can easily be made, either by using a coloured compound or by dry-colouring the material before processing.

Various semifabricated forms are available, the chief ones being rod, sheet and tube.

## PROCESSING

Because of its molecular structure acetal copolymer has a relatively sharp melting point. The melt is free-flowing and has a tendency to give off formaldehyde if overheated; the processing instructions set out in the raw material supplier's literature should therefore be closely followed. Provided these precautions are taken, however, acetal copolymer can be injection moulded easily, and it can also be extruded and blow-moulded. Sheet made from the copolymer may be heat shaped.

Welding and adhesives may both be used to join acetal copolymer parts, and components can be machined by standard techniques. Interesting possibilities exist for decorating the surfaces of acetal components; processes such as electroplating, vacuum metallizing, surface dyeing, printing and painting have proved successful.

## PROPERTIES

The physical properties of acetal copolymer are characteristic of a high molecular weight, crystalline polymer. Some basic properties are given in Table 10.1.

TABLE 10.1

| Property | Units | Typical Value |
|---|---|---|
| Density | g/cm³ | 1·410 |
| Crystalline melting point | °C | 163 |
| Coefficient of linear thermal expansion | | See Figure 10.18 |
| Thermal conductivity | cal/cm s °C | $5{\cdot}5 \times 10^{-4}$ |
| | B ThU in/ft² h °F | 1·6 |
| Specific heat | cal/g °C | 0·35 |
| | B ThU/lb °F | 0·35 |
| Flammability | | burns slowly with drips and steady blue flame, giving off formaldehyde |

### *Deformation*

The deformational behaviour at 20°C of acetal copolymer is given by the tensile creep curves of Figure 10.1 (for times of up to three years). The corresponding isochronous stress vs. strain, isometric stress vs. time, and tensile creep modulus vs. time curves are given in Figures 10.2, 10.3 and 10.4; these data are appropriate to a moulding grade of medium molecular weight acetal copolymer. It can be judged from Figures 10.1 to 10.4 that acetal plastics have a much greater rigidity than is obtainable from materials considered in earlier chapters. At the same time one more degree of complexity is added to the analysis because these materials absorb a small amount of water and their mechanical properties are affected thereby. On the other hand, they are not so strongly influenced by thermal history.

Figure 10.5 compares the 100 second isochronous stress vs. strain curve for dry material (62 days over silica gel) with that of wet material (61 days immersion in water at 20°C). These data approximate to the extremes of behaviour to be encountered. From this figure, and assuming a water content of about 0·8 % at saturation, the effect of water content can be summarized as a 1·5 % decrease in isometric stress or creep modulus for each 0·1 % increase of water, but this is an approximate guide only.

If the 100 second isochronous stress–strain relationship from Figure 10.2 is superimposed on Figure 10.5 it will be seen to fall between the dry and wet curves but not far below the former, even though the nominal rh was 55 %. This fact arises out of different storage histories of the test specimens. Different moulding conditions have little effect on, say, the 100 second isochronous stress vs. strain curve, but subsequent storage at room temperature is accompanied by a small but detectable increase in the stress–strain ratio. This change also occurs during creep tests, of course, because it is impossible to delay the start of tests until an equilibrium state has been reached. The interpolation procedures that generate the family of creep curves depicted in Figure 10.1 utilize an isochronous curve and it was decided to use one typical of a long storage time; this causes the slight discrepancy between Figures 10.2 and 10.5. The absorption of water results also in a small, but important, increase in the dimensions of a specimen or component. The experimental results presented in Figure 10.6 illustrate this concisely. The changes of length as two standard creep specimens, one initially wet and one initially dry, come into equilibrium with a 55 % rh atmosphere is a virtually foolproof method of assessment. Full equilibrium as the basis of no further change in dimensions for a $\frac{1}{8}$ in thick specimen takes roughly three months. Immersion of a dry acetal copolymer component in water at 20°C would result in an increase in linear dimensions of 0·4 % whilst the change from dry to equilibrium with air at 20°C and 55 % rh is approximately 0·15 %. Such changes might be important in precision components.

Data on recovery from tensile creep at 20°C, are given in Figure 10.7. The usual pattern of behaviour of almost immediate and complete recovery for small strains and short times, and less rapid recovery for large strains and long times is followed.

The effect of temperature on the deformational behaviour can be assessed from Figure 10.8 of 100 second tensile creep modulus (at 0·2 % strain) vs. temperature whence it can be deduced that the creep modulus of this figure is decreased by about 1 % in value for each 1°C rise in temperature above 20°C. This correction factor can be applied to the creep moduli at long times to a reasonable approximation, although it is an underestimation for temperatures higher than about 30°C and should be recalculated from Figure 10.8 for specific design purposes.

### *Limiting Stresses*

The failure of acetal plastics after long times under load is essentially of a ductile type following its behaviour in a conventional tensile test. Catastrophic failure of the specimen is preceded by some necking but the two events follow each other reasonably closely in time for constant load conditions and are not distinguished in Figure 10.9 where the material under test is a moulding grade of medium molecular weight (MMW). Again, this failure curve is likely to be markedly affected by a range of variables, including temperature, humidity, orientation, etc. and it is recommended that the 3 % strain isometric stress vs. time curves should be used instead. The failure curve is plotted in Figure 10.9 for 20°C, 65 % rh.

Illustrative data showing the effect of temperature on yield stress in a conventional tensile test for medium molecular weight material are given in Figure 10.10.

The dynamic fatigue resistance of the medium molecular weight grade acetal plastic measured under the standard experimental conditions defined in Chapter 5, i.e. fully reversed, square wave stress cycling at 30 cycles per minute, appears to be very good, at least up to 4000 lb/in$^2$. This result, plotted in Figure 10.11, is of the same general order as that derived from high frequency (30 Hz) experiments.

*Impact Behaviour*

The impact strength of notched specimens of acetal copolymer is given as a function of temperature in Figure 10.12, where it can be seen that increasing the stress concentration by decreasing the notch tip radius from 0·080 in to 0·010 in markedly reduces impact strength. The same effect is shown in Figure 10.13 where tests on sharply-notched specimens discriminated less between materials than did tests on specimens with a blunt notch.

Figure 10.13 compares the impact behaviour of three different molecular weight grades of acetal copolymer. Surprisingly, perhaps, the three curves do not follow a logical sequence in that the data show the high molecular weight grade to be inferior under these particular test conditions. A possible explanation of this anomaly, which is likely to be peculiar to mouldings similar to the test mouldings, is that increased residual orientation more than compensates for the improvement in impact strength normally associated with increased molecular weight. This observation calls for special care and consideration of the precise working conditions in specifying the optimum grade for any application.

Some work has been done on the effect of moulding conditions in the range possible for acetal copolymers: variation in these has had little effect on the impact strength of either the medium molecular weight or the high molecular weight materials, but high mould and cylinder temperatures impair the strength of the low molecular weight material.

*Friction, Abrasion and Wear*

Acetal copolymers have low coefficients of friction in the unlubricated state when mated with various metals under light applied load conditions; values of 0·15 are often quoted. The corresponding value for acetal copolymer rubbing against itself is 0·35.

These low friction values in combination with other properties make acetal copolymers suitable for a range of bearing and gear applications. As described in Chapter 6, an approximate guide to the performance of a bearing is in terms of its $PV$ value, the product of the pressure on the projected area of the bearing and the velocity of sliding. The limits for this quantity are given in Figure 10.14 as a function of velocity, but these cannot be regarded as more than an approximate guide, and should be used only as an assessment of whether a proposed application warrants the making of a prototype.

*Electrical Properties*

The electrical properties of acetal copolymers, although not in the same class as those of the polyolefines, are adequate for many applications and compare favourably with the properties of materials already found in such non-critical uses. The permittivity is low and the loss tangent reasonably constant in the room temperature region at audio frequencies (10–10,000 Hz). Data showing the general behaviour are given in Figures 10.15 and 10.16.

The apparent volume resistivity is approximately $10^{14}$ ohm cm and depends on the time of electrification in the expected manner as shown in Figure 10.17.

*Optical Properties*

Acetal plastics are not at all transparent: the direct transmission factor, even for a sheet of thickness 0·1 mm, is essentially zero.

Refraction measurements indicate a refractive index of 1·489 for 5893 Å spectral wavelength and a critical angle of 42° 51′ for a material of density 1·413 g/cm$^3$.

*Chemical Properties*

Acetal copolymer has excellent resistance to alkalis and is resistant to most organic solvents. Weak acids, including organic acids, and materials which liberate acid substances, such as moist chlorine, bleach and acetic anhydride, will sometimes cause deterioration, the extent depending on the strength of the solution, the temperature and the time of immersion. Long-term immersion tests have shown that concentrated mineral acids such as sulphuric, nitric, hydrochloric and chromic acids eventually cause the material to become unserviceable. Similarly, ethylene dichloride, acetone and ethyl acetate will attack acetal plastics after prolonged immersion, and cause a deterioration in mechanical properties.

TABLE 10.2

| | |
|---|---|
| Equilibrium water content (immersed) | 0·8% |
| Gas permeabilities (cm$^3$ cm/cm$^2$ s cm Hg) × $10^9$ | |
| Carbon dioxide | 0·004 |
| Oxygen | 0·003 |
| Nitrogen | 0·001 |
| Water vapour permeability (g/m$^2$ 24 h) for 0·001 in thickness | 42 |
| Resistance to: | |
| Mineral acids (dilute) | Fair |
| Mineral acids (conc.) | Poor |
| Alkalis | Excellent |
| Solvents: | |
| alcohols | Excellent |
| ketones | Good |
| aromatic hydrocarbons | Excellent |
| chlorinated hydrocarbons | Good |
| Detergents | Excellent |
| Greases and oils | Good |

## APPLICATIONS

Acetal copolymer is used mainly in the form of injection mouldings. Its combination of properties, particularly its excellent retention of mechanical properties at temperatures up to around 100°C and in a wide variety of chemical environments, suggests its great potential as a replacement for metal, especially die-cast zinc and aluminium, bronze and brass.

Bearings and gears made from it can be run without lubrication under light loading, but at higher loads, water or a non-acidic oil or grease should be used as a lubricant. Because of its good creep properties, acetal copolymer has good holding power for snap fits, items such as bushes, and inserts which require interference fits; self-tapping screws also can be firmly held.

*Typical examples*

Car components, e.g. instrument housings, gears for speedometers and windscreen wipers, choke controls, body trim clips.

Plumbing components, e.g. taps and pipe fittings, ball valves, valve seats and glands, shower heads. Extractor fan parts, door and window furniture, washing machine agitators, sewing machine parts.

Terminal blocks, meter components, coil formers, pump and motor housings.

Seat belt buckles, conveyor belting slats, textile machinery parts.

## LIST OF FIGURES

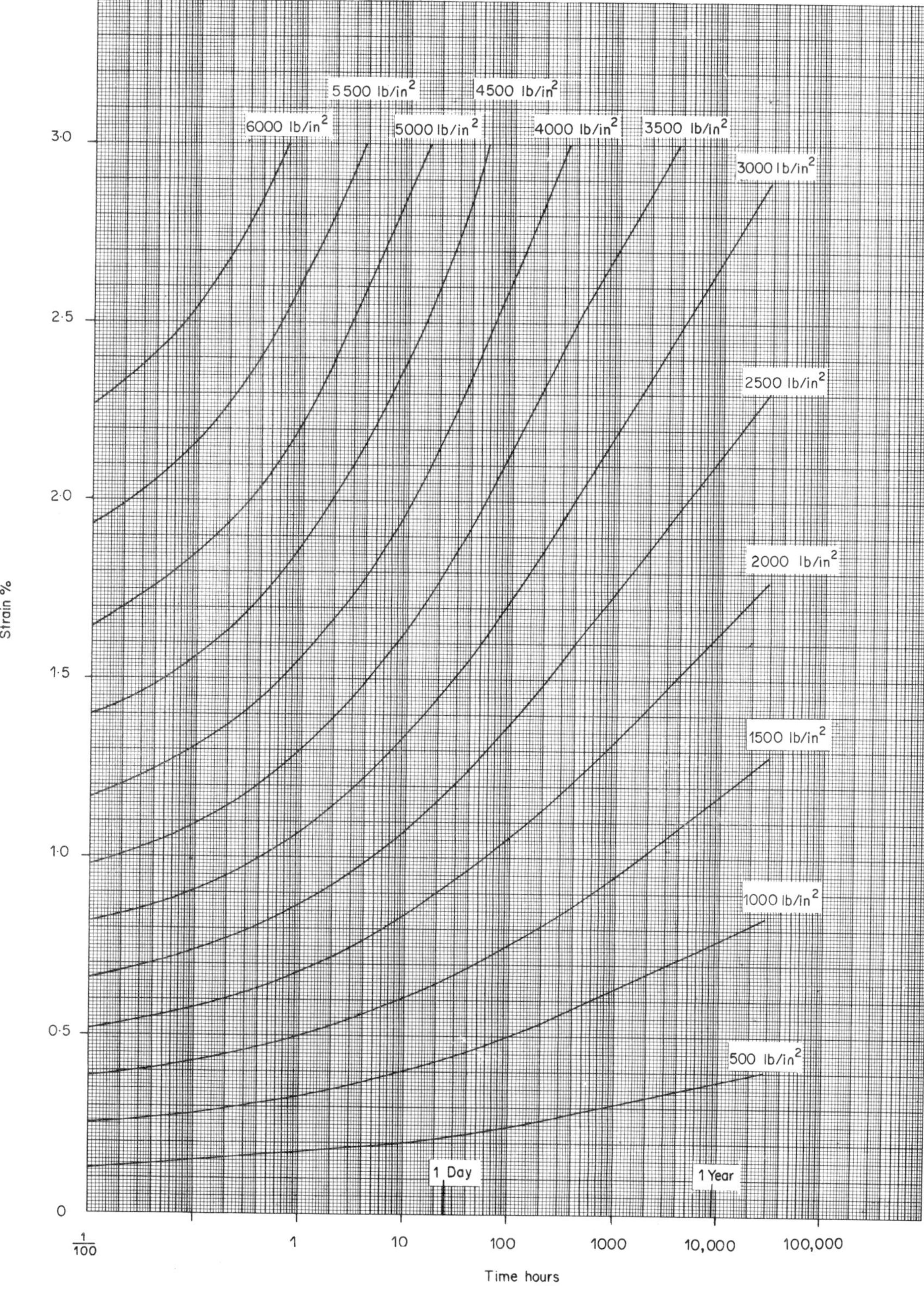


**Figure 10.1.** Creep curves in tension: 20°C, 55 RH. Acetal copolymer ('Kematal' M 90–02)

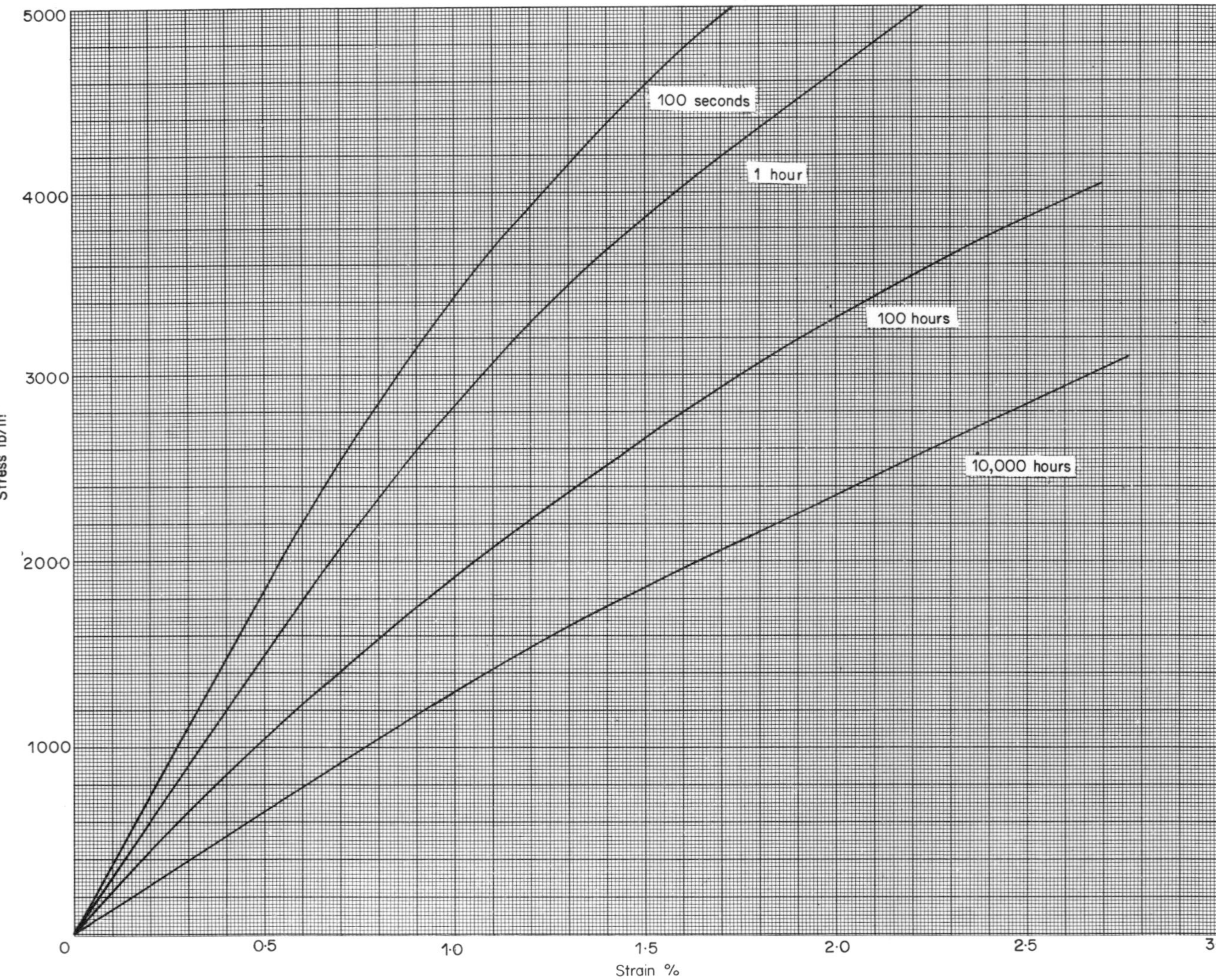


**Figure 10.2.** Isochronous stress vs strain curves: 20°C, 55 RH. Acetal copolymer ('Kematal' M 90-02)

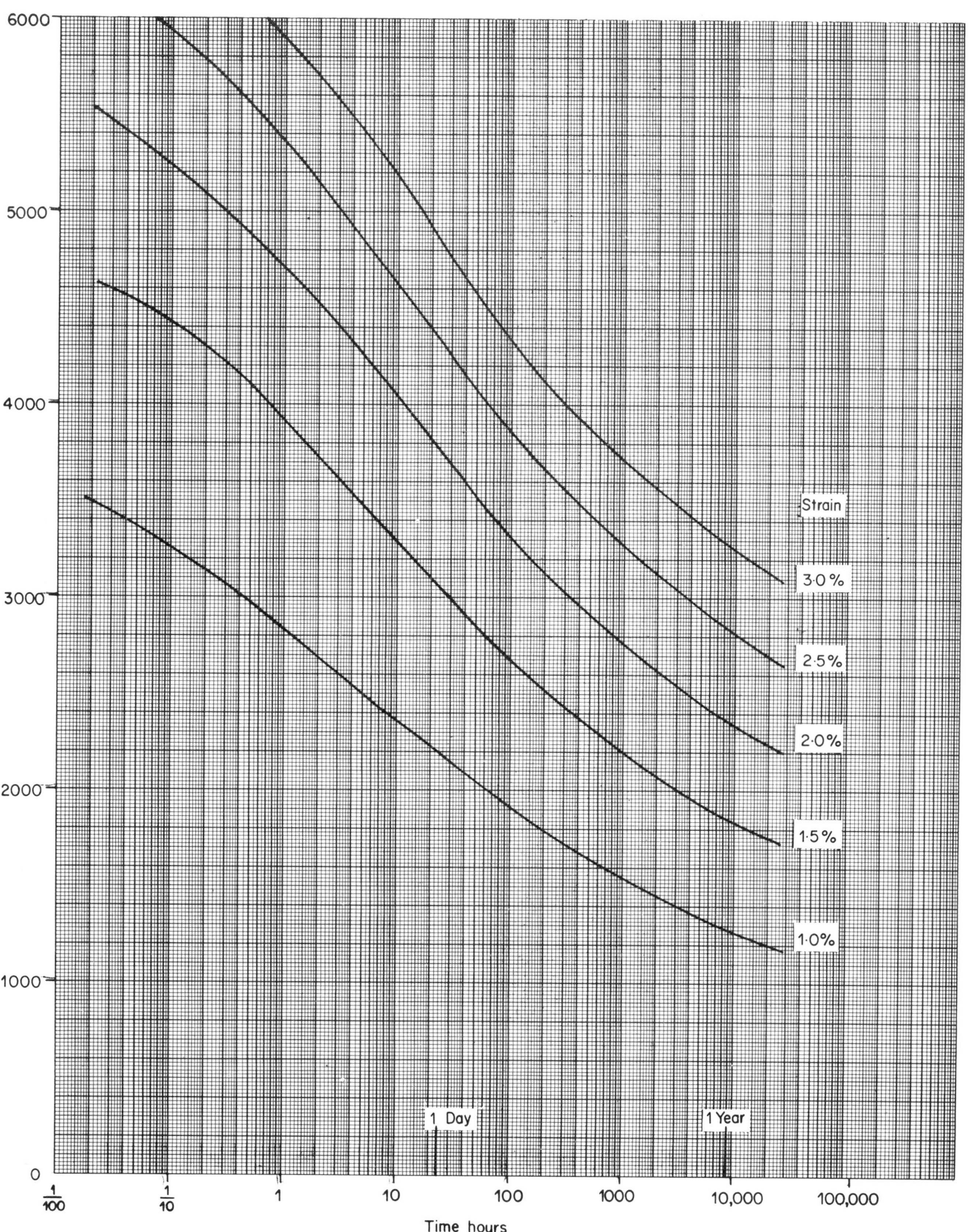


**Figure 10.3.** Isometric stress vs time curves: 20°C, 55 RH. Acetal copolymer. ('Kematal' M 90-02)

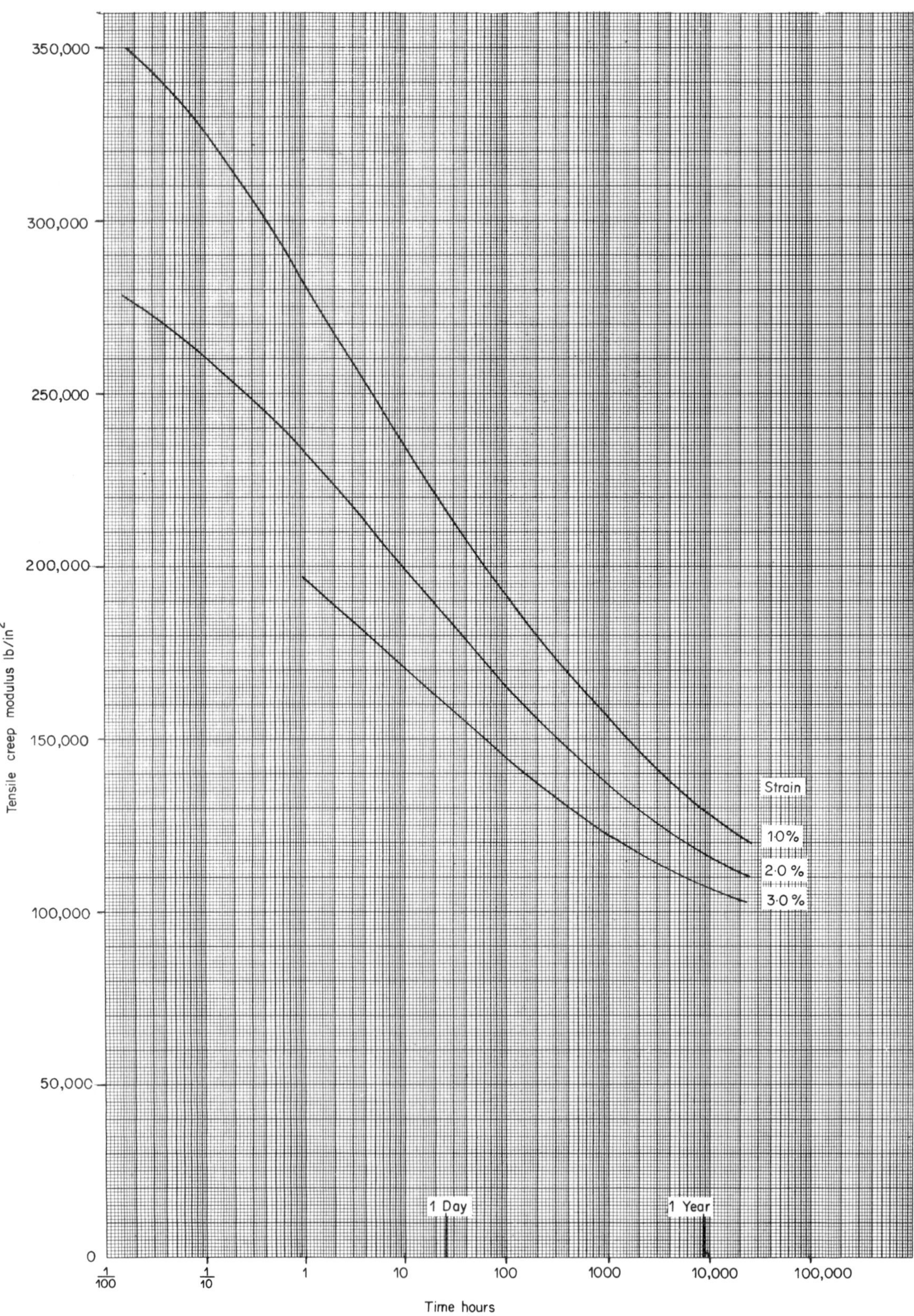


**Figure 10.4.** Tensile creep modulus vs time curves: 20°C, 55 RH. Acetal copolymer ('Kematal' M 90–02)

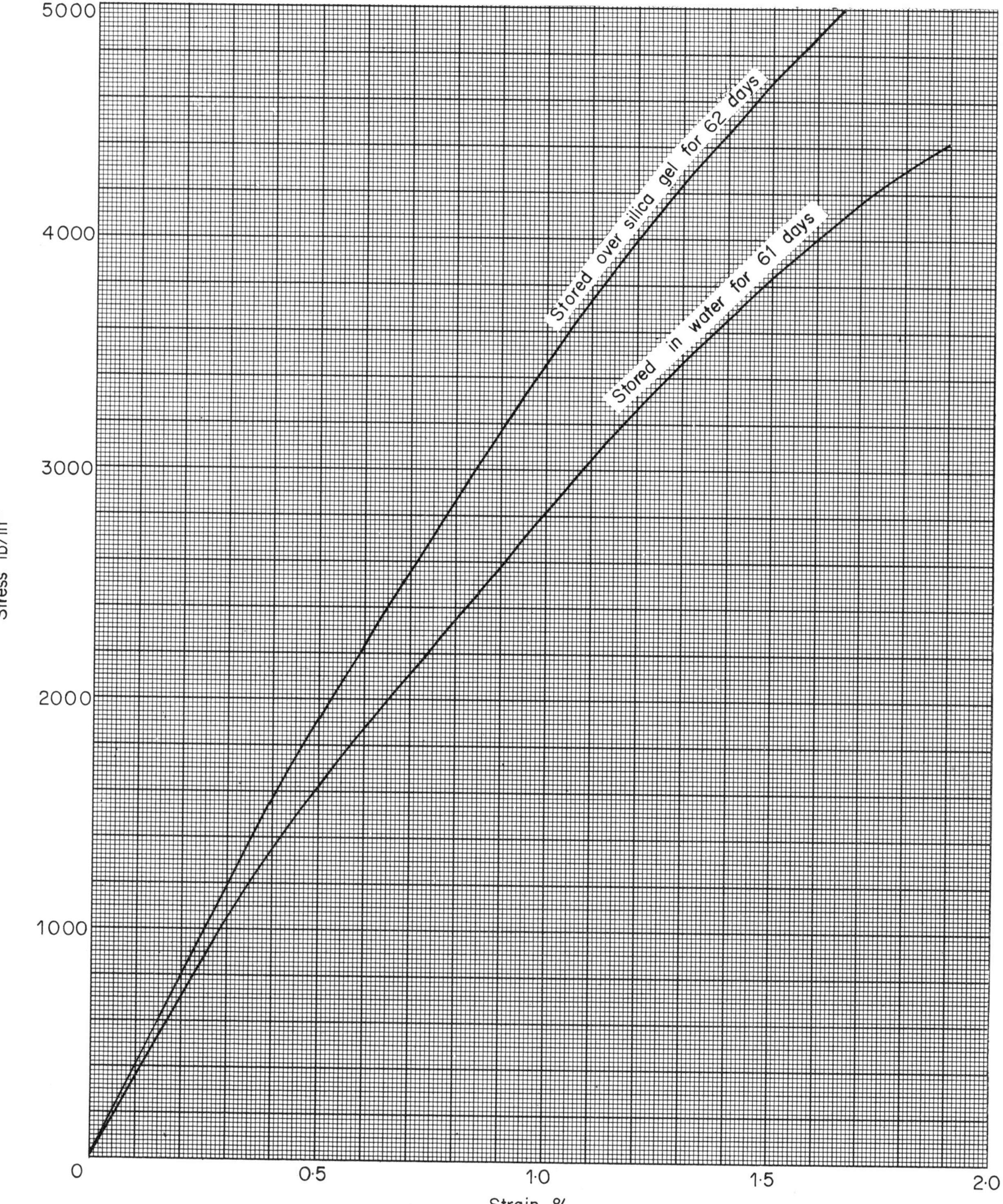


**Figure 10.5.** Isochronous stress vs strain curves: 20°C. Effect of water content on acetal copolymer. ('Kematal' M 90–02)

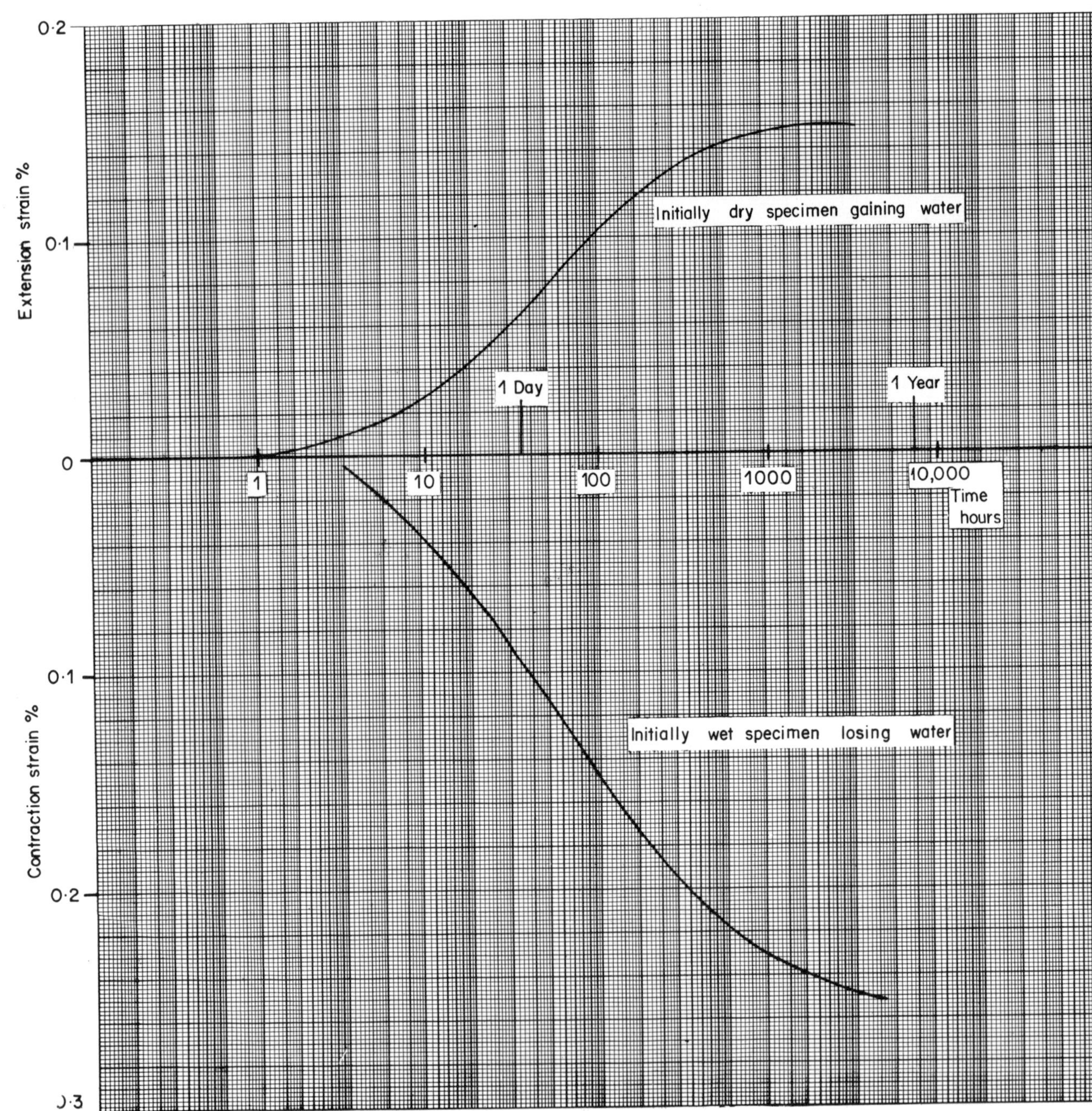


**Figure 10.6.** Change in linear dimensions with water content. Rate of equilibration of specimens initially wet and dry stored at 20°C, 55 RH. Acetal copolymer. ('Kematal' M 90-02)

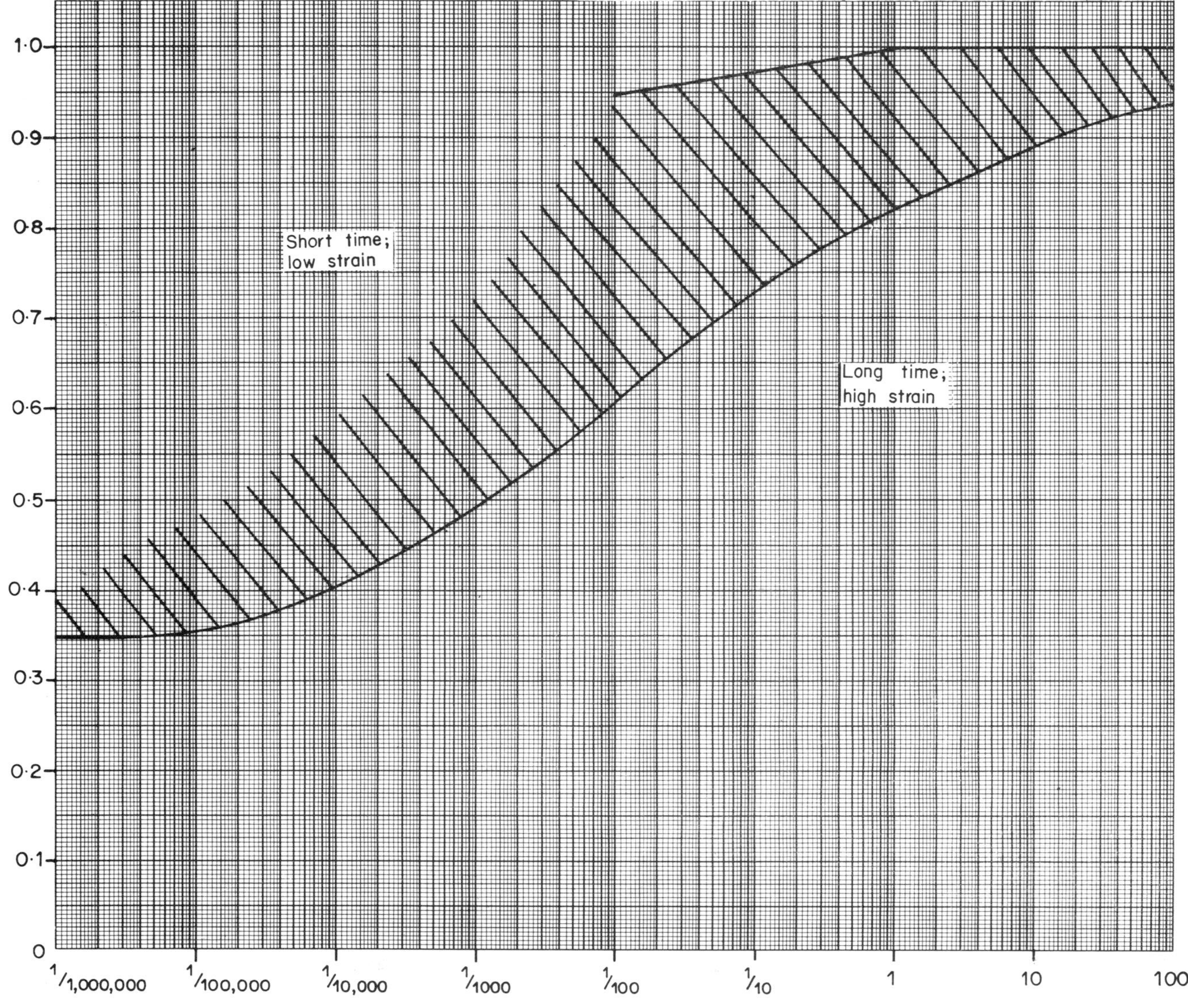


**Figure 10.7.** Recovery from creep in tension: 20°C, 55 RH. Acetal copolymer. ('Kematal' M 90-02)

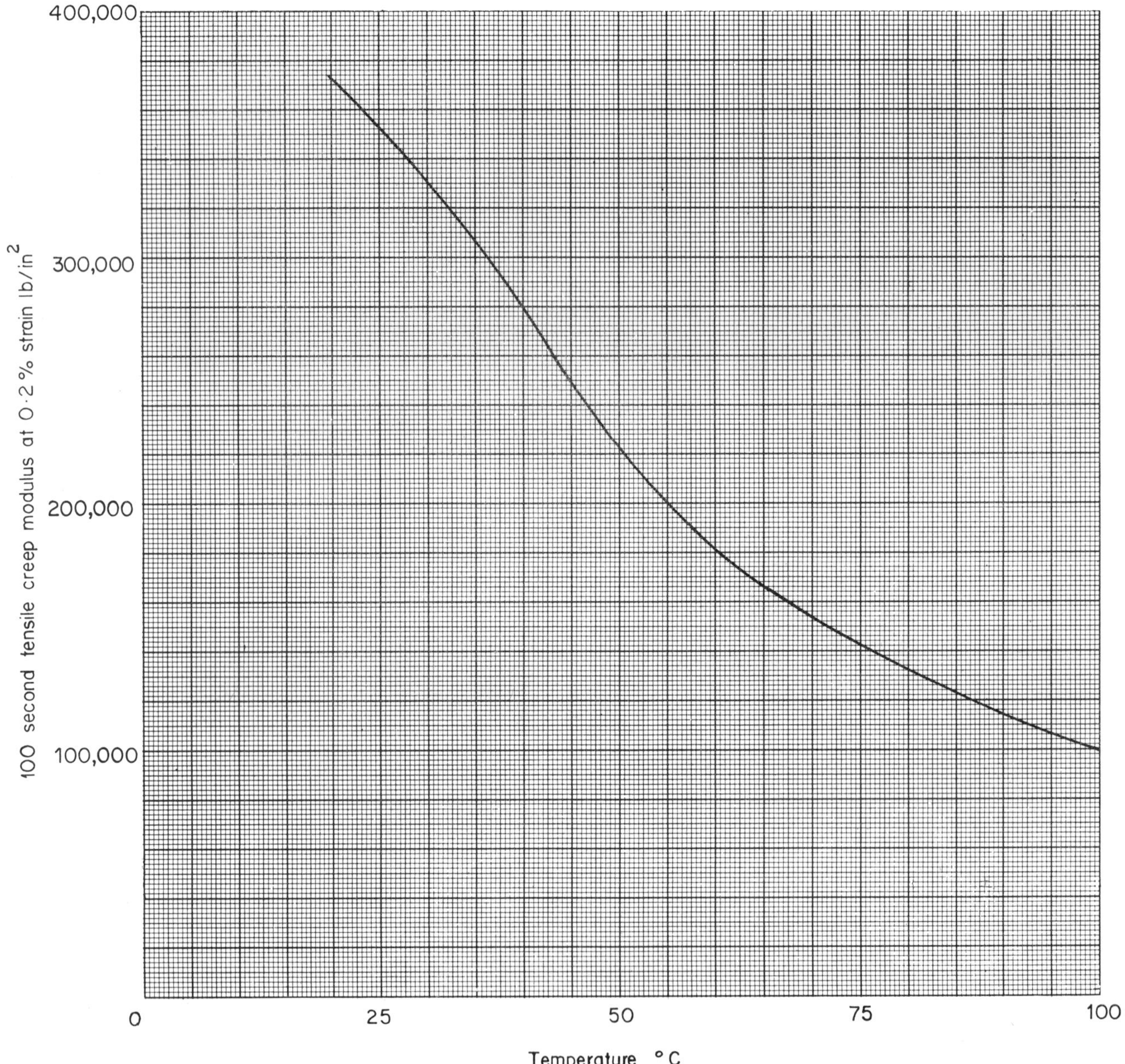


**Figure 10.8.** Tensile creep modulus (100 sec, 0·2% strain) vs temperature. Acetal copolymer ('Kematal' M 90-02)

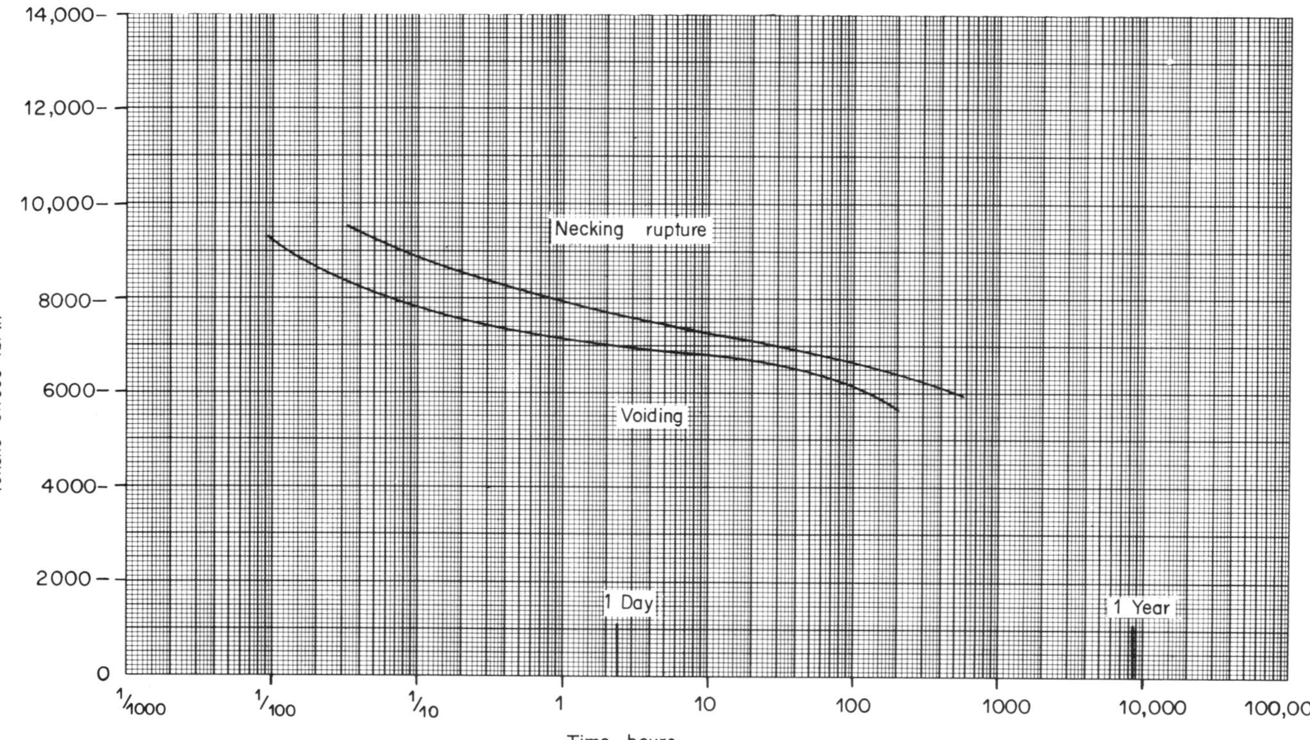


**Figure 10.9.** Creep rupture stress in tension vs time to failure: 20°C, 65 RH. Acetal copolymer. ('Kematal' M 90-04)

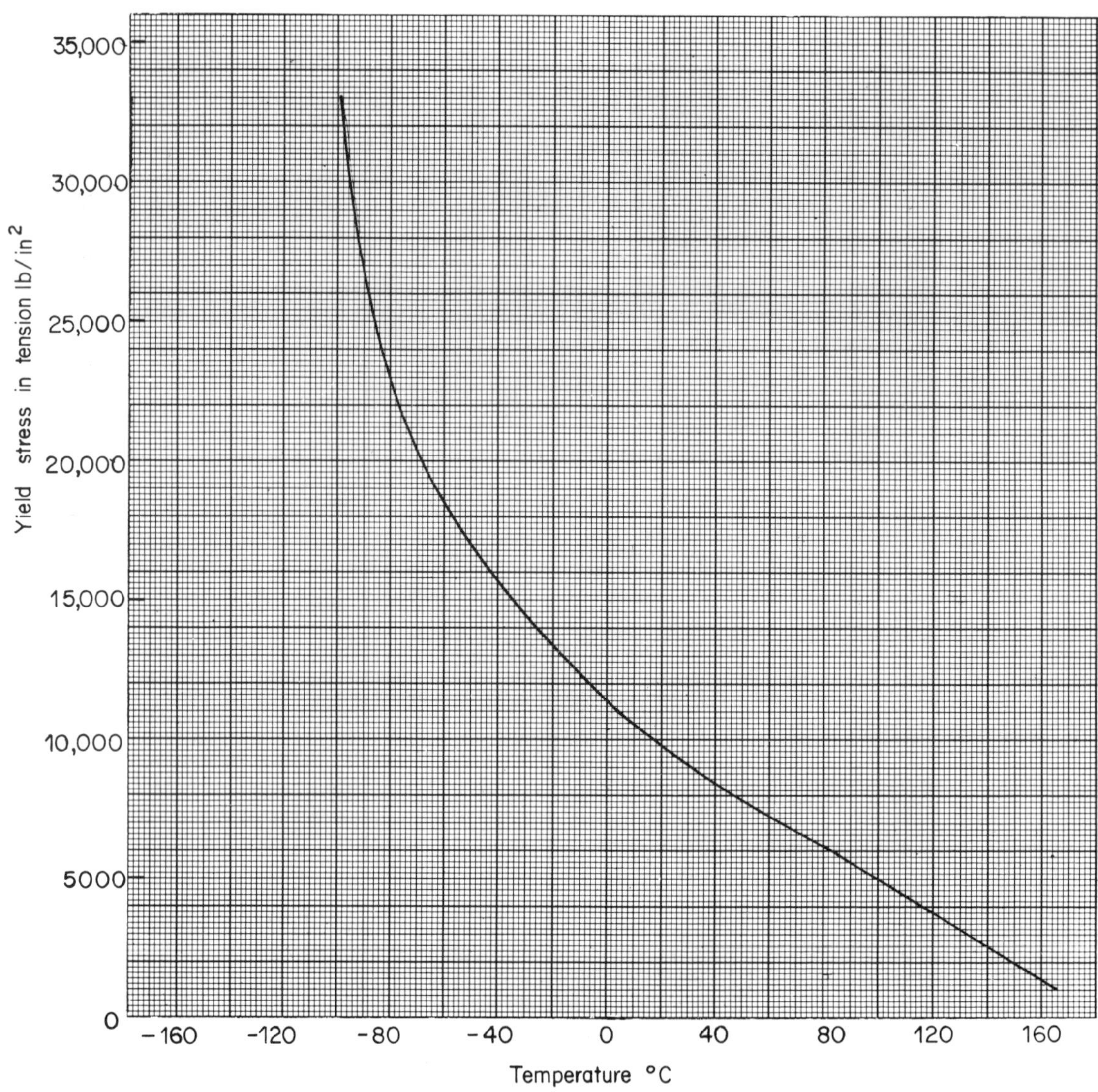


**Figure 10.10.** Yield stress in tension vs temperature: 50% per min straining rate. Acetal copolymer. ('Kematal' M 90-02)

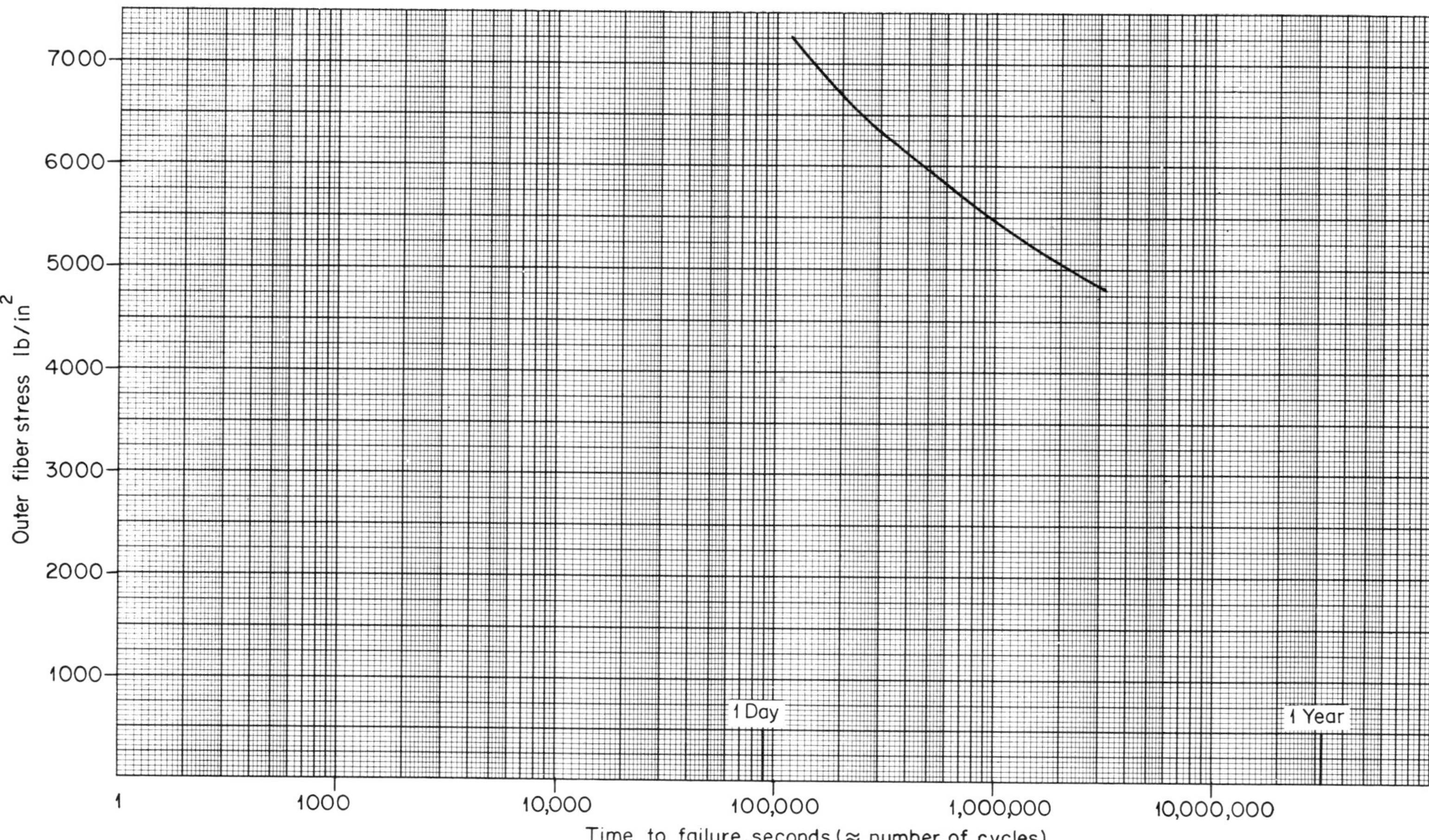


**Figure 10.11.** Dynamic fatigue in flexure: 20°C, 30 cycles/min, square wave applied stress. Acetal copolymer, medium molecular weight grade ('Kematal' M 90-04)

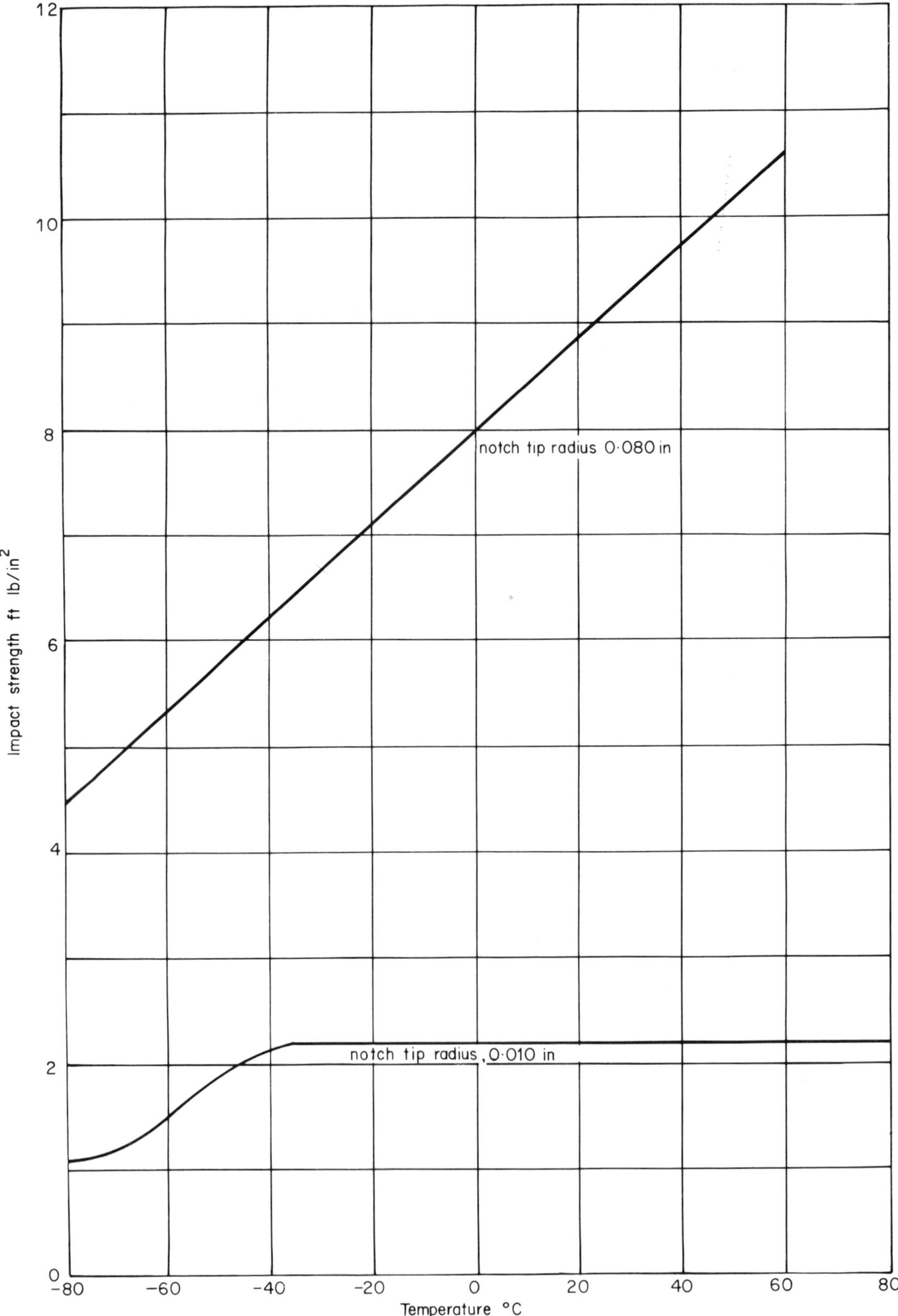


**Figure 10.12.** Impact strength vs temperature. Acetal copolymer, medium molecular weight grade ('Kematal' M 90–04)

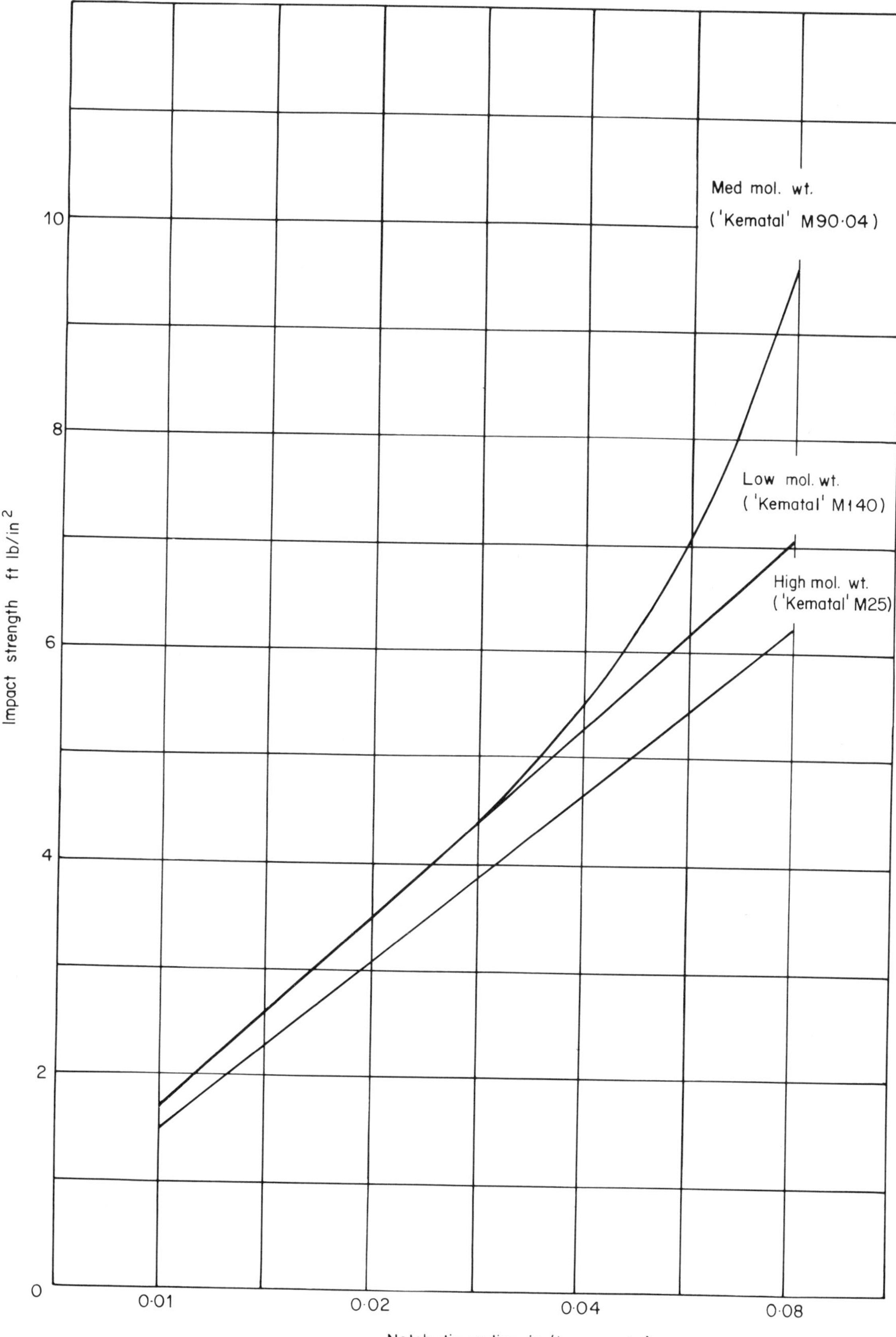


**Figure 10.13.** Impact strength vs notch tip radius: 20°C. Effect of molecular weight. (Various 'Kematal' grades)

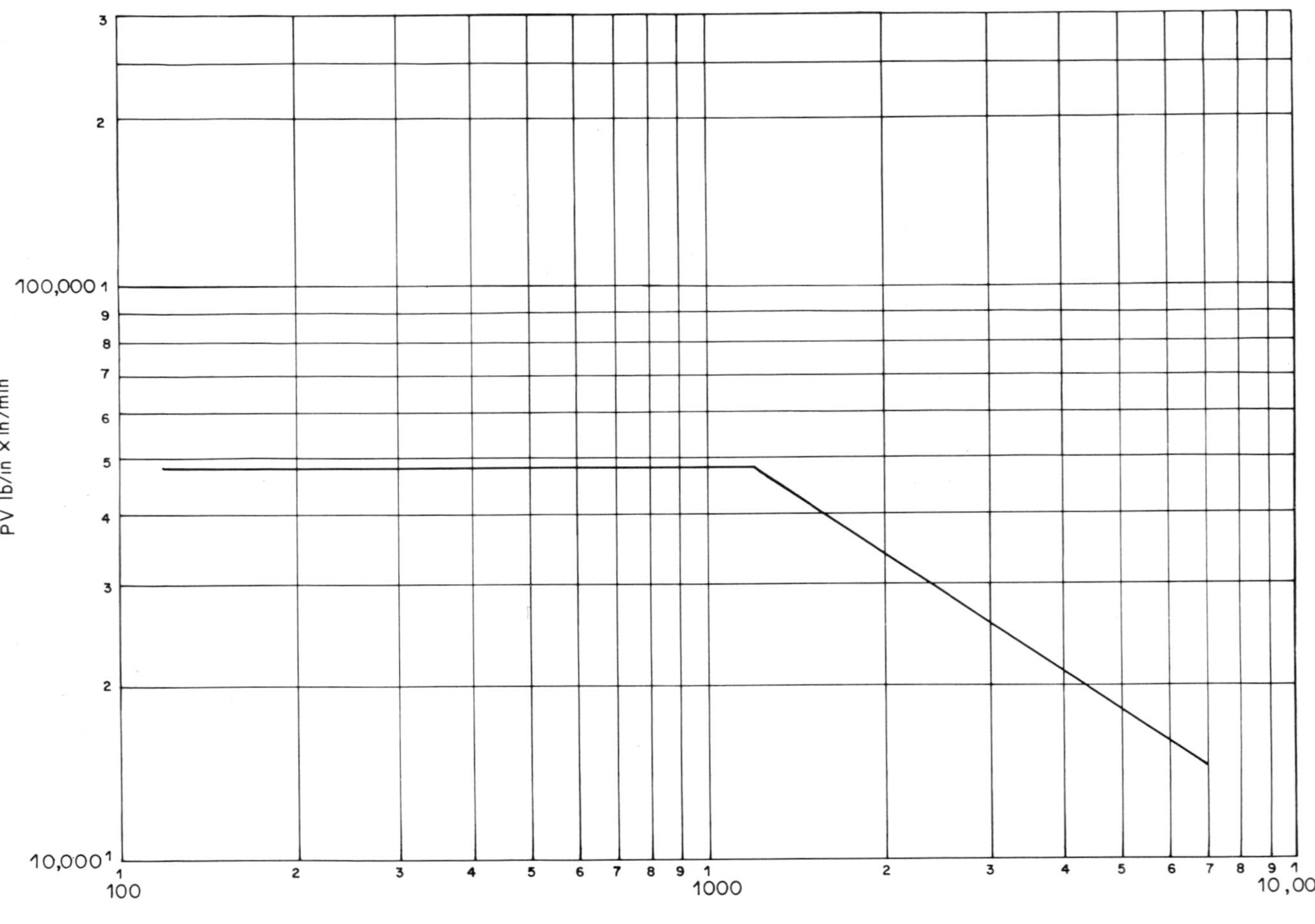


**Figure 10.14.** Limiting PV values vs velocity of sliding. Acetal copolymer. Data obtained from American Celanese Plastics Co.

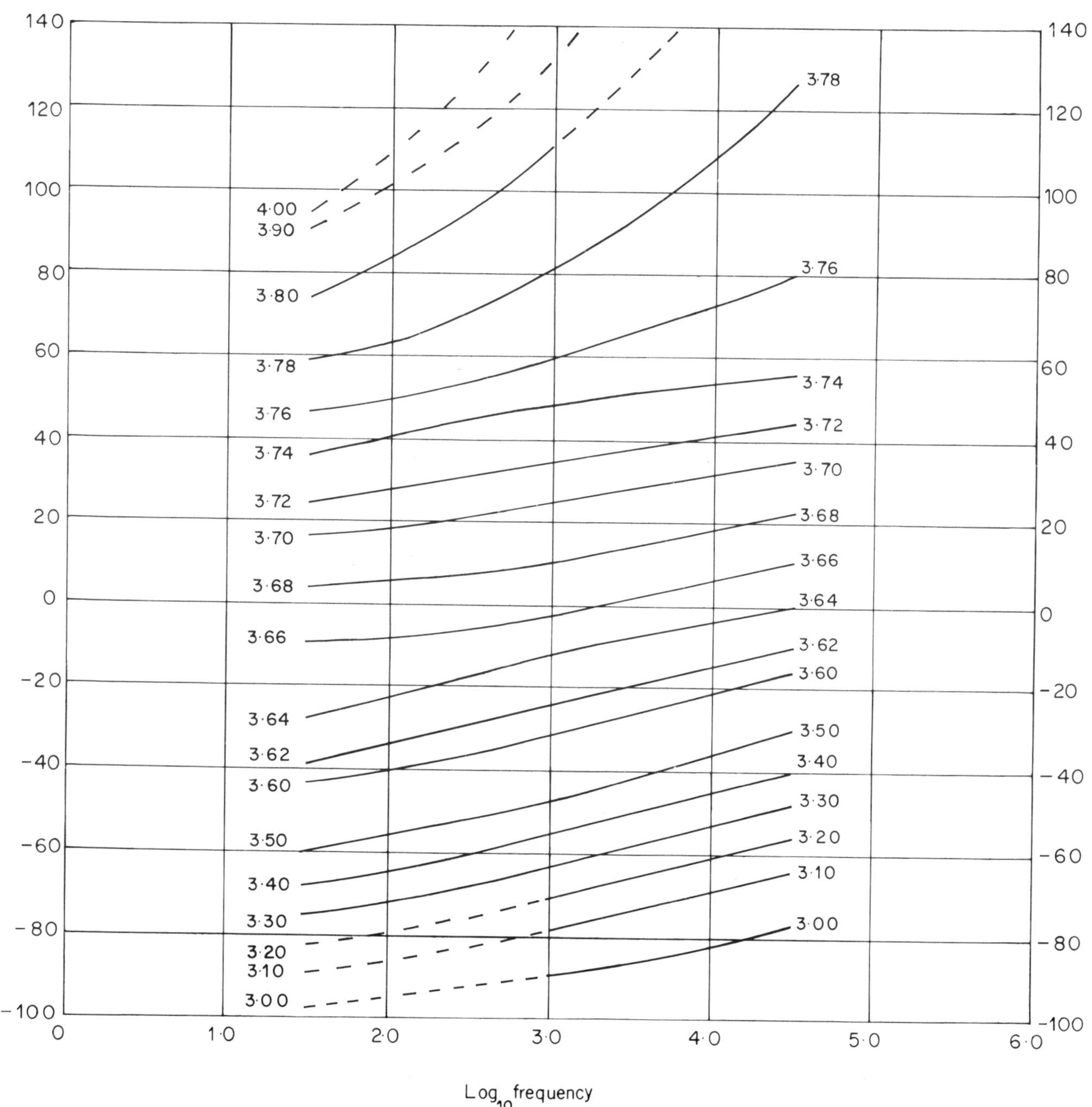


**Figure 10.15.** Permittivity vs frequency and temperature. Acetal copolymer, dry. ('Kematal' M 90-02)

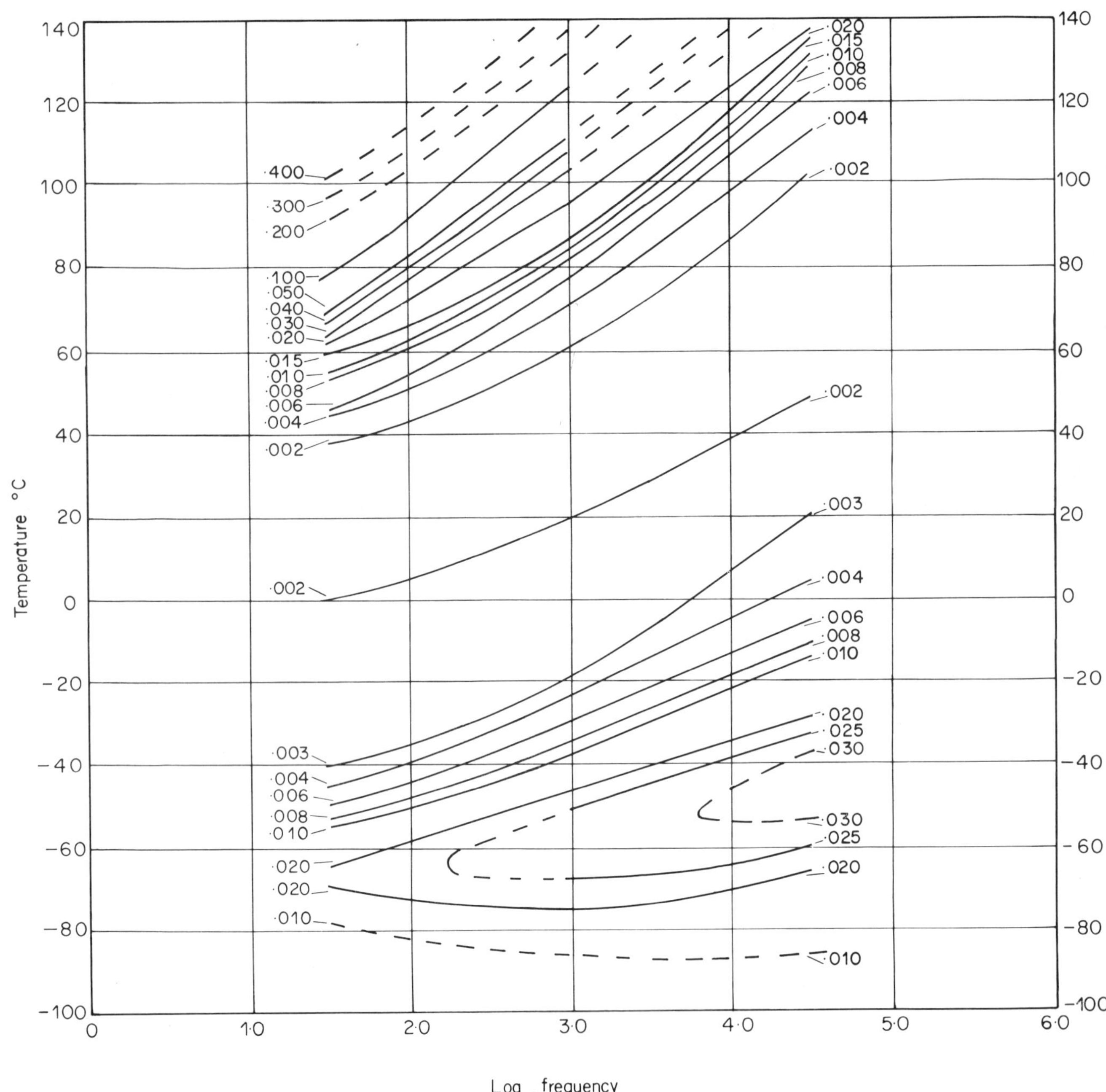


**Figure 10.16.** Loss tangent vs frequency and temperature. Acetal copolymer, dry. ('Kematal' M 90-02)

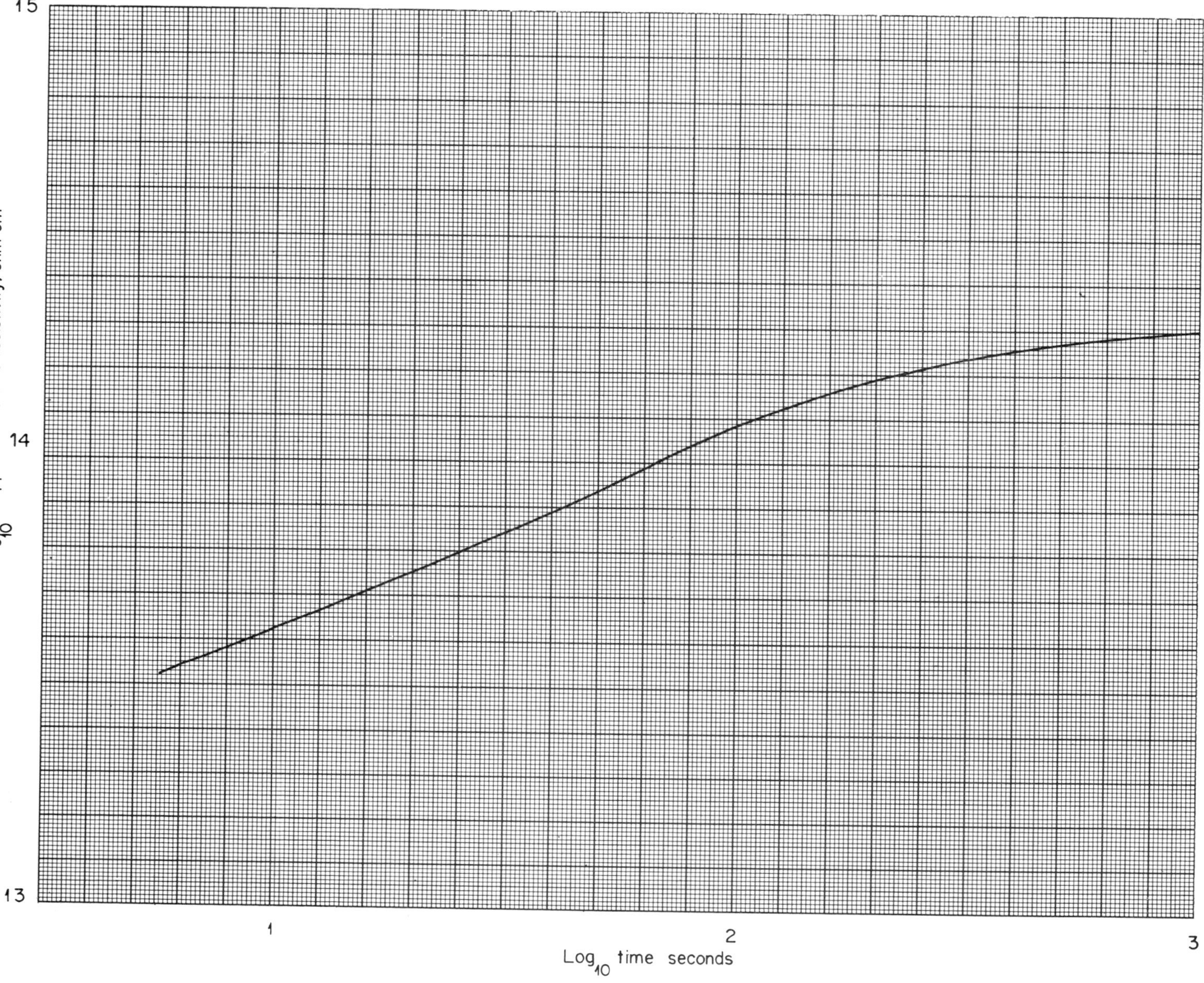


**Figure 10.17.** Apparent volume resistivity vs time of electrification: 20°C. Acetal copolymer, dry. ('Kematal' M 90-02)

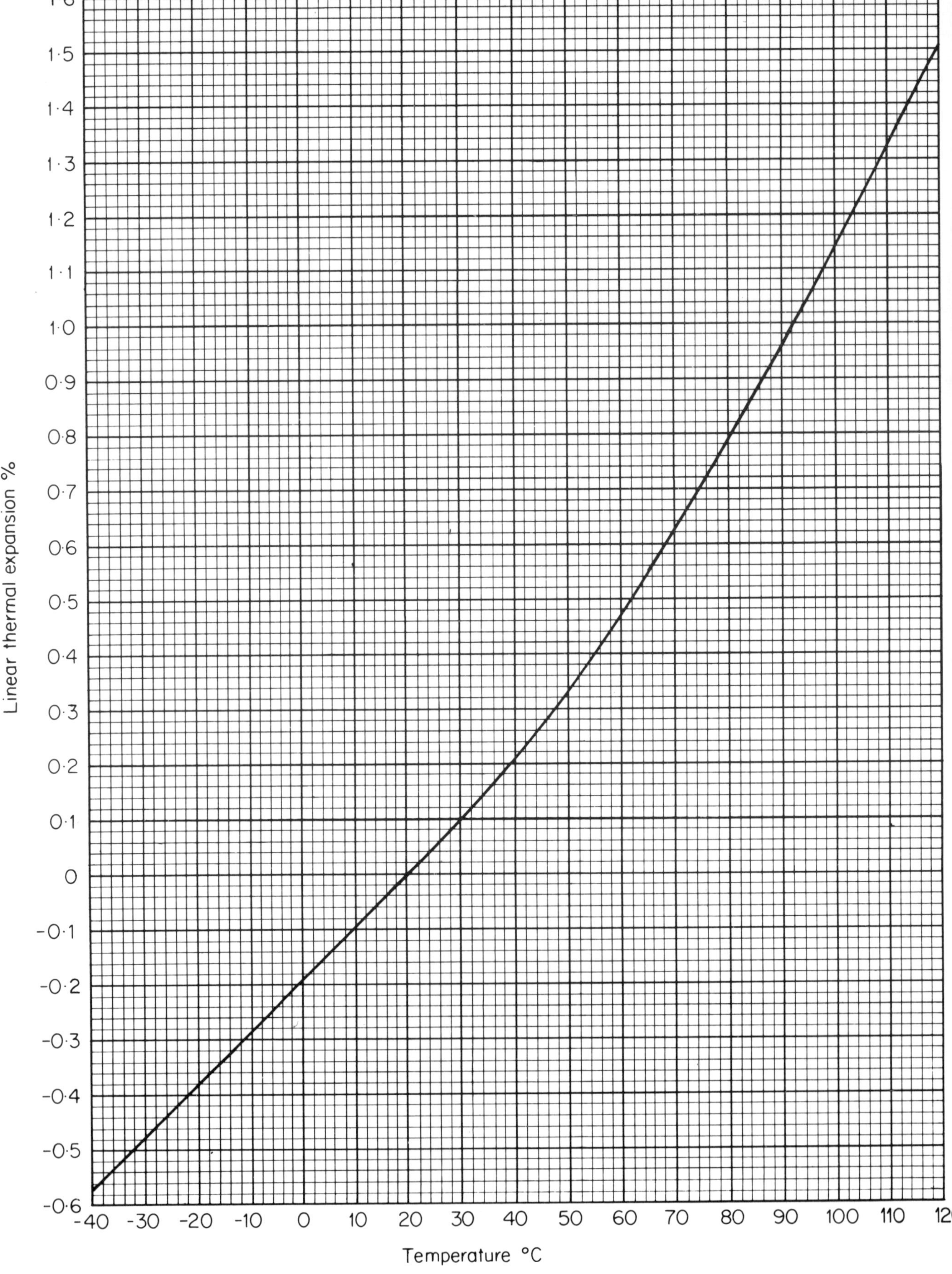


**Figure 10.18.** Linear thermal expansion vs temperature. Acetal copolymer. ('Kematal')

# 11

# NYLON

## INTRODUCTION

Nylons (polyamides) form a family of thermoplastics. All members are structurally interrelated and show a general pattern of behaviour, but each nylon is a distinct chemical entity and there are significant differences in properties and processing behaviour from one type of nylon to another. Nylons are characterized by high strength, good toughness, and good abrasion and chemical resistance, particularly to non-acidic substances, fuels and oils. Mechanical components made from them have excellent wear resistance and can often be used without lubrication. Glass-fibre filled nylons possess several advantages over the unfilled grades: they are stronger and stiffer, have an increased working temperature range and greater dimensional stability. Thus the usefulness of the nylons is extended still further, notably into applications where metal die castings have previously been used.

All the nylons in their equilibrium state contain a small amount of water: the more water it absorbs, the more flexible, soft and tough does a nylon become. The absorbed water also causes nylons to swell slightly, although this is often unimportant because the accompanying increases in flexibility and resilience allow the component to deform slightly, thus tending to cancel the effect of small dimensional changes. A further point is that the absorption of water is a very slow process and therefore day-to-day variations in humidity have a negligible effect on the performance of any nylon part.

Nylons were developed as a result of work carried out in the 1930's by W. H. Carothers on 'condensation polymerization', i.e. the formation of large molecules by combining smaller molecules containing reactive groups, with the elimination of a simple substance such as water, as opposed to 'addition polymerization' in which molecules are combined by breaking double bonds and joining the resulting activated molecules together. All other polymers dealt with in this book are 'addition polymers'.

Nylons are formed from reactions between molecules containing amino ($-NH_2$) and carboxylic acid ($-COOH$) groups. The first to be made resulted from the condensation of diamines and dibasic acids, e.g.

$$\underset{\text{diamine}}{NH_2(CH_2)_xNH_2} + \underset{\text{dibasic acid}}{COOH(CH_2)_yCOOH} \longrightarrow \underset{\text{nylon salt}}{NH_2(CH_2)_xNHCO(CH_2)_yCOOH} + H_2O$$

$$\underset{\text{nylon salt}}{NH_2(CH_2)_xNHCO(CH_2)_yCOOH} \longrightarrow$$

$$\underset{\text{polyamide}}{-NH(CH_2)_xNHCO(CH_2)_yCONH(CH_2)_xNHCO(CH_2)_yCO-}$$

More recently, nylons have been produced from one type of molecule (amino acid) only, one containing both the reactive groups necessary, e.g.

$$\underset{\text{amino acid}}{NH_2(CH_2)_xCOOH} \longrightarrow \underset{\text{polyamide}}{-NH(CH_2)_xCONH(CH_2)_xCO-}$$

Nylons were first used in the form of filaments and fibres, but in 1941 the first nylon moulding materials were introduced. Since then, improvements in materials and processing techniques have led to great expansion in the use of nylons and although they will continue to remain small tonnage plastics compared with polyethylene, PVC and polystyrene, they are now firmly established as 'engineering plastics'.

## NATURE OF THE NYLONS

Because of the large number of acids and amines available, many different nylons can be made; each is named after the number of carbon atoms in the acid and amine or the amino acid from which they are derived. The four of commercial importance are: type 66 (six-six), derived from hexamethylene diamine and adipic acid, each of which contains six carbon atoms; type 610 (six-ten), derived from the same diamine and sebacic acid, which contains ten carbon atoms; type 6, made from caprolactam which is derived from an amino acid containing six carbon atoms, and type 11, derived from the $C_{11}$ amino acid manufactured from castor oil. Copolymers can also be made. Further discussion in this chapter is limited to types 66, 610 and 6 nylons.

These three nylons are highly crystalline polymers, the individual chains of which are rigidly held by attractive forces ('hydrogen bonds') between the amine hydrogen atoms and the oxygen atoms of neighbouring molecules. These bonds influence the properties of the nylons and further differences are caused by the frequency with which the amide groups occur along the chains. These differences will be discussed further in the later section on 'Properties'.

## GRADES AND FORMS

All three types of nylon are available in various grades differing in melt viscosity and thus suitable for moulding, extrusion and extrusion blow-moulding. Many special grades have been developed, e.g. to give faster moulding cycles, improved heat and light stability, better lubricating properties (these contain molybdenum disulphide and graphite) and greater strength and stiffness (glass-filled grades).

Nylon moulding and extrusion materials are sold in the form of granules, packed dry (i.e. without their natural water content) in sealed tins. They are easily dry-coloured, but coloured compounds are also available.

Several semifabricated forms of nylon, including rod, tube, film and sheet are available, and nylon monofilament is supplied either in continuous form or in chopped lengths.

## PROCESSING

Nylons are usually injection moulded, although they may also be extruded and extrusion blow-moulded. All three nylons have low melt viscosities and some form of shut-off nozzle is required when they are injection moulded. They also have sharp melting points and screws in both injection moulding and extrusion machines should therefore have short compression zones; that recommended for type 66 and 610 nylons is only half a turn of the screw.

Nylons must be dry when processed; the tins in which they are supplied should therefore be kept tightly sealed when not in use. If the nylon becomes damp, it should be dried in a vacuum oven. After processing, nylons will gradually pick up their natural water content from the surrounding atmosphere. If dimensional stability is required in a component, this process of taking in water can be accelerated by 'humidifying' the nylon, that is by immersing the component in boiling water until it has absorbed an amount of water equivalent to that which it would absorb under service conditions. In this way, any swelling likely to be caused by water absorption during service occurs substantially before the part is used. This treatment may, however, cause some slight deterioration in mechanical properties.

Type 6 nylon can be polymerized *in situ* in a mould by adding a catalyst to the monomer. This is a particularly useful process for making very large parts, but the tolerances obtained are not as good as those obtained with injection moulding.

Nylons are easily machined by conventional techniques and they can also be cemented with a limited range of adhesives.

## PROPERTIES

Although nylons share certain characteristics, such as toughness, temperature resistance and strength, there is quite a wide variation in actual values of these and other properties from one type of nylon to another. The differences are, in fact, greater than those between polythene and polypropylene. An example of this is the melting point, as Table 11.1 shows.

TABLE 11.1

| Property | Units | Typical Value | | | |
|---|---|---|---|---|---|
| | | Nylon 66 | Nylon 610 | Nylon 6 | Glass-filled nylon 66 |
| Density | g/cm³ | 1·14 | 1·09 | 1·13 | 1·39 |
| Crystalline melting point | °C | 264 | 222 | 225 | 264 |
| Coefficient of linear thermal expansion | $°C^{-1}$ | $9{\cdot}3 \times 10^{-5}$ | $10 \times 10^{-5}$ | $9{\cdot}4 \times 10^{-5}$ | $3$ to $7 \times 10^{-5}$[a] |
| −20 to 80°C | $°F^{-1}$ | $5{\cdot}2 \times 10^{-5}$ | $5{\cdot}6 \times 10^{-5}$ | $5{\cdot}2 \times 10^{-5}$ | $1{\cdot}7$ to $3{\cdot}9 \times 10^{-5}$ |
| Thermal conductivity | cal/cm s °C | $6{\cdot}0 \times 10^{-4}$ | $5{\cdot}3 \times 10^{-4}$ | $6{\cdot}0 \times 10^{-4}$ | $5{\cdot}5 \times 10^{-4}$ |
| | B ThU in/ft³ h °F | 1·7 | 1·54 | 1·7 | 1·6 |
| Specific heat, 20 to 70°C | cal/g °C | 0·5 | 0·4 | 0·4 | 0·3 |
| | B ThU/lb °F | 0·5 | 0·4 | 0·4 | 0·3 |
| Flammability[b] | | | | | |

[a] Value depends on orientation; for unoriented material it is $5\text{–}6 \times 10^{-5}$/°C.

[b] Both unfilled and filled grades of nylon can be made to burn. While burning, unfilled nylons drip and excess of heat is carried away. Provided dripping can continue, burning stops as soon as the flame is removed, and unfilled nylons can therefore be classed as self-extinguishing. With glass-filled nylon, however, the glass holds the nylon together and the nylon will continue to burn when the flame is removed. This does not necessarily preclude the use of glass-filled nylon where there is a fire hazard. For hinges on fire-proof doors for example, unfilled nylon cannot be used because of its low softening temperature, but glass-filled nylon would be suitable.

When a specially selected grade of glass fibre is incorporated in the nylons the filled materials obtained retain many of the characteristics of their unfilled counterparts, because the glass is inert. Here again, a good example is provided by the melting point. The glass does not melt under the conditions under which the nylons are used or processed and the melting points can, therefore, be regarded as the same as those for the unfilled types. The advantages of filling with glass fibre are that the strength of the material, which is already more than twice that of the unfilled material, is retained at a useful level to much higher temperatures, and the dimensional stability of the material is considerably increased.

### *Deformation*

Under this heading data are given for nylon 66, glass-filled nylon 66 (33% glass by weight), nylon 6, and glass-filled nylon 6 (25% glass by weight). From their creep curves it can be seen that dry nylons are materials of high rigidity, and that the rigidity of nylons is enhanced by glass fibre reinforcement. It should be added that of the range of plastics described in this book nylons have the greatest affinity for water and their physical properties are considerably affected by the absorption of moisture. Amongst the nylons, nylon 66 and nylon 6, the types most commonly used in engineering, have the highest absorption of water. In this section data are given on the effect of water absorption on deformational properties; data on the rate of water absorption and the change in dimensions consequent to water absorption are left to the following sections.

*Nylon 66*

The deformational behaviour at 20°C of dry nylon 66 is given by the tensile creep curves of Figure 11.1. The corresponding isochronous stress vs. strain, isometric stress vs. time, and tensile creep modulus vs. time curves are given in Figures 11.2, 11.3 and 11.4. Recovery data are given in Figure 11.5. In Figure 11.6 the 100 second isochronous stress vs. strain curves are shown for nylon 66 at different states of water absorption. It is to be noted that the effect of water content on deformational behaviour is considerable; for wet material the stress for any particular strain is less than one quarter of the value for dry material.

The deformational behaviour at 60°C of dry nylon 66 is given by the tensile creep curves of Figure 11.7 and the corresponding 100 second isochronous stress vs. strain curve of Figure 11.8.

The temperature dependence of the deformation of nylon 66 is given in Figure 11.9 in the form of the 100 second tensile creep modulus (at 0·2% strain) vs. temperature curve.

*Nylon 66 + Glass Fibre*

The deformational behaviour at 20°C of glass-filled nylon 66 is given by the tensile creep curves of Figure 11.10 and the 100 second isochronous stress vs. strain curves of Figure 11.11. Recovery data are given in Figure 11.12. Again the considerable influence of absorbed water on the deformational behaviour is demonstrated although it is significantly less than for the unreinforced material.

The temperature dependence of the deformation for glass-filled nylon 66 is given by the 100 second tensile creep modulus (at 0·2% strain) vs. temperature curve shown in Figure 11.9 and compared with the data for unreinforced nylon 66.

*Nylon 6*

The deformational behaviour at 20°C of dry nylon 6 is given by the tensile creep curves in Figure 11.13. The two grades depicted differ in that one is a crystal nucleated grade, a material for optical applications. The corresponding 100 second isochronous stress vs. strain curves are given in Figure 11.14 and the 100 second tensile creep modulus (at 0·2% strain) vs. temperature curves in Figure 11.15.

*Nylon 6 + Glass Fibre*

The 100 second isochronous stress vs. strain curve at 20°C for dry glass-filled nylon 6 is given in Figure 11.14 and compared with the curves for unfilled nylon 6. The 100 second tensile creep modulus (at 0·2% strain) vs. temperature curve is given in Figure 11.15, with the data for unfilled nylon 6.

*Water Absorption*

The effect of water absorption on the deformational behaviour has been noted but no mention was made of the time required to reach saturation nor of the change in dimensions consequent on absorption or desorption of moisture. Some information on these points is now given.

For each grade of nylon there is a maximum water content for material in contact with an atmosphere of any given relative humidity. This is known as the equilibrium water content. Figure 11.16 shows equilibrium water contents for the most important types of nylon for different relative humidities.

The rate at which water is absorbed depends on the relative humidity, the type of nylon and the dimensions of the component and above all, on the temperature. For example, a nylon 66 disc $\frac{1}{8}$ in thick takes 400–500 days to attain equilibrium at 20°C in an atmosphere at 65% rh, whilst equilibrium is attained in boiling water in a matter of 30 hours. The thickness effect is illustrated by the data in Table 11.2.

TABLE 11.2

| Thickness of nylon 66 (in) | Time to attain equilibrium In water, 20°C | In water, 100°C |
|---|---|---|
| $\frac{1}{4}$ | >900 days | 100 h |
| $\frac{1}{8}$ | ~150 days | ~30 h |
| $\frac{1}{16}$ | ~45 days | ~6 h |
| $\frac{1}{32}$ | — | ~2 h |

A simplifying feature of water absorption is that the equilibrium content appears to be largely independent of temperature, exemplified by the data in Table 11.3 for water-immersed samples at 20°C and 100°C.

TABLE 11.3

| Type of nylon | Water immersed At 20°C | At 100°C |
|---|---|---|
| Nylon 66 | 8·5% | ~8·2% |
| Nylon 66+glass fibre | 5·6% | 5·6% |
| Nylon 6 | 11% | ~9% |
| Nylon 610 | 3·5% | ~3·3% |

The only exception appears to be nylon 6. Equilibrium values of moisture content for 20°C, 65% rh are given below:

| | |
|---|---|
| Nylon 66 | 3% |
| Nylon 66+glass fibre | 2·5% |
| Nylon 6 | 4·5% |
| Nylon 610 | 1·5% |

The effect of moisture on specimen dimensions is of importance in design and relevant information is given in Figures 11.17 and 11.18: the dimensional change of glass fibre-filled nylon 66 is depicted as a band, not as a line, because the dimensional change in any direction depends on the degree of orientation of the glass fibres. In a strongly oriented specimen, the dimensional change will be represented by the lower region of the band for the direction of orientation; perpendicular to this direction the higher region of the band will apply.

*Limiting Stress*

In the absence of stress concentrations, nylons are ductile materials which fail by yielding, as they do for example, in a conventional tensile test. Data on long term strength of these materials are limited and the preferred course is to work from the 3% isometric stress vs. log time curve as a failure curve because more information on the effect of variables such as fabrication, moisture content and temperature is available at the isometric stresses. However, some yield stress data at 20°C, 65% rh are presented in Figure 11.19 for nylon 66 and the same material with glass fibre.

The effect of temperature on the stress at failure may be judged approximately from the variation of yield stress in a conventional test, shown in Figure 11.20 for several grades of nylon.

*Impact Behaviour*

The impact strength of dry nylon 66 is plotted as a function of temperature for sharp and blunt notches in Figure 11.21. The pronounced effect of sharpness of notch is obvious from these results and is further confirmed by the fact that un-notched nylon 66 specimens do not break at −20°C. Similar data for the glass fibre-filled nylon 66 are given in Figure 11.22, where it can be seen that the impact behaviour of this material is not nearly so markedly dependent on the severity of the notch. Further consideration of Figure 11.22 reveals that the actual level of impact strength of the glass-filled material is maintained at a reasonably high level, even at low temperatures with sharp notches. This is clearly a feature supporting the widespread satisfactory practical use of this material. Comparison of dry unfilled and glass-filled nylon 66 is made in Figures 11.23 (sharp notch) and 11.24 (blunt notch).

It is well known that the impact strength of nylon components is markedly affected by moisture content, and because nylon is not normally used as a dry material, data on the effect of moisture content are required. There are numerous ways in which the moisture content may be increased:

1. By boiling in water.
2. By soaking in hot potassium acetate solutions.
3. By immersion in cold water.
4. By conditioning at the required humidity.

The first two techniques have been discarded in this work because there is some indication that surface oxidation following these treatments leads to spuriously low results. Impact data for 0·010 in notch radius specimens after various periods of water immersion are given in Figures 11.25 (nylon 66) and 11.26 (glass fibre-filled nylon 66). The rapid improvement in impact strength with water immersion compared with the slow attainment of equilibrium water content on a weight basis might have been expected in that the outer skin of the specimen controls the impact behaviour and this is wetted rapidly. This has been proved by measuring the impact strength as equilibrium is attained at 65% rh: the result is similar to that achieved after only three days immersion. A further confirmation was obtained by machining away the skin (0·02 in) from specimens immersed for three days. Impact strength data similar to those of the dry material were obtained.

In some preliminary design considerations, particularly where the humidity of the environment is likely to change within wide limits, the concept of a brittleness temperature might be helpful; a similar concept was introduced in Chapter 8. Again, if one defines the brittleness temperature as the temperature at which the 0·010 in notch radius impact strength attains 5 ft-lb/in$^2$, plots of brittleness temperature for the two grades of material vs. nominal water content (water content % w/w, assumed evenly distributed) can be made (Figure 11.27).

*Friction*

Data concerning the frictional properties of nylon 66 have been obtained from a modified Bowden Leben machine, the measurements being made between a loaded hemisphere (the slider) and a flat surface (the plate). A load of 2 kg was employed and the measurements were made at 20°C at various sliding speeds.

It can be seen from Table 11.4 that both water and liquid paraffin have a considerable lubricating effect on nylon 66.

TABLE 11.4

| Lubricant | Material | | Coefficient of friction Sliding speed cm/s | | | |
|---|---|---|---|---|---|---|
| | Slider | Plate | 0·003 | 0·04 | 0·30 | 1·06 |
| None | Nylon (as moulded) | Nylon (as moulded) | 0·63 | 0·69 | 0·70 | 0·65 |
| | Nylon (machined) | Nylon (machined) | 0·42 | 0·44 | 0·46 | 0·47 |
| | Mild steel | Nylon (machined) | 0·33 | 0·33 | 0·30 | 0·30 |
| | Nylon (machined) | Mild steel | 0·39 | 0·41 | 0·40 | 0·40 |
| Water | Nylon (machined) | Nylon (machined) | 0·27 | 0·24 | 0·21 | 0·19 |
| | Mild steel | Nylon (machined) | 0·23 | 0·20 | 0·19 | 0·17 |
| | Nylon (machined) | Mild steel | 0·20 | 0·23 | 0·22 | 0·18 |
| Liquid paraffin | Nylon (machined) | Nylon (machined) | 0·22 | 0·15 | 0·11 | 0·08 |
| | Mild steel | Nylon (machined) | 0·16 | 0·11 | 0·08 | 0·08 |
| | Nylon (machined) | Mild steel | 0·26 | 0·15 | 0·07 | 0·04 |

The coefficient of friction is affected by other variables too; for example the combination of steel and machined nylon gives greater friction at elevated temperatures and increasing the moisture content increases the friction.

Nylons are used both in gears and bearings. For the former, conventional formulae applicable to metal gears may be used although in practice the permissible torques derived are conservative figures because the greater resilience of the nylon enables the loads to be spread over a larger surface area.

Some guidance on bearing design may be obtained from *PV* values, the product of the pressure on the bearing area and the peripheral speed for which the units are usually lb/in$^2$ and ft/min respectively. Maximum and typical *PV* values are given in Table 11.5 (the latter in brackets) in these units.

These values are indicative of behaviour in a specific test only and practical trials should always be carried out under the specific conditions of surface, clearance and wall thickness. Maximum *PV* values for bearings lubricated continuously are certainly higher than those indicated in the Table.

For nylon bearings of 1–2 in internal diameter and wall thickness 0·1–0·2 in a clearance of the order of 0·005–0·010 in will normally be suitable. It is usual to allow greater clearances on the

TABLE 11.5

| Lubricant | Type of operation Continuous | Intermittent |
|---|---|---|
| None | 1000 (230) | 3000 (690) |
| Water | 1500 (350) | 4000 (920) |
| Oil (initially) | 2000 (460) | 8000 (1840) |

shaft than those used with metal bearings, particularly if it is not possible to ensure an environment of constant humidity. Small tolerances are not essential with nylon bearings to ensure smooth running and a satisfactory life.

### *Electrical Properties*

The permittivities of nylons compared with those of other plastics are high only at low frequencies. Also the power losses increase with the water content of the specimen and consequently with relative humidity. Comprehensive data on dielectric properties in the form of contour maps are not available but Figure 11.28 shows the considerable effect of relative humidity on the dielectric losses for glass-filled nylon 66 and nylon 610. One important practical advantage of the high power losses of nylons is that they can be conveniently welded by radio frequency dielectric heating.

Nylons are reasonably good electrical insulators at low temperatures and humidities but their insulating properties deteriorate with increasing temperature and humidity. These effects are shown in Figures 11.29 and 11.30 which give the dependence of steady state volume resistivity on relative humidity and temperature for several types of nylon.

### *Optical Properties*

Like acetal copolymers, nylons are not at all transparent in massive form as can be seen from the following data for a film specimen of nylon 66 of thickness 0·05 mm illuminated at 5461 Å. The results are corrected for surface defects and reflection.

| | |
|---|---|
| Direct transmission factor | 48 % |
| Direct transmission factor corrected to 1 mm thickness | ~0 |
| Scattering coefficient | $>100$ cm$^{-1}$ |
| Forward scattering fraction (ASTM D1003) | 26·5 % |

Although the transparency of nylon is poor the clarity of components rapidly quenched from the melt is sufficient to be of use in some instances, for example the viewing of a liquid level through the material.

Refraction measurements indicate a refractive index value of 1·448 and a critical angle of 55° 21′ for a material of density 1·489 g/cm$^3$ at a spectral wavelength 5893 Å.

*Chemical Properties*

One of the most useful properties of the nylons is that they are resistant to organic solvents and in particular to oils and fuels.

Nylons are inert to most inorganic reagents except hydrogen peroxide and bleaching agents. Some aqueous solutions may have a plasticizing effect because the nylons will absorb water from the solution. Concentrated mineral acids attack nylon, but the effect of dilute acids is much less marked, type 66 and 610 nylons having better resistance than type 6.

All nylons are dissolved by phenols, cresols and similar materials, but their dilute aqueous solutions cause only very slight deterioration. Benzyl alcohol is a solvent at high temperatures as are certain other substances such as nitrobenzene and nitroalcohols. The effect of these materials at room temperature is negligible.

TABLE 11.6

| | Type 66 | Type 610 | Type 6 |
|---|---|---|---|
| Equilibrium water content (immersed) | 8·5% | 3·5% | 11% |
| Gas permeabilities ($cm^3$ $cm/cm^2$ s cm Hg) × $10^9$ at 20°C and 0% rh | | | |
| Carbon dioxide | 0·05 | | 0·06 |
| Oxygen | 0·03 | | 0·009 |
| Nitrogen | 0·002 | | 0·003 |
| Water vapour permeability ($g/m^2$ 24 h) at 38°C and 90% rh for 0·001 in thickness | 110 | | 125–300 |
| Resistance to: | | | |
| Mineral acids (dilute) | Good | Good | Poor |
| Mineral acids (conc.) | Poor | Poor | Poor |
| Alkalis | Good | Good | Excellent |
| Solvents: | | | |
| alcohols | Fair, but may be absorbed to give a plasticizing effect | | Excellent |
| ketones | ,, | | Excellent |
| aromatic hydrocarbons | ,, | | Excellent |
| chlorinated hydrocarbons | ,, | | Good |
| Detergents | Fair to good | Fair to good | Good |
| Greases and oils | Fair to good | Fair to good | Excellent |

## APPLICATIONS

The nylons are used principally in light engineering. Gears and bearings made from nylon have a low noise level, good abrasion resistance and the ability to run under light loads without lubrication. Nylon gears are used in such diverse industries as food processing, where the absence of conventional lubricants is a great advantage, to agricultural machinery where wear resistance, light weight and resistance to corrosive fertilizers are important.

Where resistance to shock loading is a prime requirement unfilled nylon is the preferred material. Where wear resistance is most important, e.g. in abrasive environments, evidence is accumulating to show that glass-filled nylon performs better. Glass-filled nylon offers also better dimensional stability and greater strength, although the increase in strength is gained at the expense of some loss of resilience.

Although the nylons are not in the same class as polythene and PTFE as electrical insulators, they are nevertheless adequate insulators for a number of products such as drill-housings, switch gear mechanisms, coil formers and terminal blocks. Glass-filled nylon possesses the useful added advantage of being able to withstand soldering temperatures, for example, during the manufacture of telephone relay bobbins.

*Typical examples*

Light engineering components—gears, bushes, cams, bearings.

Chains, belts, conveyor rollers, pulleys, wheels.

Car components—speedometer drive gears, ball joints, door lock mechanisms, windscreen wiper motor drives, tubing for high pressure lubrication, bulb sockets, floats and other carburettor components.

Domestic appliance housings, power tool housings.

Desk drawer rollers, office machine parts, motor housings.

Ship and boat components—stabilizers, propellers, fairleads, cleats and rowlocks.

Coil formers, terminal blocks, connectors, brush holders, relay bobbins, switch actuators, insulating clips, cable sheathing.

Textile machinery parts—threaders, guides and ring travellers.

## LIST OF FIGURES

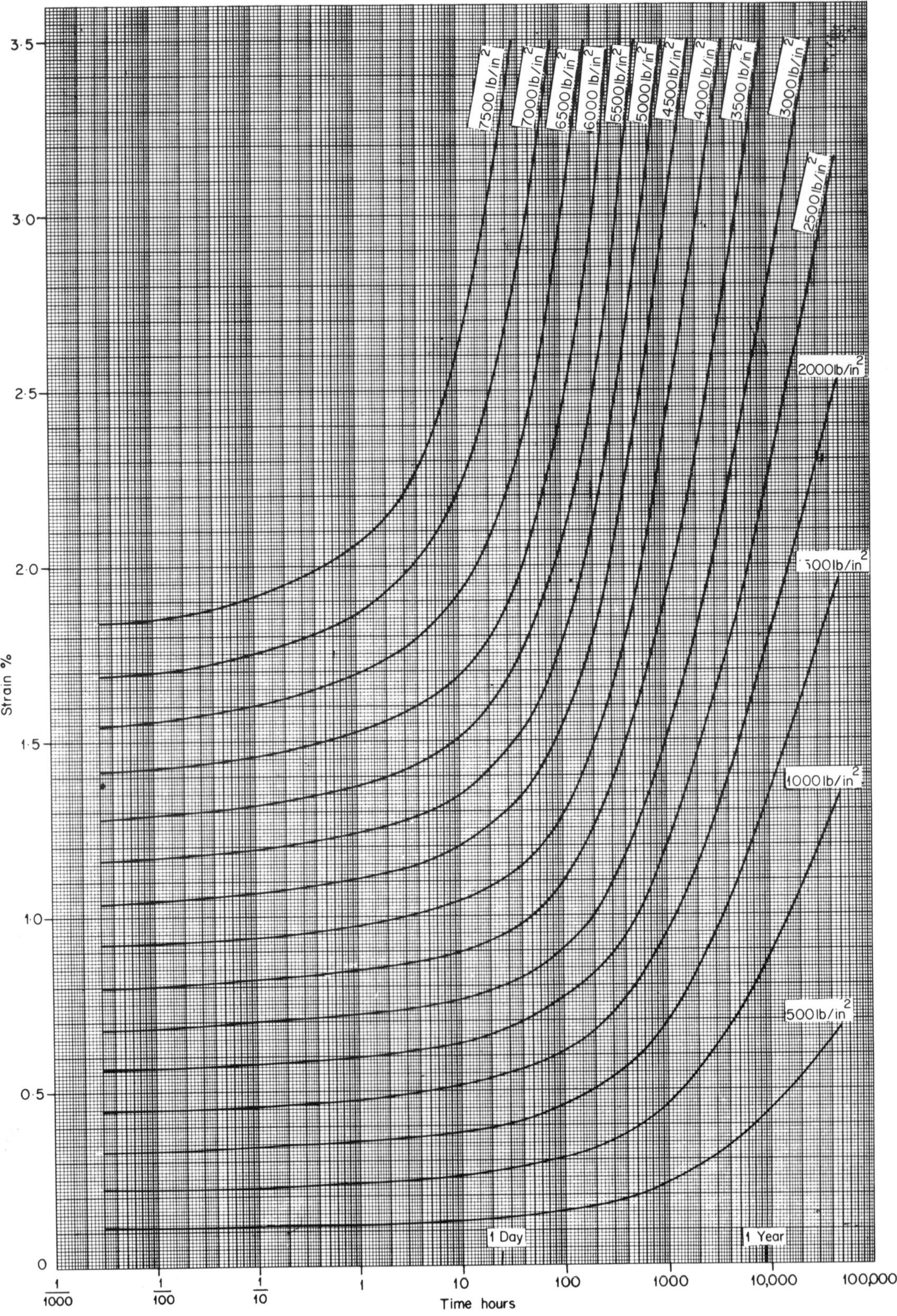


**Figure 11.1.** Creep curves in tension: 20°C, dry. Nylon 66 ('Maranyl' A101)

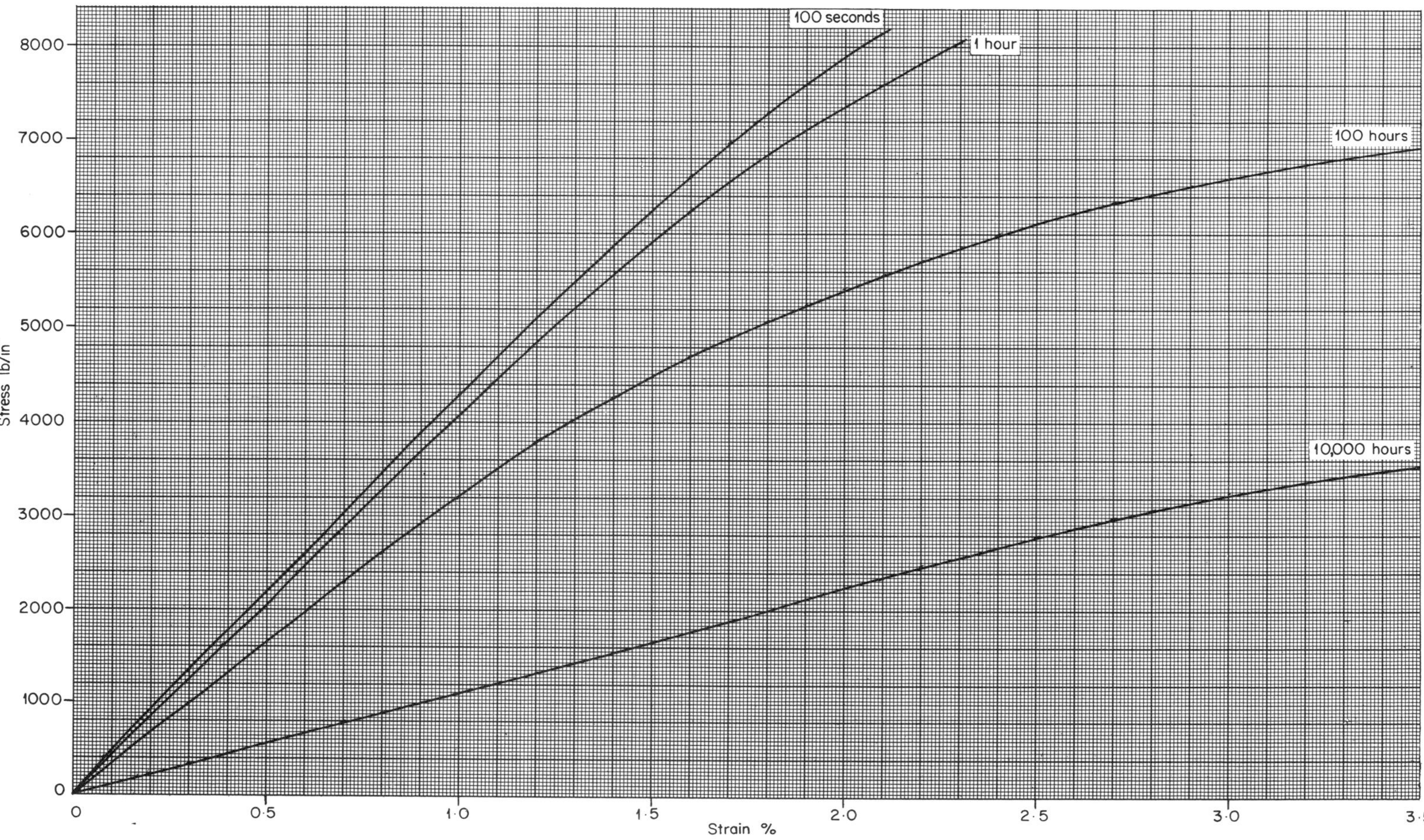


**Figure 11.2.** Isochronous stress vs strain curves: 20°C, dry. Nylon 66 ('Maranyl' A101)

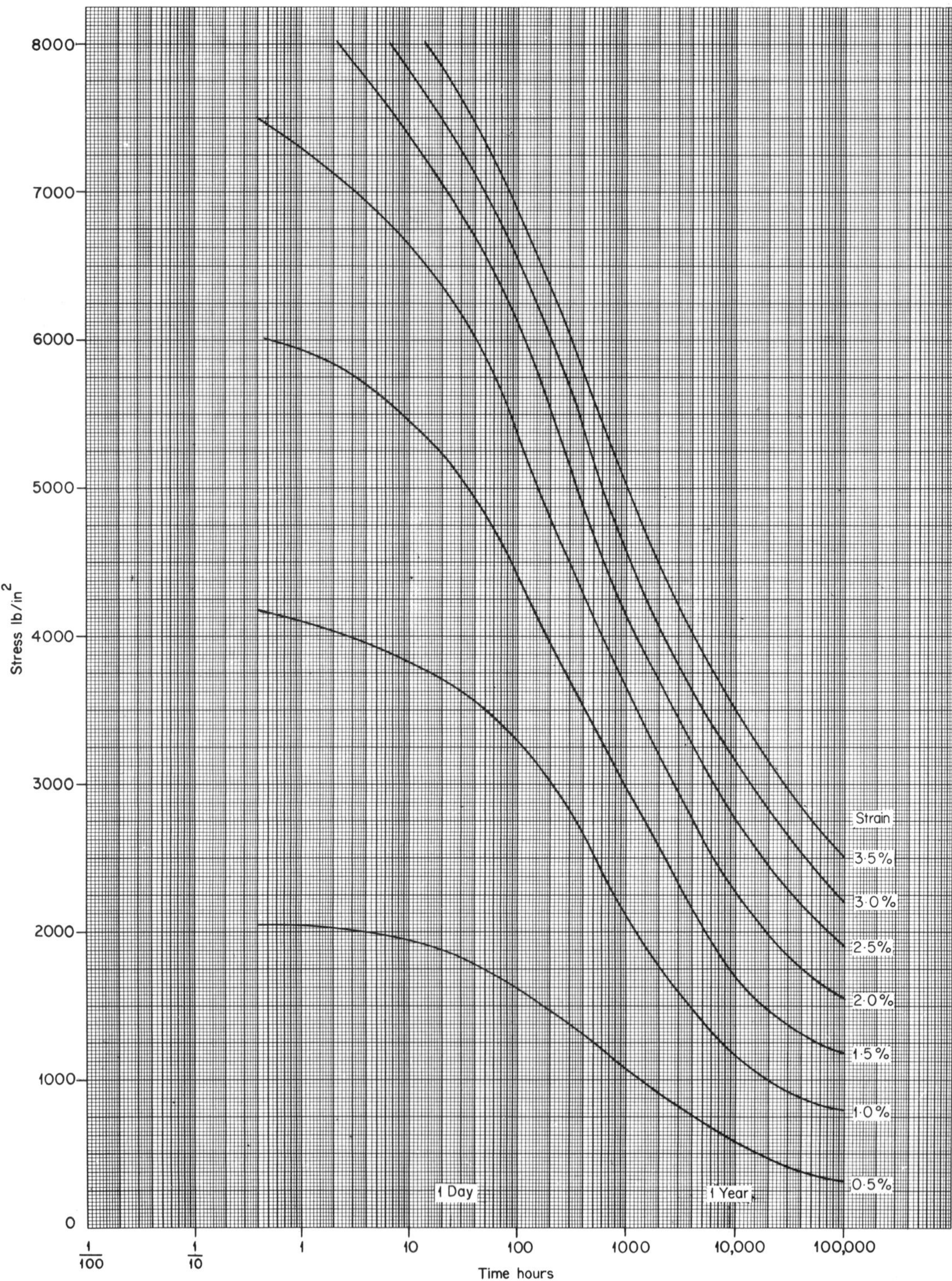


**Figure 11.3.** Isometric stress vs time curves: 20°C, dry. Nylon 66 ('Maranyl' A101)

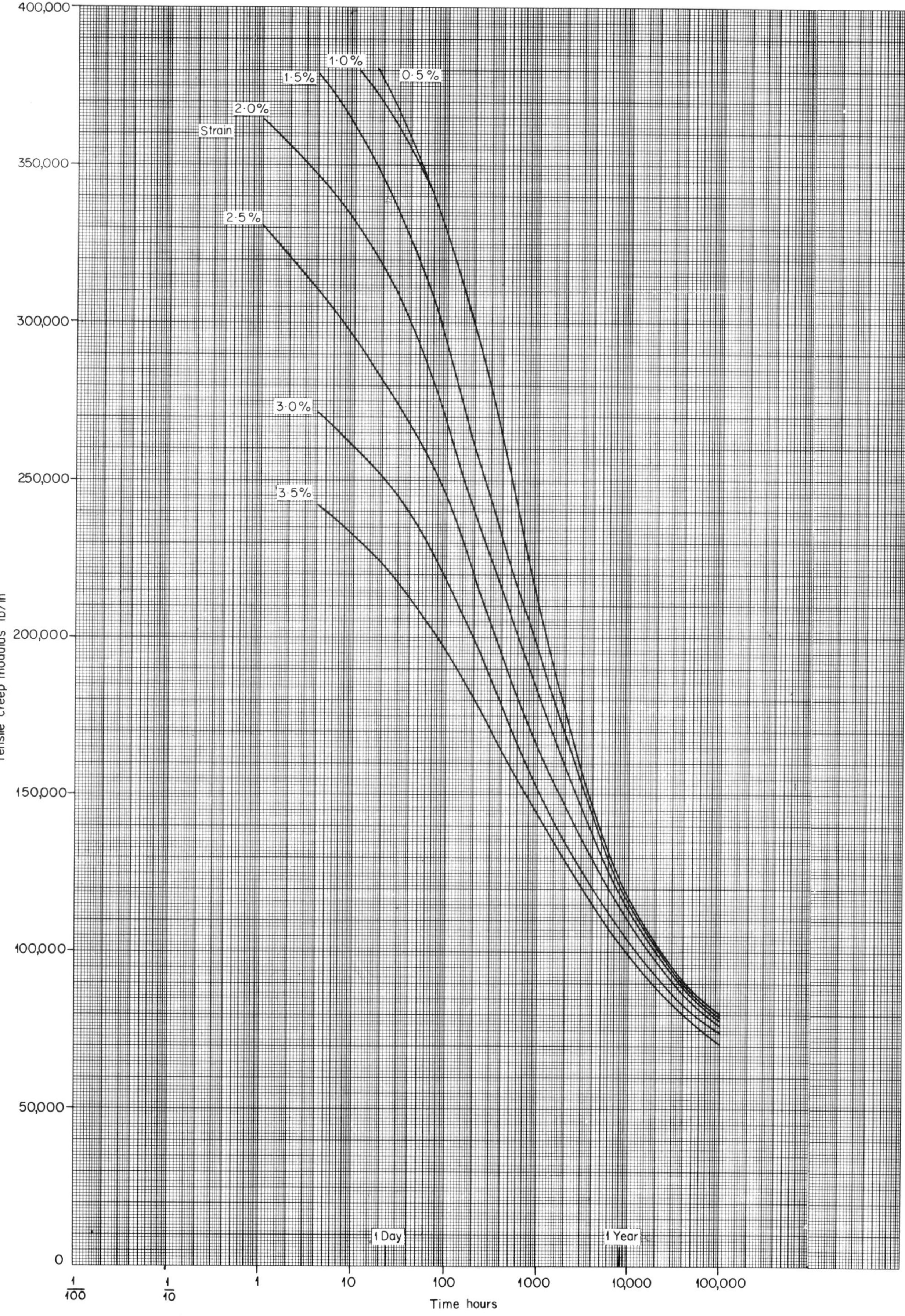


**Figure 11.4.** Tensile creep modulus **vs time curves**: 20°C, **dry.** Nylon 66 ('Maranyl' A101)

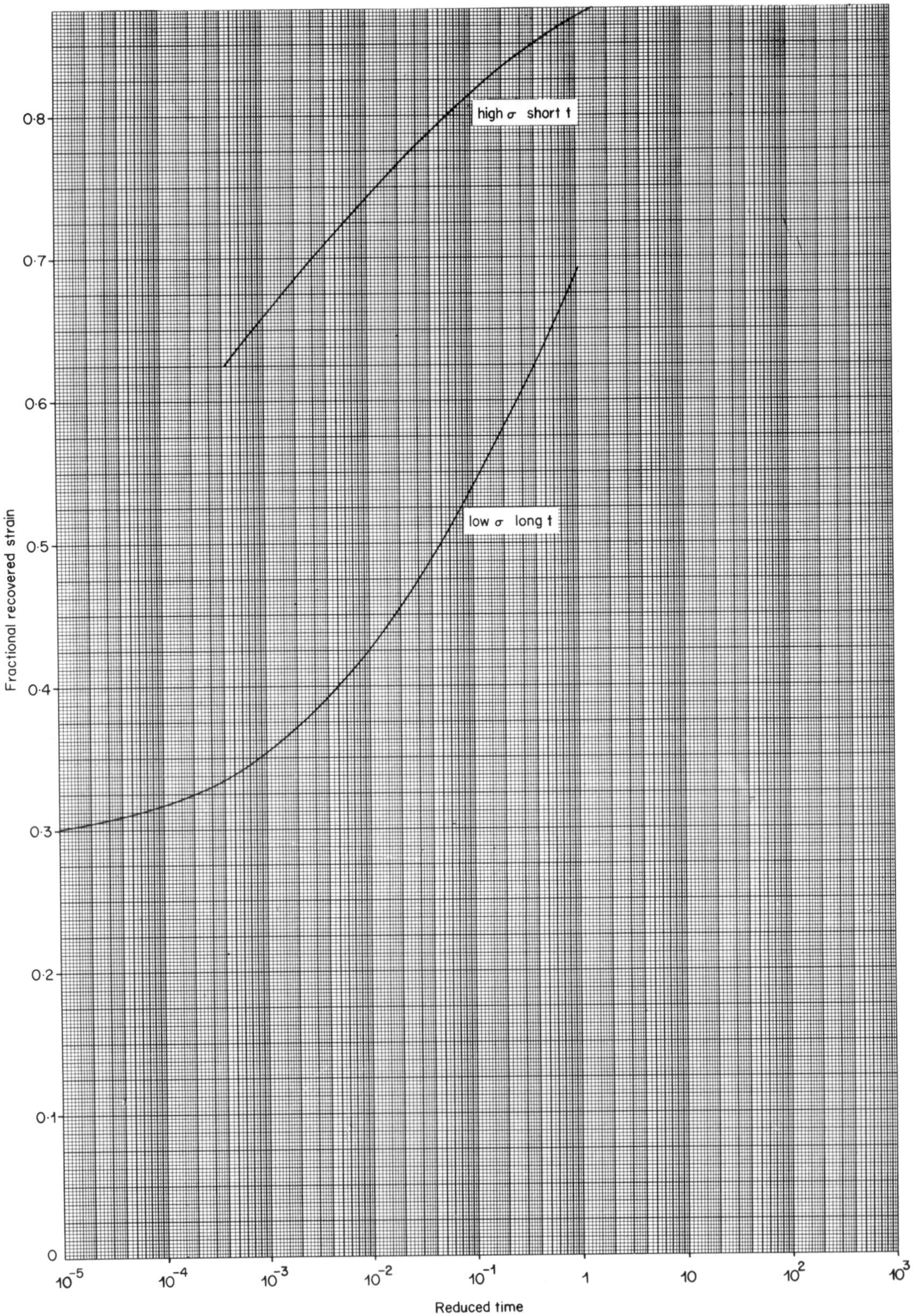


**Figure 11.5.** Recovery from creep in tension: 20°C, dry. Nylon 66 ('Maranyl' A101)

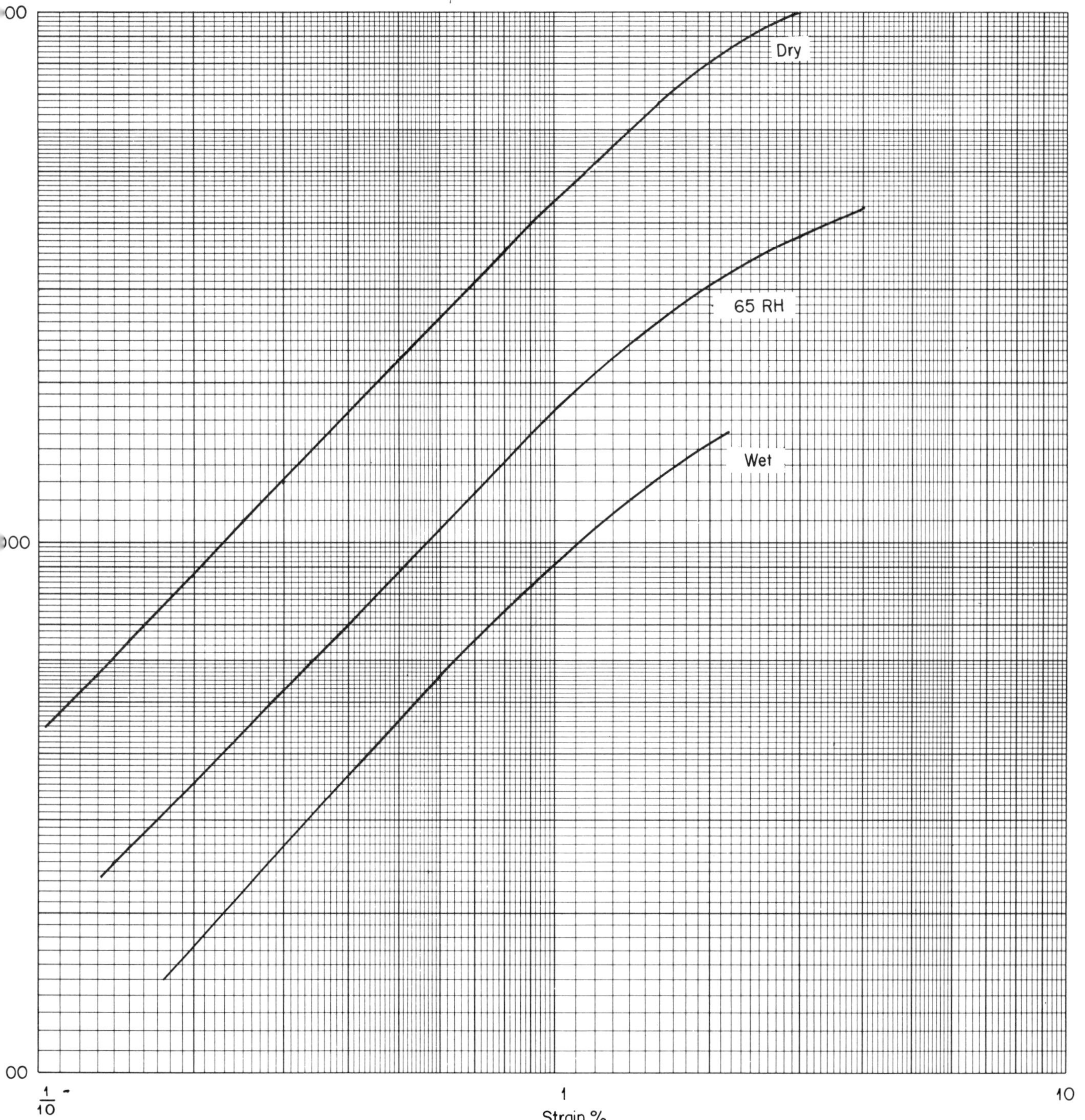


**Figure 11.6.** Isochronous stress vs strain curves: 20°C, 100 sec: logarithmic axes. Effect of water content. Nylon 66 ('Maranyl' A100, A101)

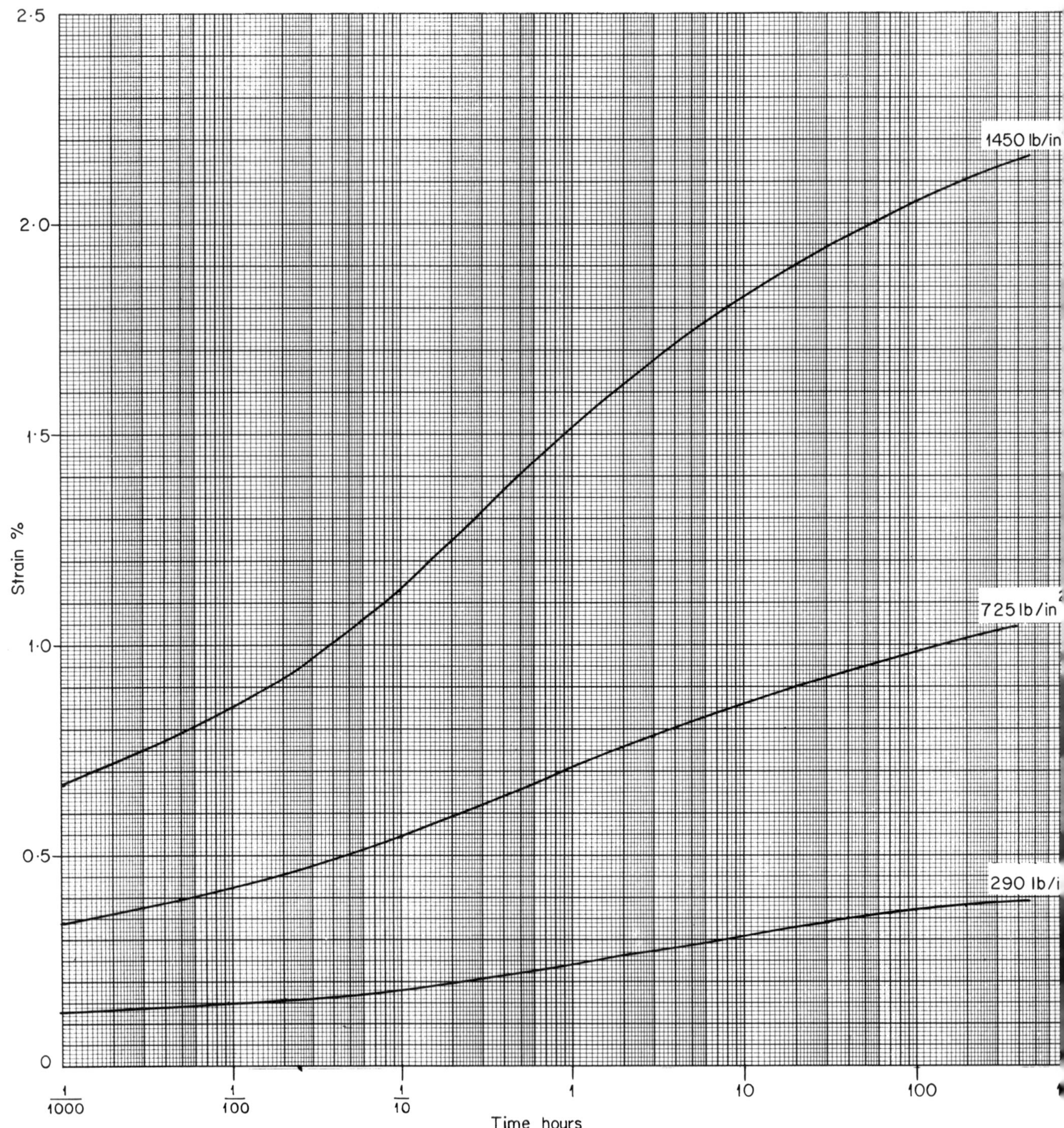


**Figure 11.7.** Creep curves in tension: 60°C, dry. Nylon 66 ('Maranyl' A100)

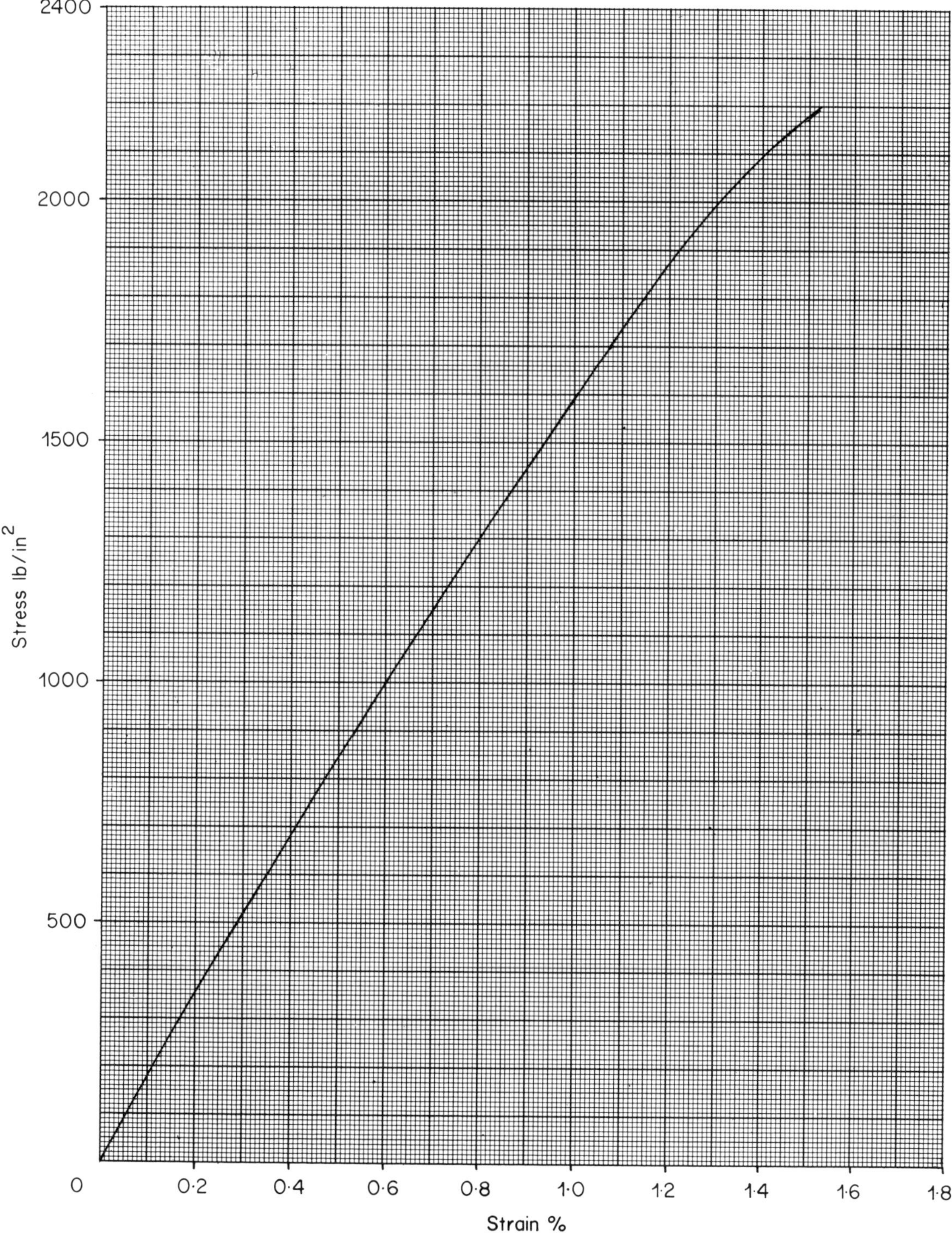


**Figure 11.8.** Isochronous stress vs strain curve: 60°C, 100 sec, dry. Nylon 66 ('Maranyl' A100)

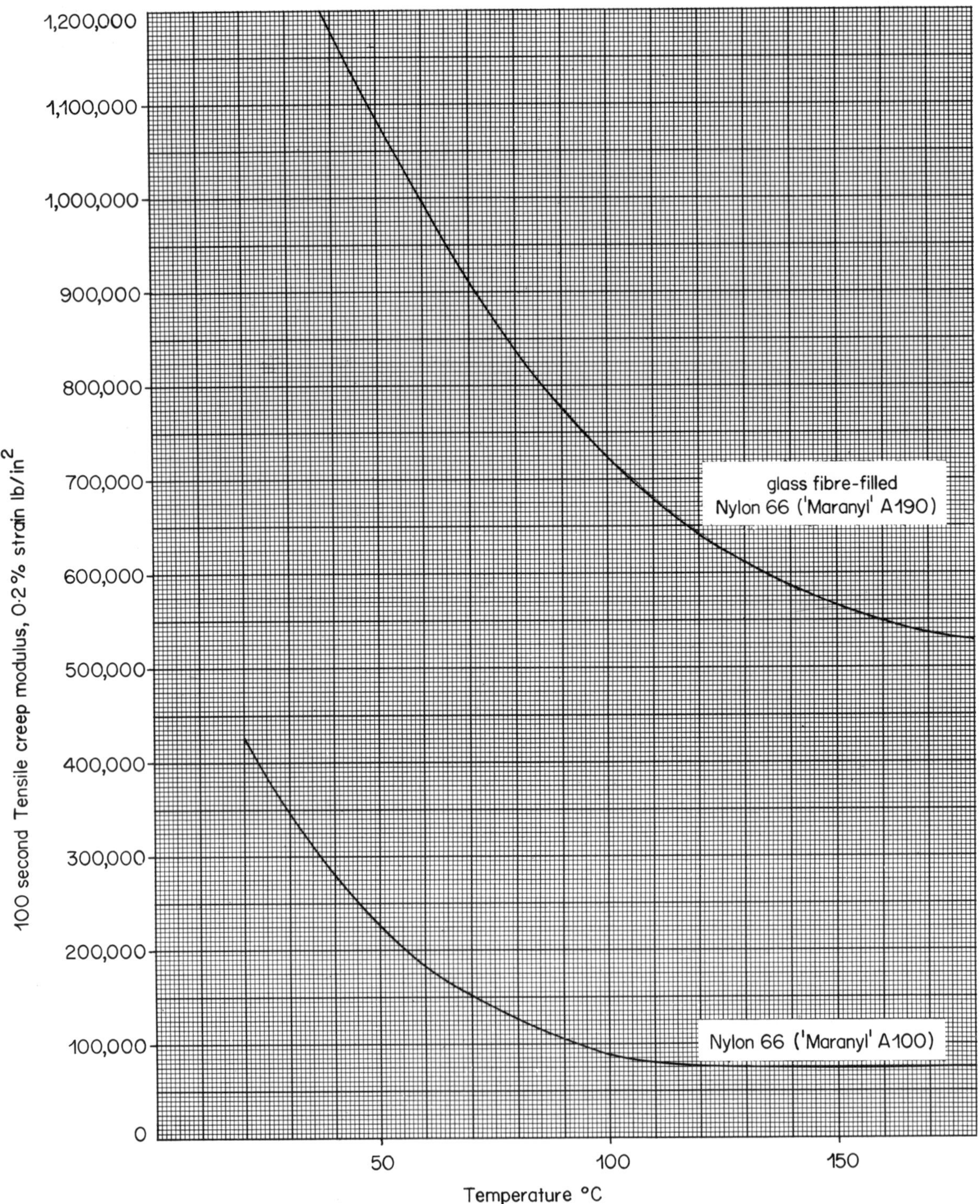


**Figure 11.9.** Tensile creep modulus (100 sec, 0·2% strain) vs temperature. Nylon 66 and glass fibre-filled nylon 66, dry specimens ('Maranyl' A100, A190)

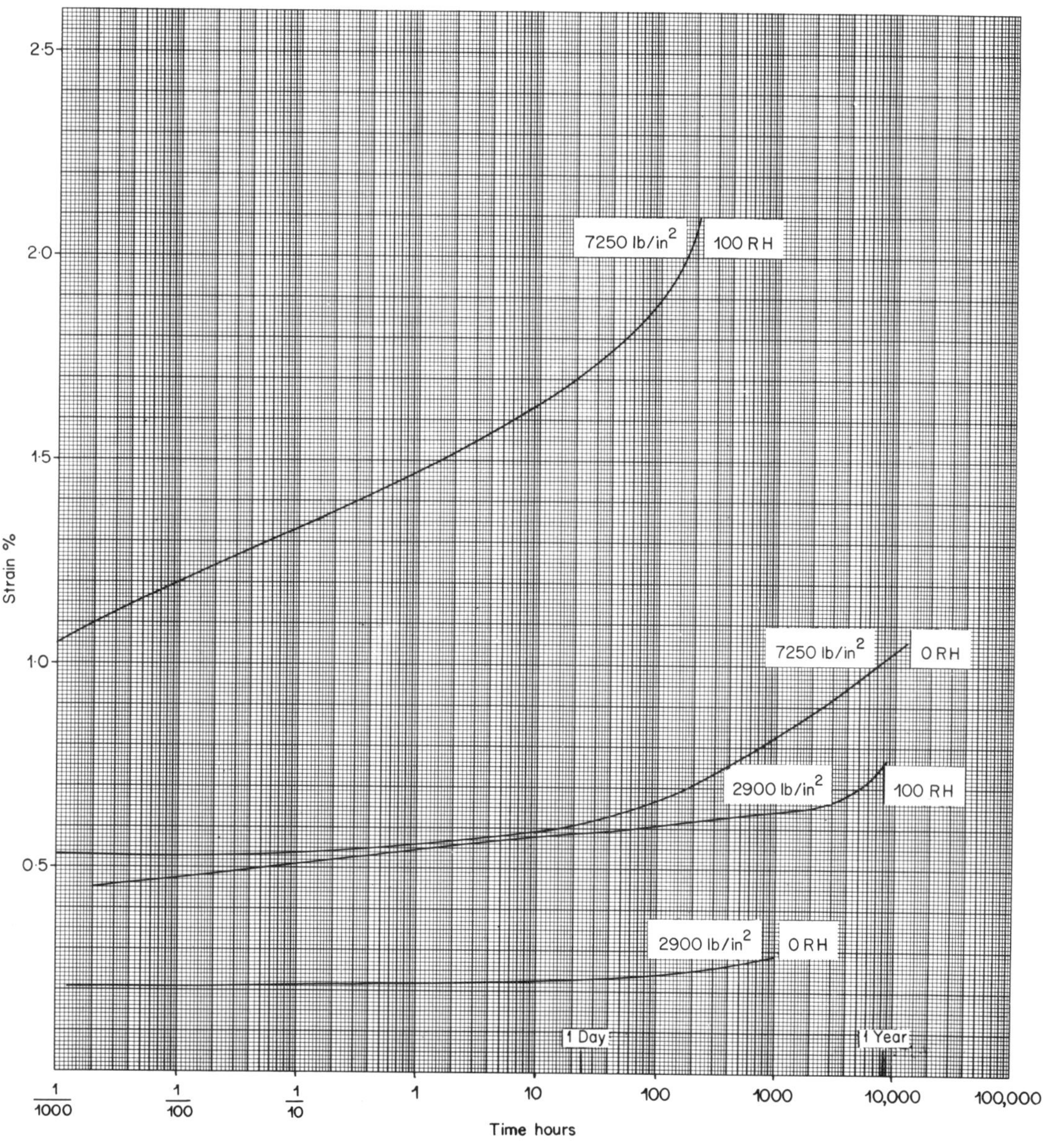


**Figure 11.10.** Creep curves in tension: 20°C. Comparison of wet and dry specimens. Glass fibre-filled nylon 66 ('Maranyl' A190)

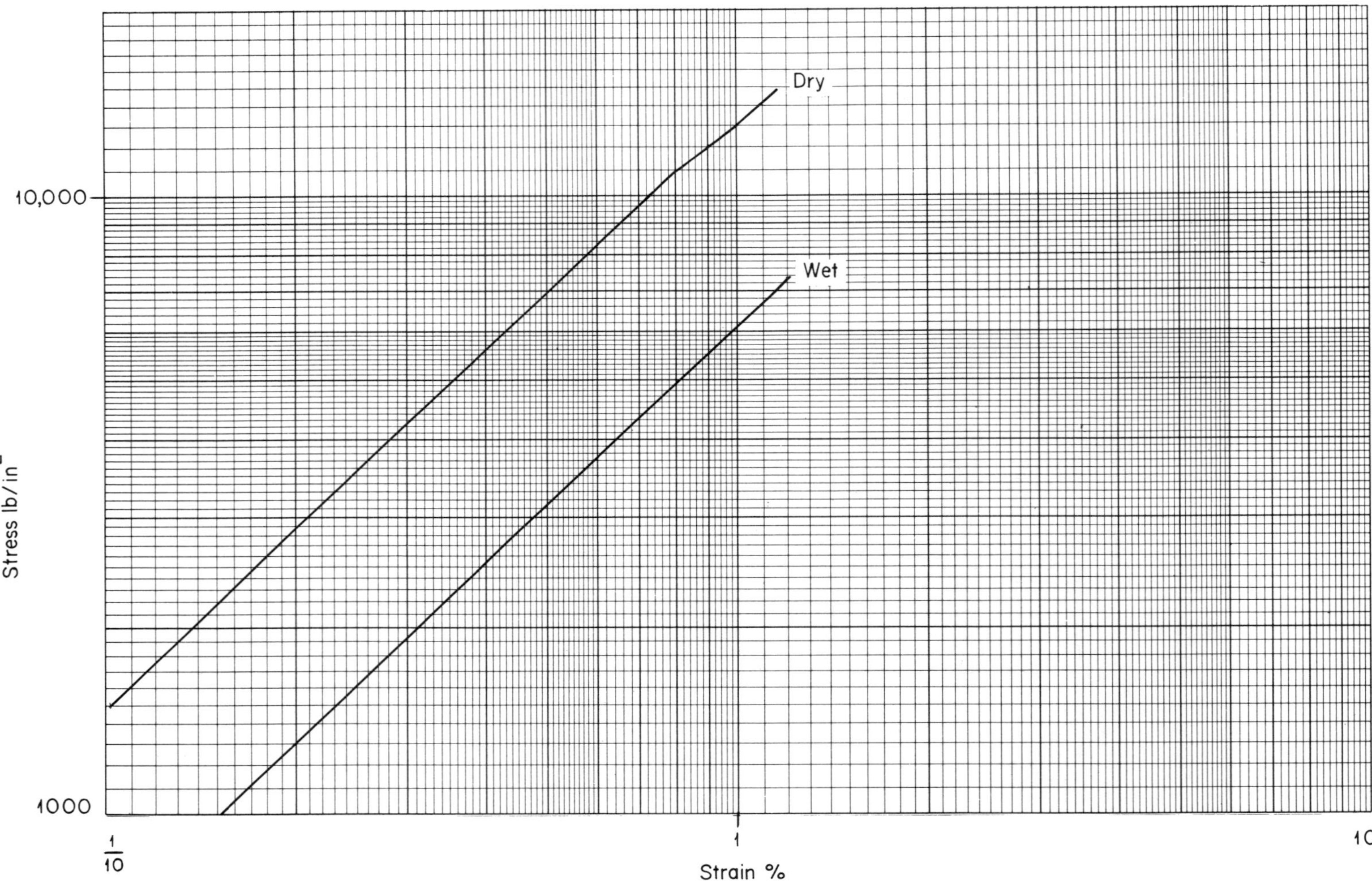


**Figure 11.11.** Isochronous stress vs strain curves: 20°C, 100 sec, logarithmic axes. Effect of water content. Glass fibre-filled nylon 66 ('Maranyl' A190)

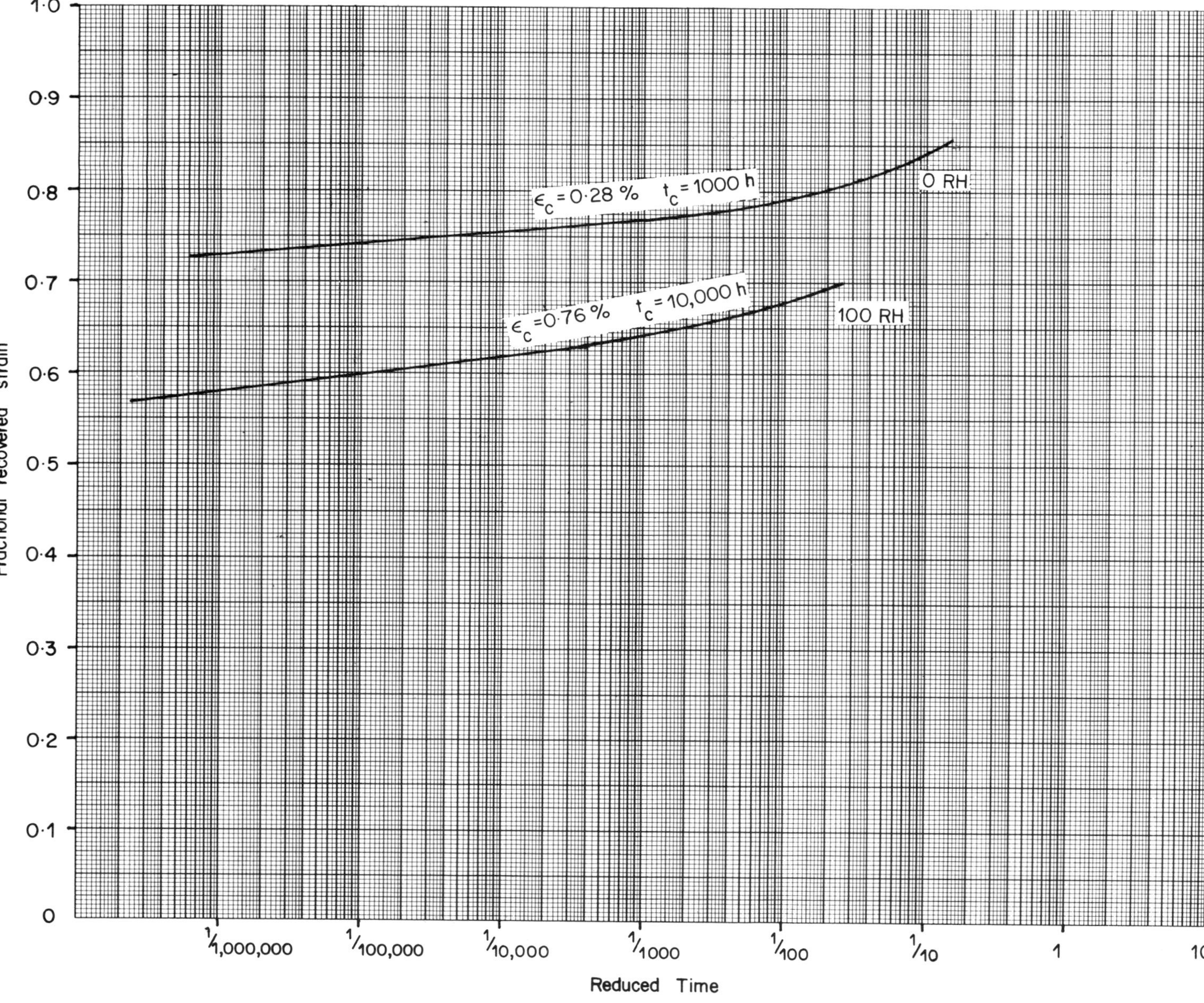


**Figure 11.12.** Recovery from creep in tension: 20°C. Wet and dry specimens. Glass fibre-filled nylon 66 ('Maranyl' A190)

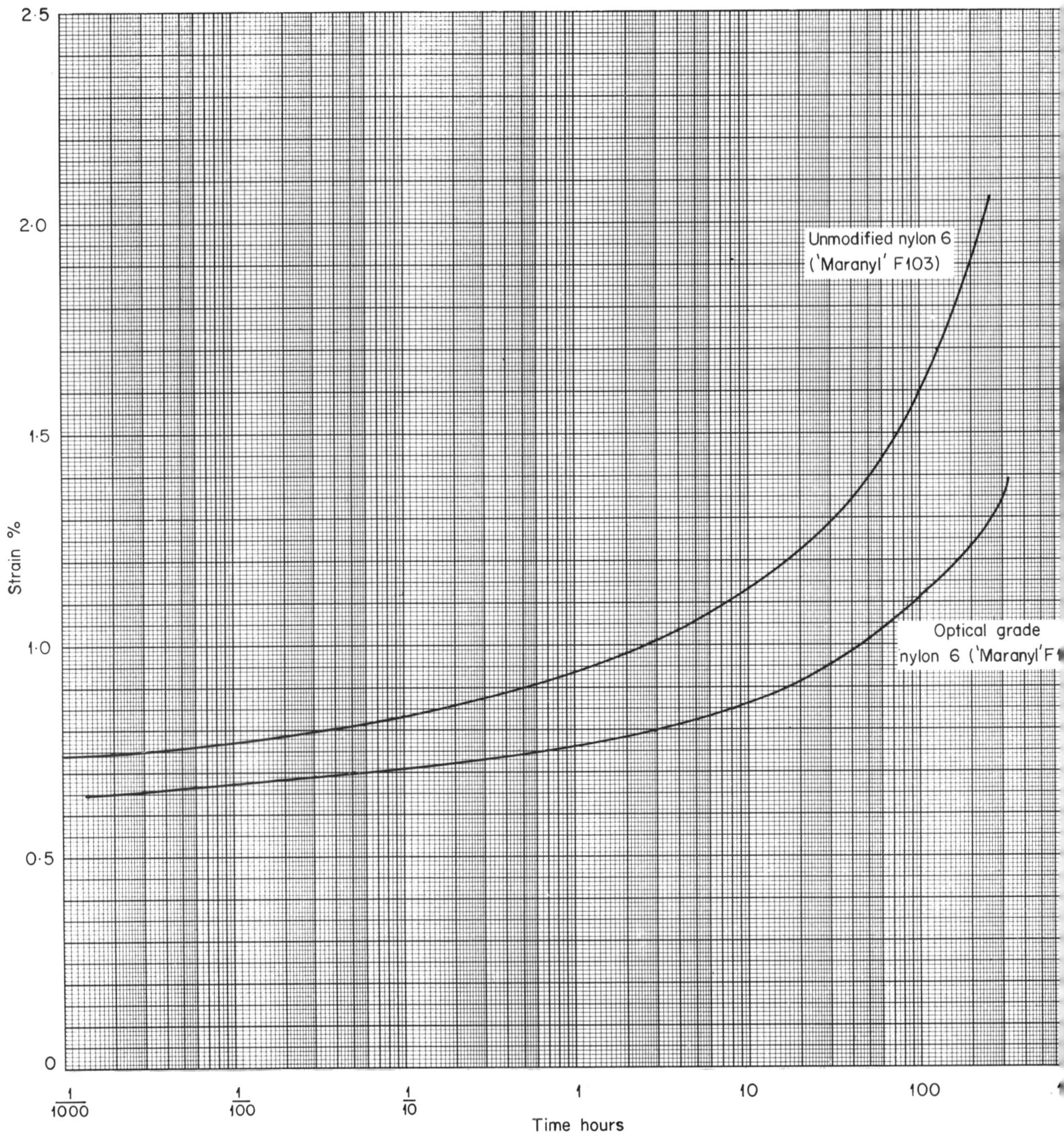


**Figure 11.13.** Creep curves in tension: 2900 lb/in$^2$ 20°C, dry. Nylon 6, unmodified and optical grades ('Maranyl' F103, F113)

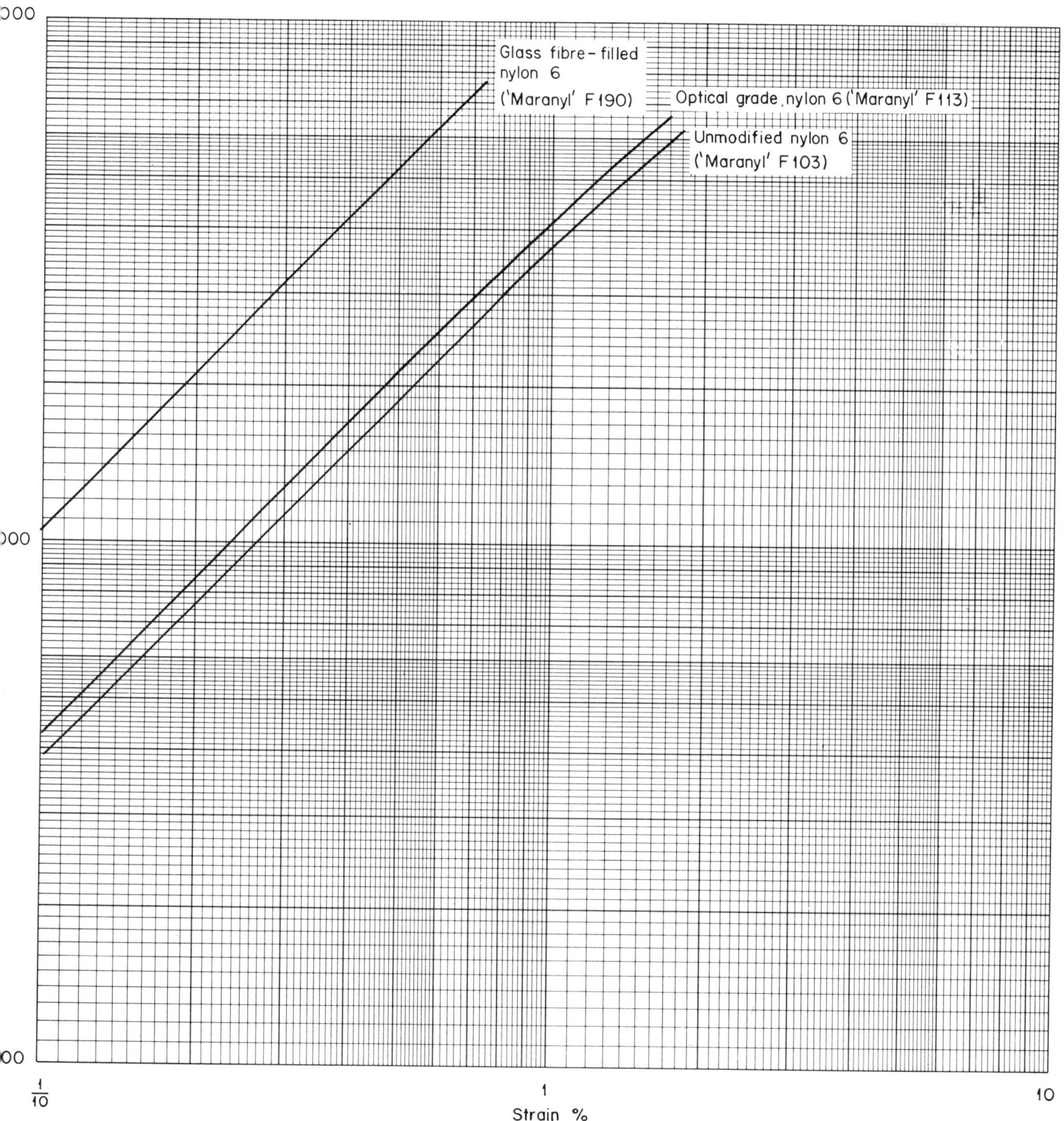


**Figure 11.14.** Isochronous stress vs strain curves: 20°C, 100 sec, dry, logarithmic axes. Nylon 6; unmodified, optical and glass fibre-filled grades ('Maranyl' F103, F113, F190)

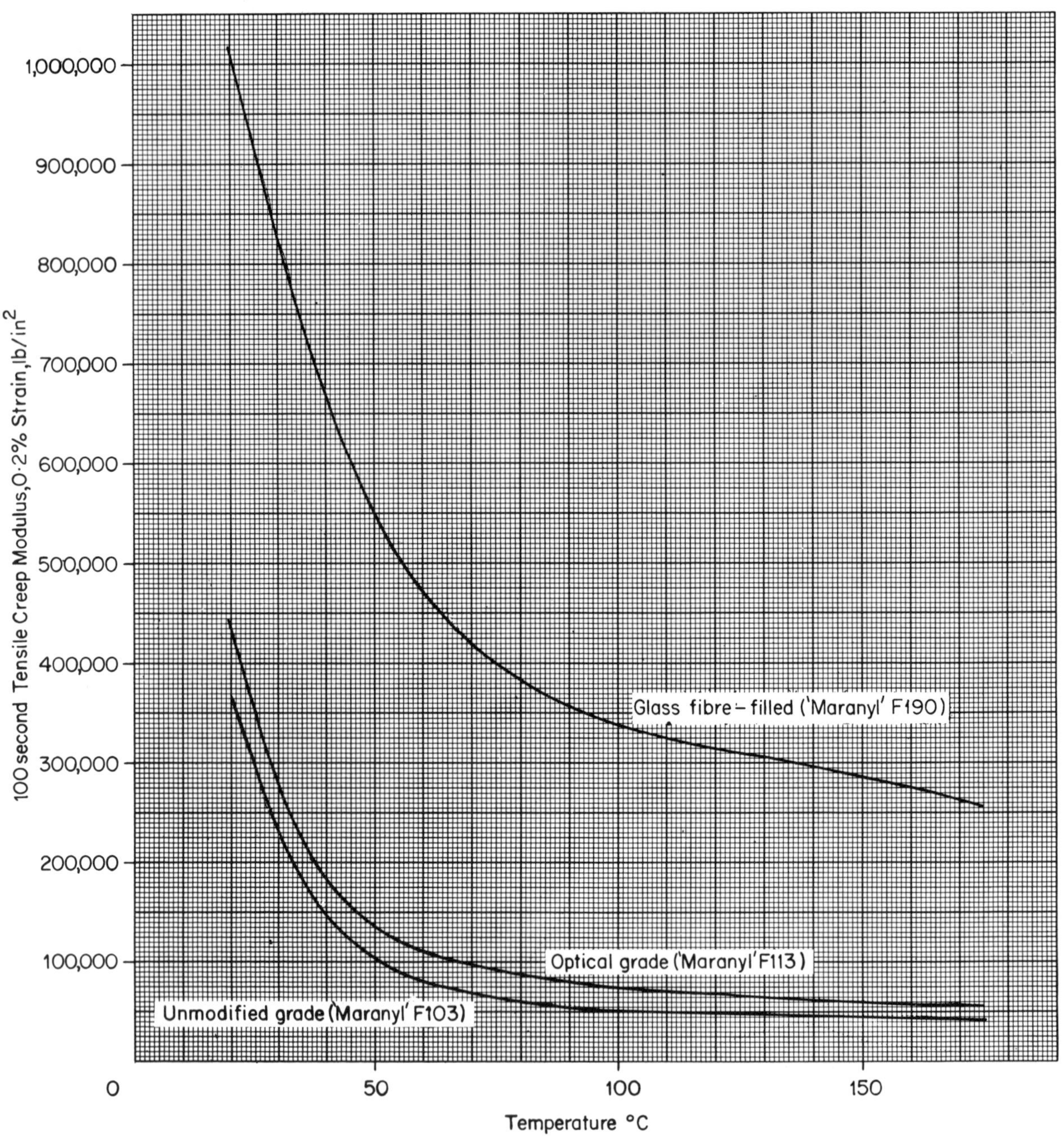


**Figure 11.15.** Tensile creep modulus (100 sec, 0·2% strain) vs temperature: dry. Nylon 6, unmodified, optical and glass fibre-filled grades ('Maranyl' F103, F113, F190)

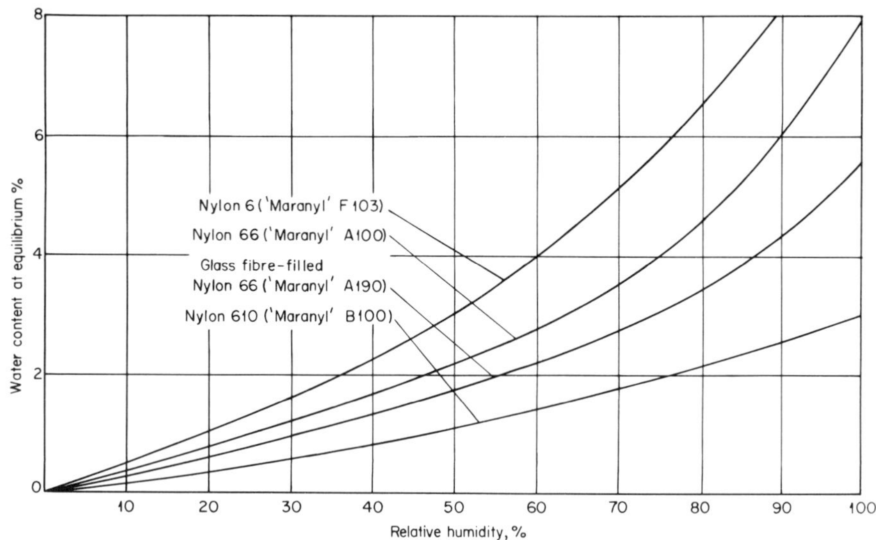


**Figure 11.16.** Equilibrium water content vs relative humidity of environment. Nylon 6, 66, 610 and glass fibre-filled nylon 66 (various 'Maranyl' grades)

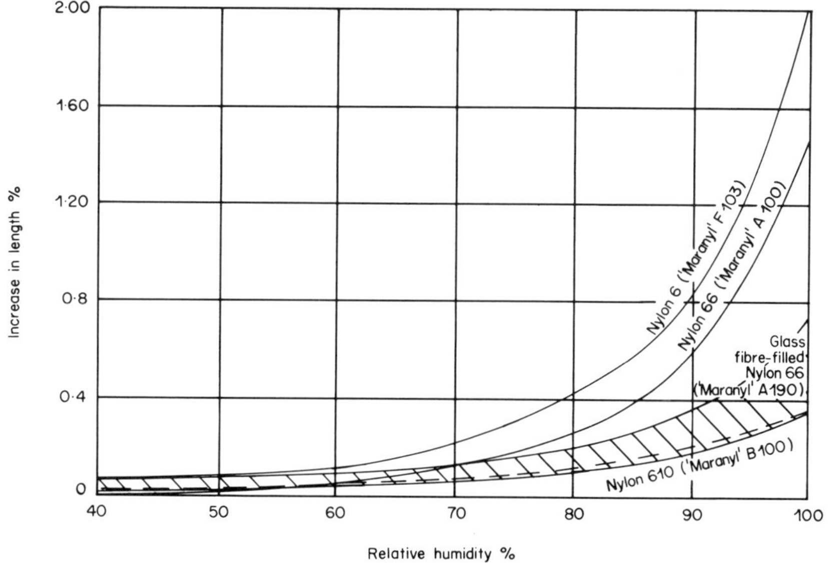


**Figure 11.17.** Equilibrium change in linear dimensions vs relative humidity of environment. Nylon 6, 66, 610 and glass fibre-filled nylon 66 (various 'Maranyl' grades)

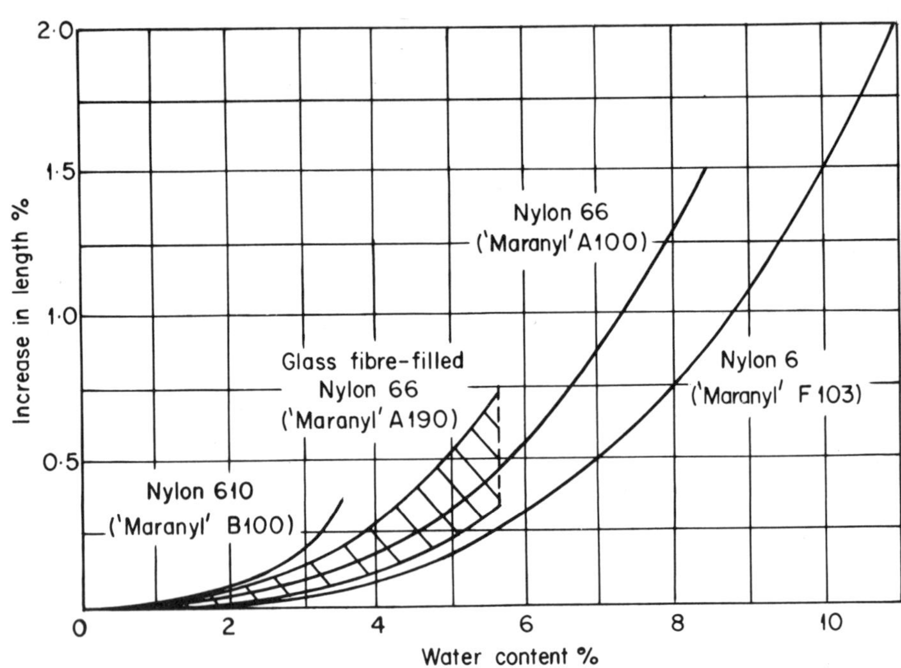


**Figure 11.18.** Equilibrium change in linear dimensions vs water content. Nylon 6, 66, 610 and glass fibre-filled nylon 66 (various 'Maranyl' grades)

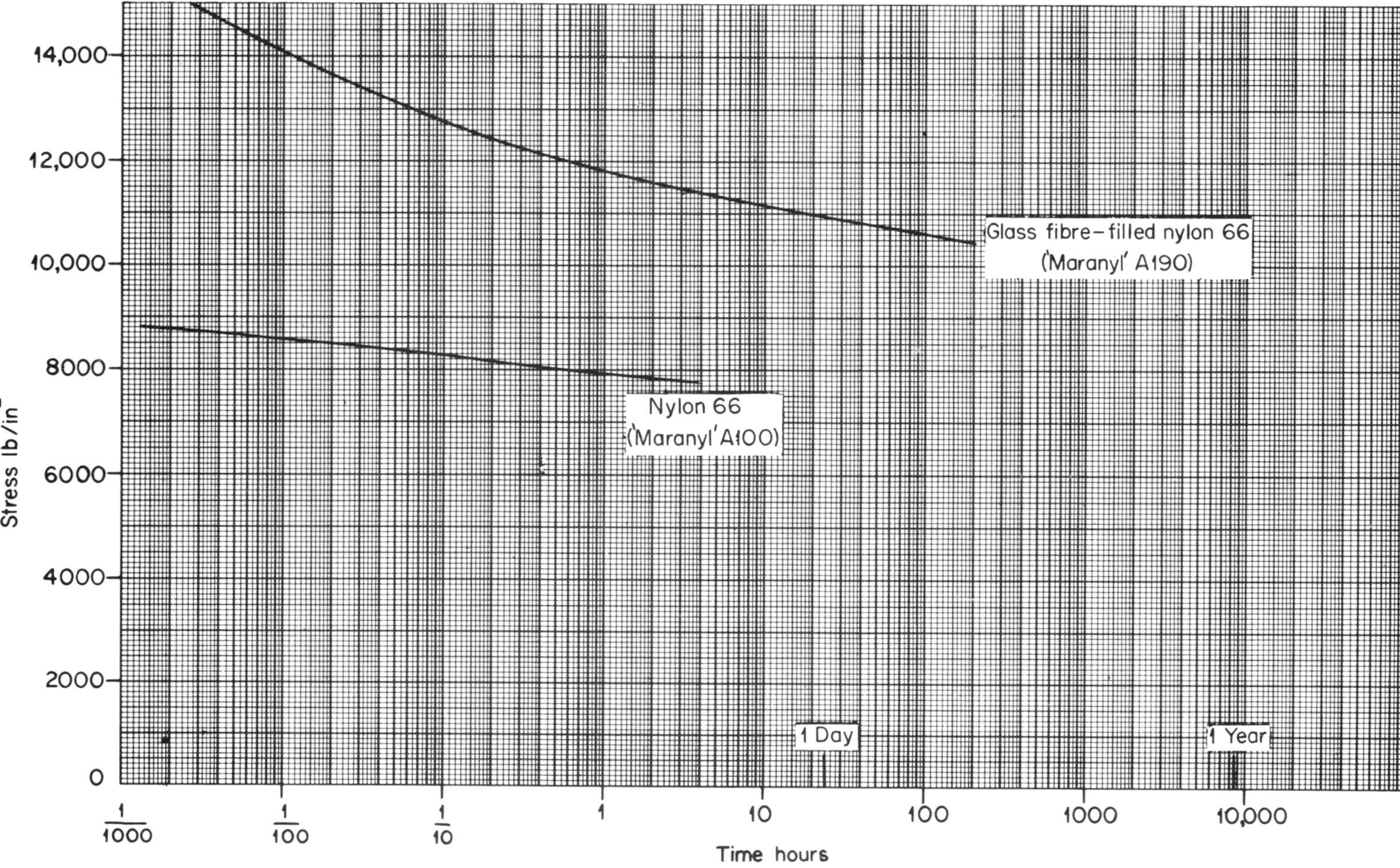


**Figure 11.19.** Creep rupture stress in tension vs time to failure: 20°C, 65RH. Nylon 66 and glass fibre-filled nylon 66 ('Maranyl' A100, A190)

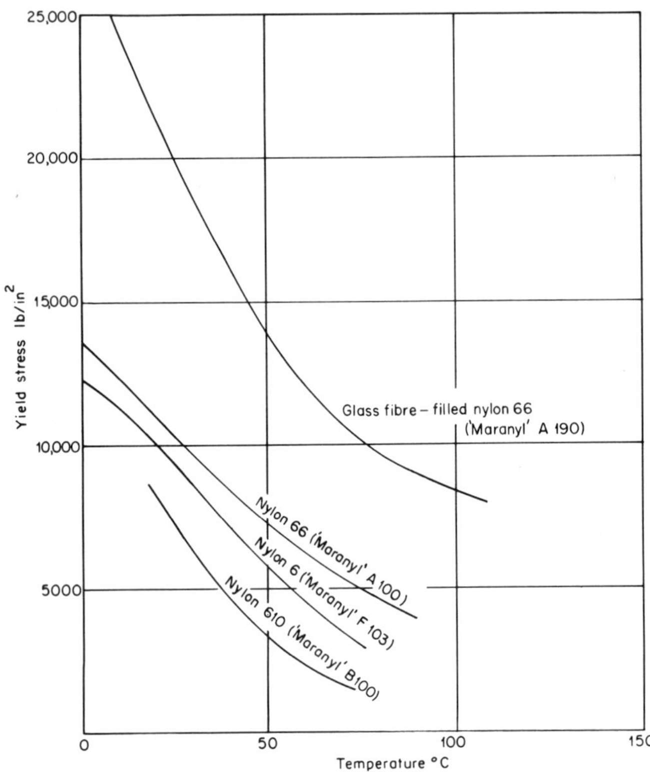


**Figure 11.20.** Yield stress in tension vs temperature: dry, 50% per min. straining rate. Nylon 6, 66, 610 and glass fibre-filled nylon 66 (various 'Maranyl' grades)

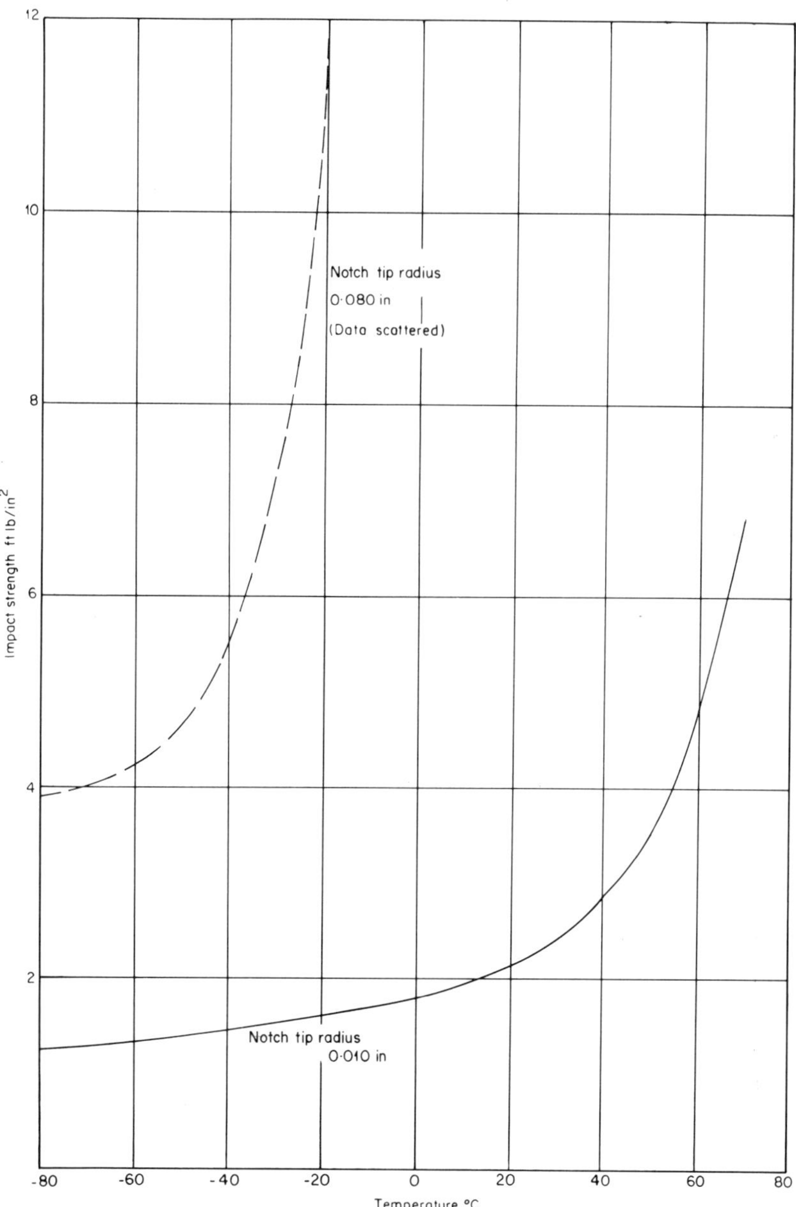


**Figure 11.21.** Impact strength vs temperature: dry. Nylon 66 ('Maranyl' A100)

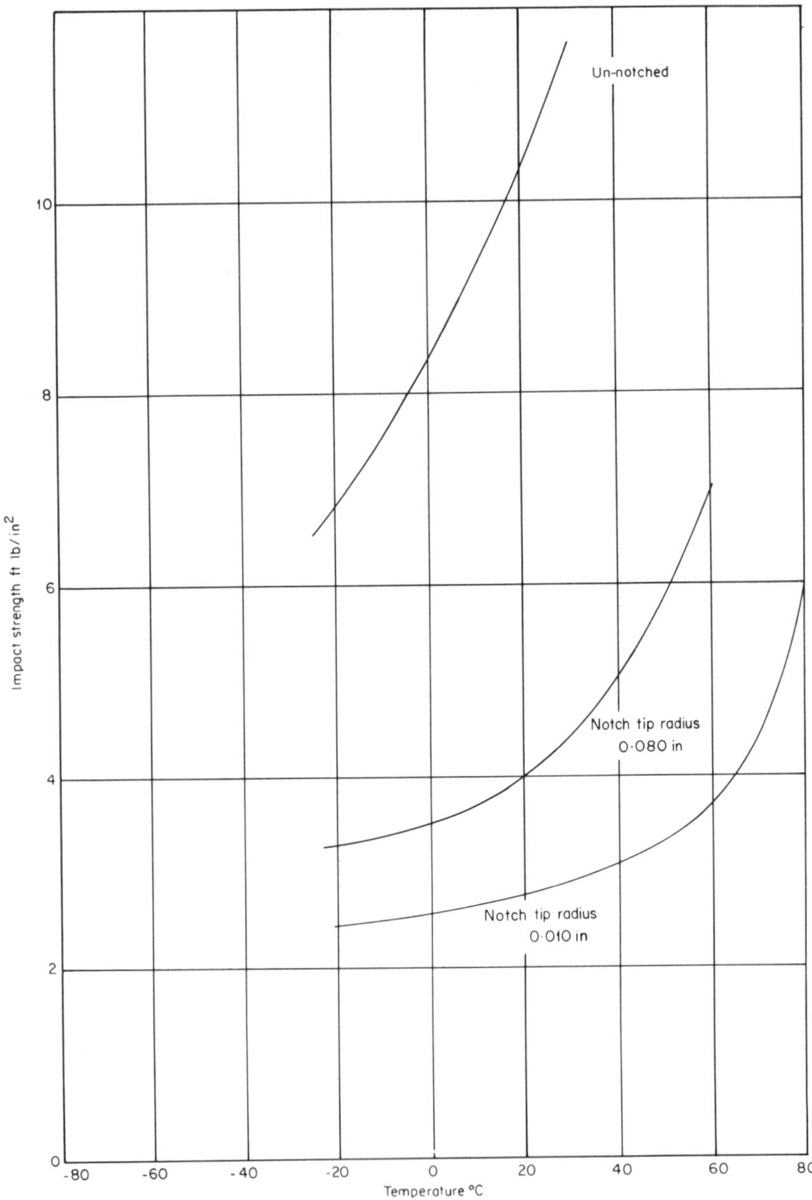


**Figure 11.22.** Impact strength vs temperature: dry. Glass fibre-filled nylon 66 ('Maranyl' A190)

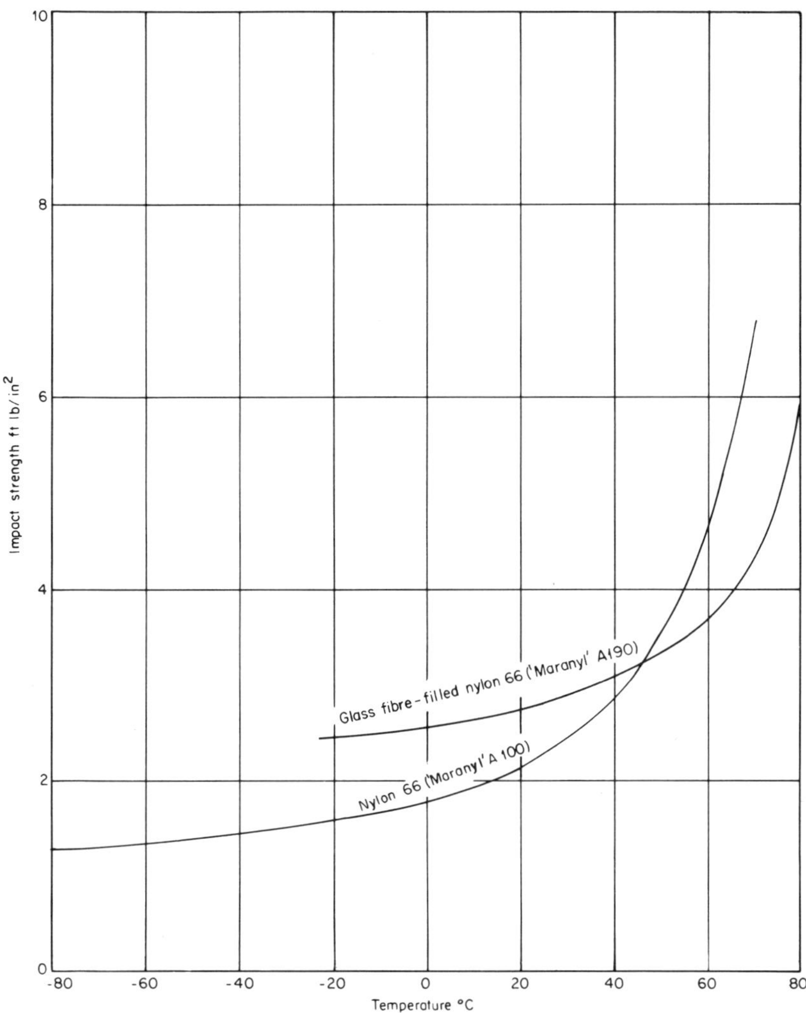


**Figure 11.23.** Impact strength vs temperature: 0·010 in notch tip radius, dry. Nylon 66 and glass fibre-filled nylon 66 ('Maranyl' A100, A190)

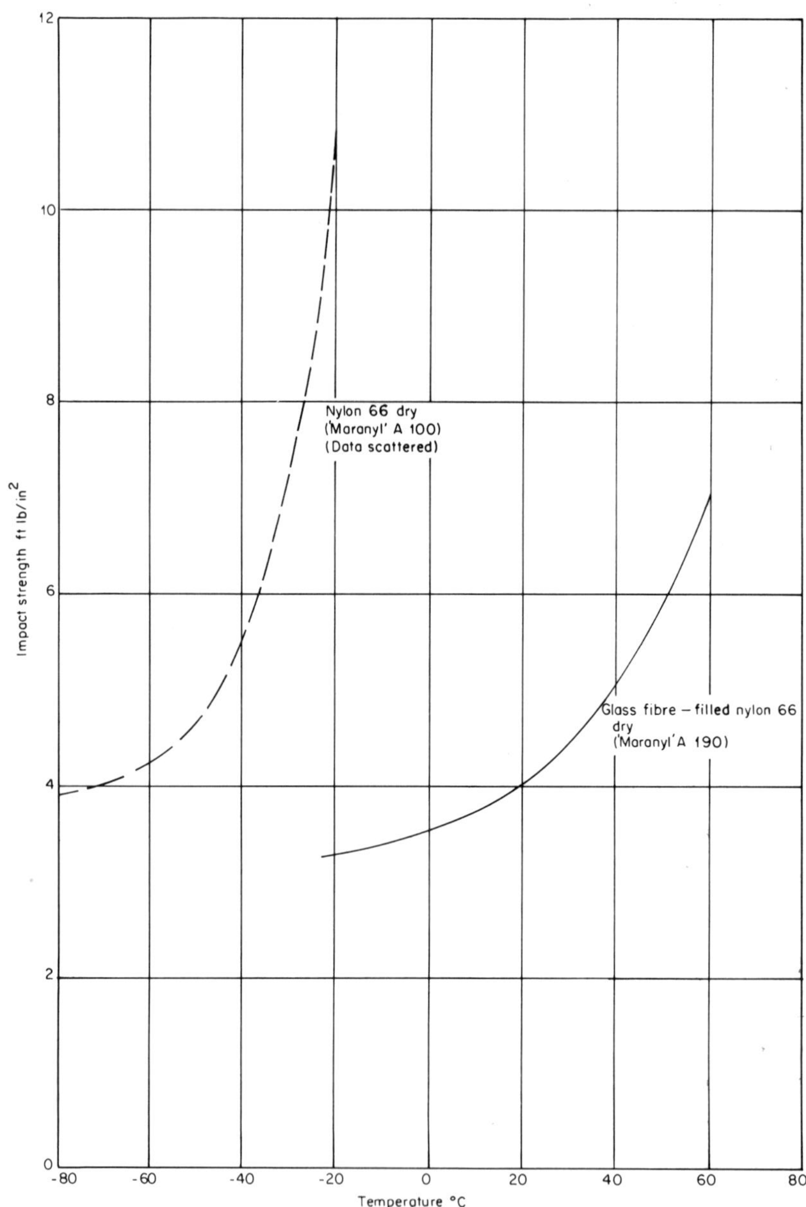


**Figure 11.24.** Impact strength vs temperature: 0·080 in notch tip radius, dry. Nylon 66 and glass fibre-filled nylon 66 ('Maranyl' A100, A190)

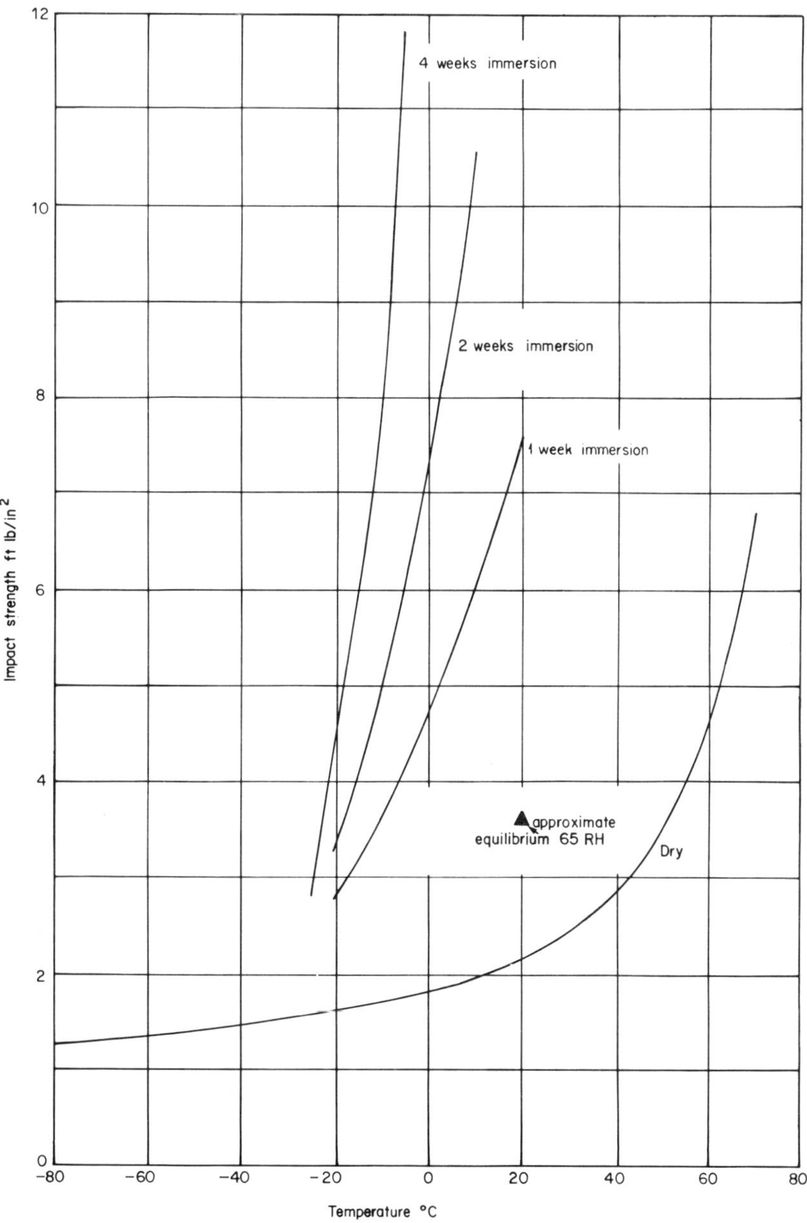


**Figure 11.25.** Impact strength vs temperature: 0·010 in notch tip radius. Effect of water immersion. Nylon 66 ('Maranyl' A100)

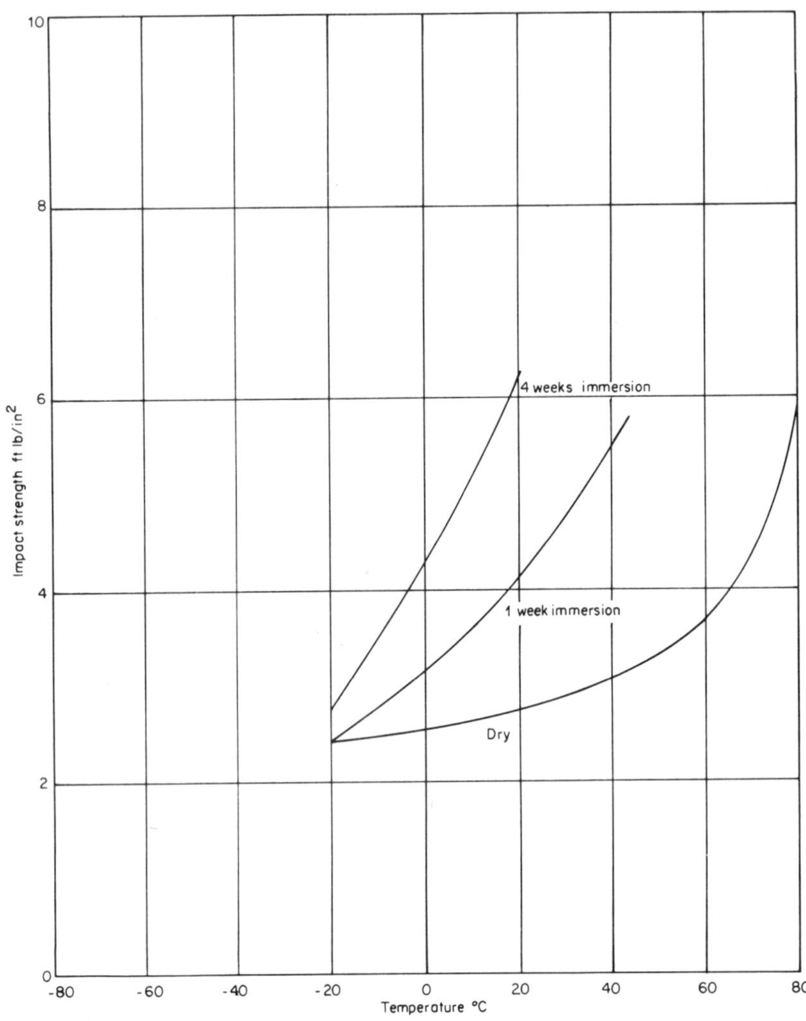


**Figure 11.26.** Impact strength vs temperature: 0·010 in notch tip radius. Effect of water immersion. Glass fibre-filled nylon 66 ('Maranyl' A190)

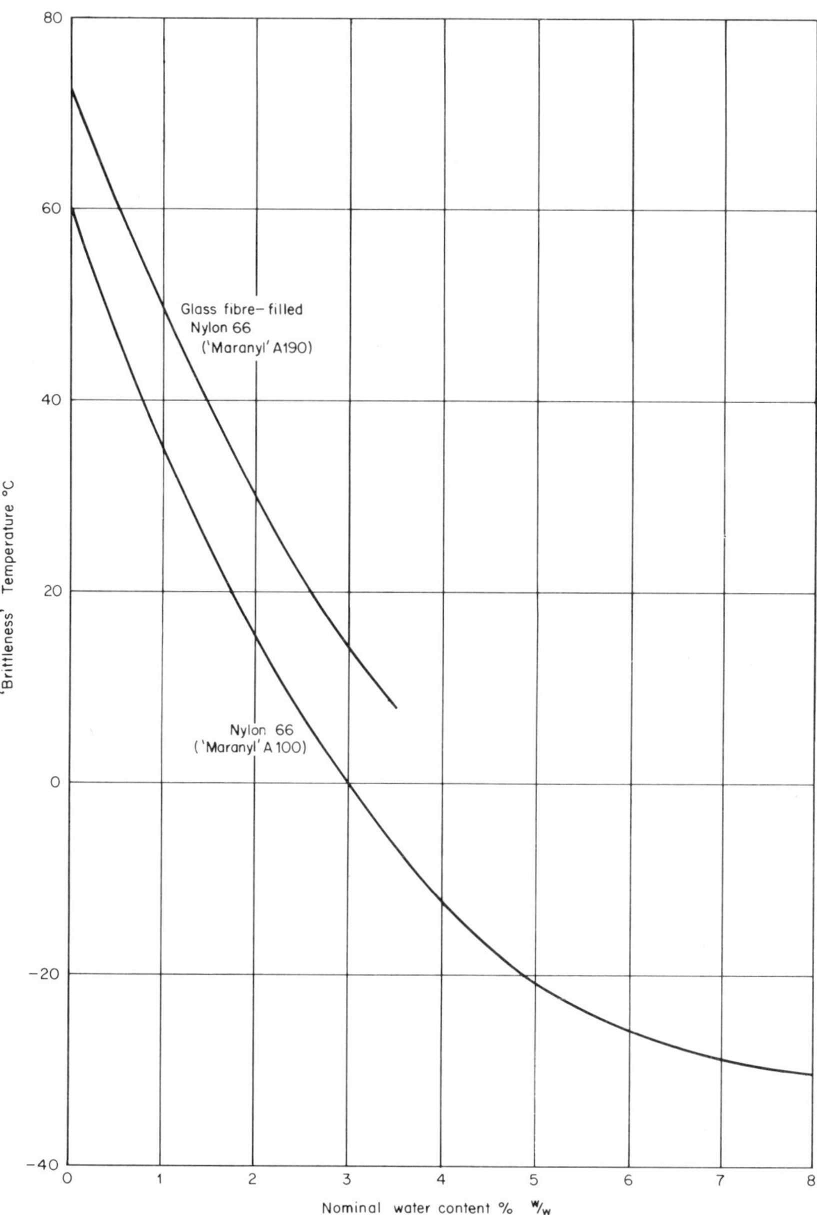


**Figure 11.27.** 'Brittleness' temperature (0·010 in. notch tip radius, 5 ft lb/in$^2$) vs nominal water content. Nylon 66 and glass fibre-filled nylon 66 ('Maranyl' A100, A190)

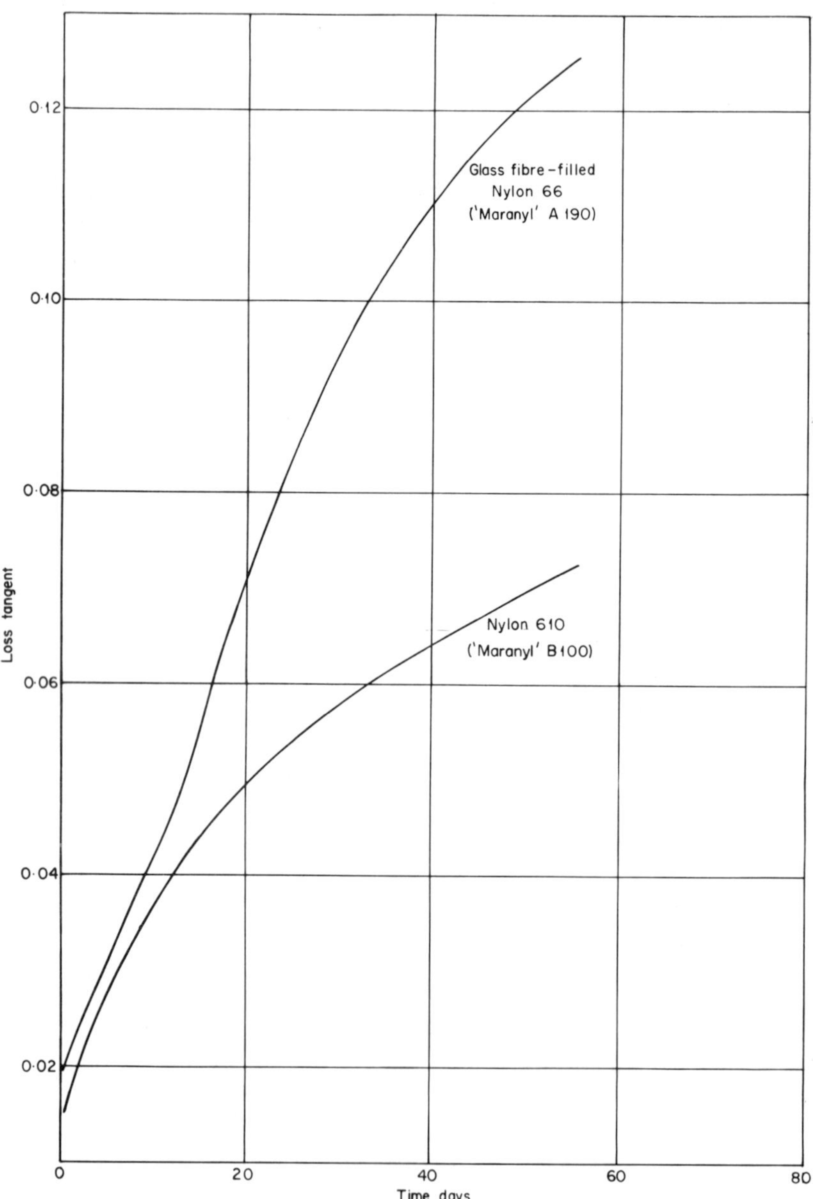


**Figure 11.28.** Loss tangent (1000 Hz) vs time of storage at 95 RH. Nylon 610 and glass fibre-filled nylon 66 ('Maranyl' B100, A190)

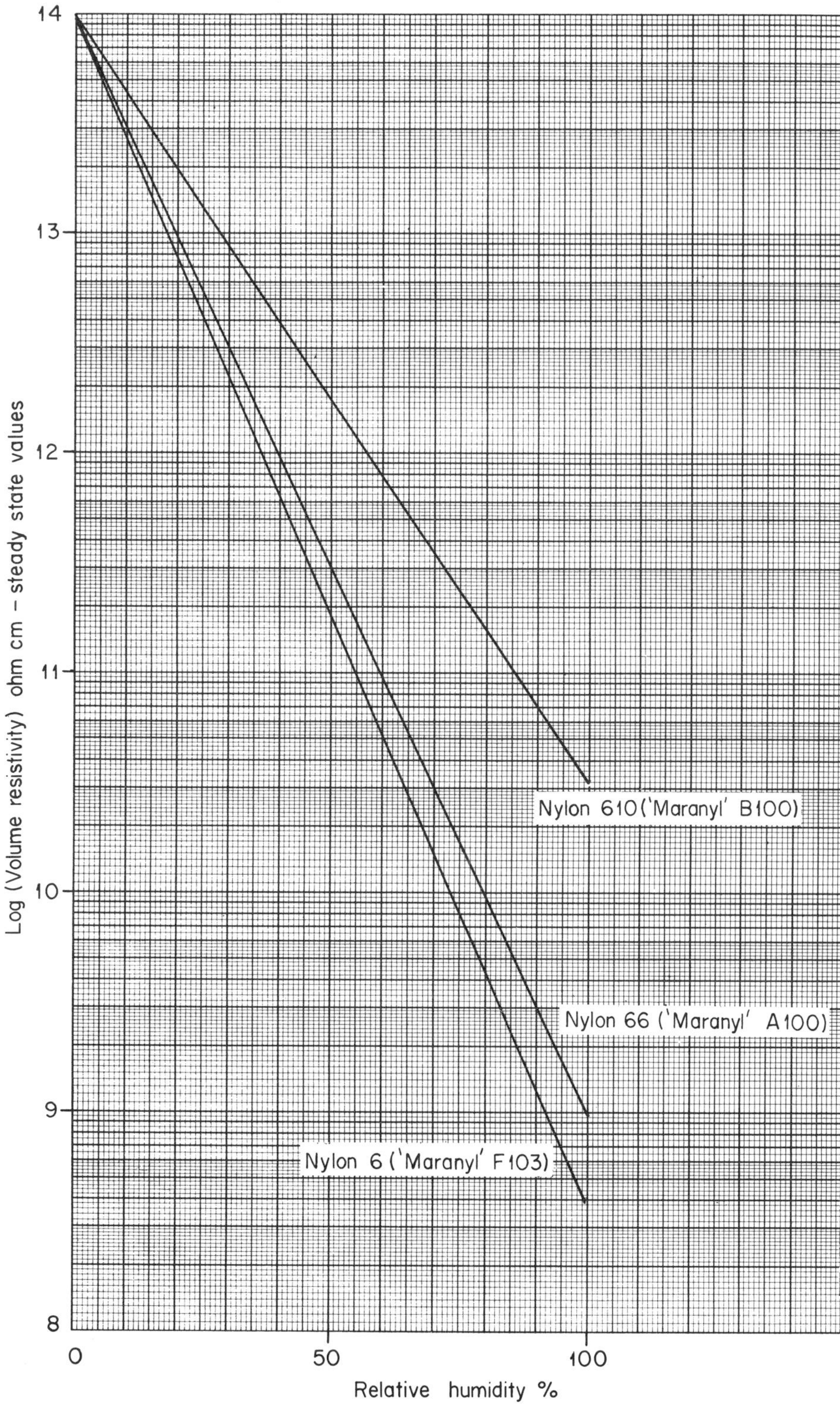


**Figure 11.29.** Steady state volume resistivity vs relative humidity of environment. Equilibrium data. Nylon 6, 66, 610 ('Maranyl' F103, A100, B100)

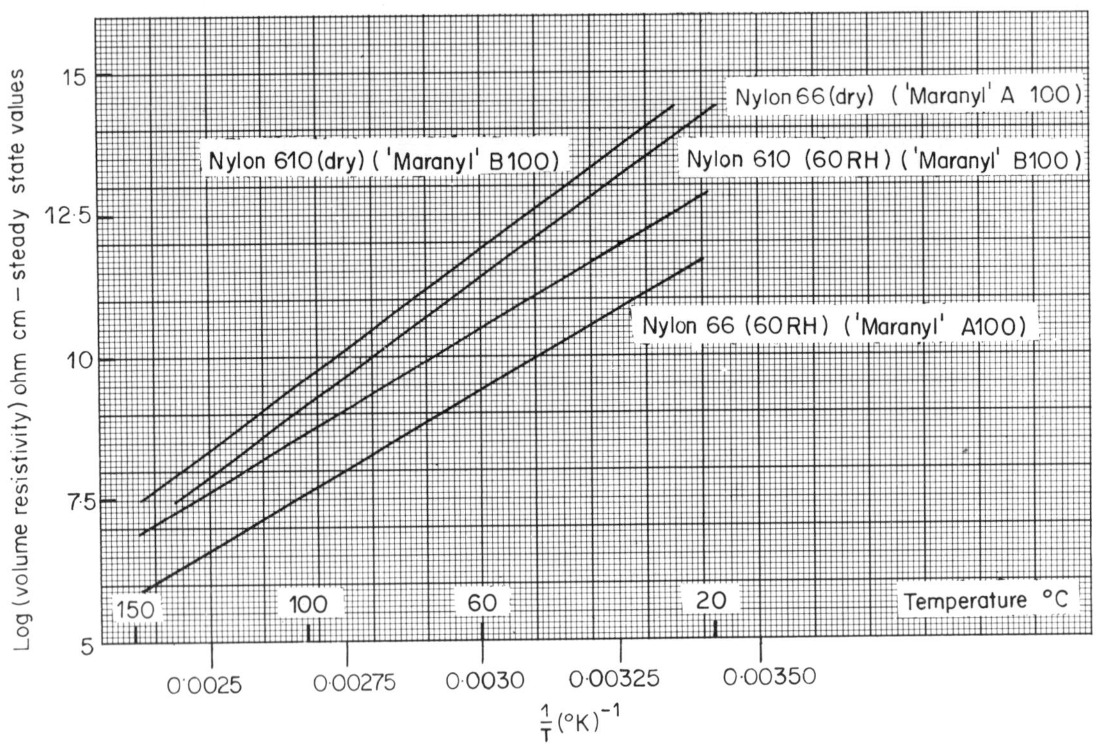


**Figure 11.30.** Steady state volume resistivity vs temperature: various relative humidities. Nylon 66, 610 ('Maranyl' A100, B100)

# 12

# ACRYLICS

## INTRODUCTION

Glass-like polymers of acrylic esters were first reported in the 1870's, but it was not until 1901 that Otto Rohm in Germany began the systematic study that was to lay the foundation for the development of acrylic plastics. Rohm was concerned mainly with soft materials with potential as lacquers, although by about 1933 the production of small articles from poly(methyl methacrylate) had begun.

British interest in poly(methyl methacrylate) was quite independent and arose from the desire to find a laminated safety glass interlayer to replace cellulose nitrate which yellowed in sunlight. These attempts were unsuccessful, but during the work Hill and Crawford of I.C.I. realized the possibilities of acrylic plastics in other fields, and in 1932 Crawford devised an economic way of making the monomer. This discovery marked the true beginning of poly(methyl methacrylate) as an important plastic and Crawford's process is still the basis of the manufacture of acrylic plastics throughout the world.

In 1933 the first acrylic sheet was cast and shortly afterwards acrylic moulding materials became available. During the 1939–45 war acrylic sheet was used in vast quantities for aircraft glazing and since then both it and the powder form have continued to expand into markets such as lighting, signs and display, and sanitary ware.

## NATURE

Poly(methyl methacrylate) consists of polythene-like molecules in which both hydrogen atoms on alternate carbon atoms have been replaced with two different groups. Both these groups are bulky; thus crystallization is effectively prevented and the material is wholly amorphous.

## GRADES AND FORMS

Acrylic sheet may be either cast or extruded. It may be clear, opal or coloured in a very wide range of transparent, translucent or opaque colours; flat, profiled or surface-patterned. Flat and patterned sheets are available in a wide range of thicknesses and sizes, and profiled sheets in a smaller range. Clear block and rod are also available.

Acrylic moulding and extrusion materials are supplied in various grades, differing in flow properties, heat resistance and craze resistance; all are of lower molecular weight than cast sheet. Each is normally available in the form of small transparent spheres, which may be easily dry-coloured, but some grades are also supplied as coloured compounds in the form of regular-cut granules.

## PROCESSING

Acrylic sheet may be shaped by air, vacuum, mechanical pressure or by combinations of these methods. The sheet should be heated to a temperature of 150–160°C in an oven maintained at 150–170°C. Pressures of up to 60 lb/in$^2$ are required, depending on the size and complexity of the mould. Because of its high molecular weight cast acrylic sheet is not suitable for vacuum forming and ground-up scrap cannot be used for injection moulding. Extruded acrylic sheet has a lower molecular weight and can be used on conventional vacuum forming machines at temperatures of 140–160°C. Both types of sheet are machined on ordinary wood-working machines, but special tool shapes and rake angles are recommended for optimum results.

On first heating to shaping temperature, cast acrylic sheet undergoes a linear shrinkage of 1·2–1·8% but thereafter no further changes in dimension occur.

Granular acrylics may be processed by extrusion or injection moulding but because of the high melt viscosity of the material somewhat higher than average pressures are required. Profiled extrusions can be obtained by postforming an extruded tube by slitting it while hot and passing the slit profile over formers placed in a cooling bath.

Acrylics may be cemented to themselves and to many other materials by means of either pure solvents or acrylic-based cements. They may be decorated with special paints and inks, and metallized using standard equipment.

## PROPERTIES

Acrylic plastics are noted for their clarity, hardness, rigidity, superb appearance and outstanding weathering resistance. Some properties are listed in Table 12.1.

TABLE 12.1

| Property | Units | Typical Value |
|---|---|---|
| Density | g/cm³ | 1·18–1·19 |
| Softening region | °C | 100–120 |
| Coefficient of linear thermal expansion, −30 to 30°C | $°C^{-1}$ | $6·0 \times 10^{-5}$ |
| | $°F^{-1}$ | $3·3 \times 10^{-5}$ |
| Thermal conductivity | cal/cm s °C | $4·7 \times 10^{-4}$ |
| | B ThU in/ft² h °F | 1·4 |
| Specific heat, 20 to 80°C | cal/g °C | 0·35 |
| | B ThU/lb °F | 0·35 |
| Flammability | | Difficult to ignite; burns slowly with drips |

*Deformation*

The deformational behaviour at 20°C, 65% rh of cast acrylic sheet is given by the tensile creep curves of Figure 12.1. The corresponding isochronous stress vs. strain, isometric stress vs. time, and tensile creep modulus vs. time curves are given in Figures 12.2, 12.3 and 12.4, and it can be judged from these data that acrylic plastics have high rigidity. Also the deformational behaviour of acrylic plastics can be approximated to that of linear viscoelasticity over greater ranges of stress, strain, and time than can that of any other plastic described in this book. Even so, the limitations of describing the behaviour as linear viscoelastic can be judged by estimating the deviations of the isochronous stress vs. strain curves from straight lines—a result implicit in linear viscoelastic theory.

The differences in deformational behaviour between the various grades of acrylic plastics are small. From the 100 second isochronous stress vs. strain curves it has been found that up to strains of 1% there is little difference between cast sheet and moulding grades.

The effect of water content on the above results is small compared with that for nylon, and is comparable with that for acetal copolymer: the total decrease in stress at constant strain, corresponding to a change of the material from a dry to a wet state, is approximately 12%. Thus the correction is approximately a 5% decrease in stress for an increase in water content of 1%. The corresponding change from dry to 0·7% water content by weight, the equilibrium value for contact with an atmosphere of 20°C and 65% rh, is a 3 to 4% decrease in isometric stress or tensile creep modulus value.

Data on recovery from tensile creep at 20°C and 65% rh are given in Figure 12.5.

The deformational behaviour at 60°C, for cast acrylic sheet is given by the tensile creep curves of Figure 12.6. The corresponding isochronous stress vs. strain, isometric stress vs. time, and tensile creep modulus vs. time curves are given in Figures 12.7, 12.8 and 12.9. Recovery data at 60°C are given in Figure 12.10. An assessment of the effect of temperature on deformational behaviour can be made by comparing creep data at 20°C and 60°C. The general trend of deformational behaviour is shown by curves such as that in Figure 12.11.

*Limiting Stress*

It is in the failure properties rather than in deformational behaviour that considerable differences between cast sheet and moulding grades manifest themselves. These differences are shown at all stages, including crazing and final failure. As with other plastics, the amount of data available is limited but such results as there are show the superiority of the higher molecular weight sheet material. Failure data for 20°C, 65% rh are given in Figure 12.12. Also included in Figure 12.12 are curves showing the onset of stress crazing in air at 20°C, 65% rh, because this must be regarded as the end point for serviceability in optical applications: these data are relevant to tensile loading. At short times, failure is by brittle fracture but ductility increases, particularly for the high molecular weight material, at long times and higher temperatures. Another aspect of the effect of time on the stress corresponding to ductile failure is shown by results from conventional load vs. elongation tests at various straining rates. Data are given in Figure 12.13.

At higher temperatures, brittle fracture gives way to failure by yielding and necking; the dependence of yield stress on temperature determined from conventional load vs. elongation tests and from tests in uniaxial compression is given by the curve in Figure 12.14.

*Dynamic Fatigue and Stress Solvent Crazing*

A limited number of measurements have been made of the effect of dynamic stresses on the failure of acrylic plastics. As far as these measurements have progressed (up to 10 days) the results are surprising in that there is no decrease beyond the strength at constant load curve for the cast acrylic sheet whereas the moulding grade falls away from the corresponding curve at comparatively short times. The results are shown in Figure 12.15. In different terms, cast acrylic sheet appears to be comparatively free from dynamic fatigue effects at the low frequency studied.

Another phenomenon to which acrylic plastics are susceptible is stress solvent crazing, that is, crazing under the combined effects of stress and a solvent for the material. This has to be distinguished from environmental stress cracking experienced with polythene where non-solvents are particularly effective. Crazing, in common with all other forms of failure, is time dependent so that it is difficult to quote detailed data on the effectiveness of various solvents in promoting crazing. As for stress crazing the resistance to stress–solvent crazing of the cast sheet is greater than that of the moulding and extrusion grades, even of the improved craze-resistant forms, which are designed to give satisfactory performance in contact with detergent solutions in sanitary ware applications. It should be remembered that moulded-in strains can be of such a magnitude as to promote crazing in moulded components.

*Impact Behaviour*

The impact strength of cast acrylic sheet vs. temperature is given in Figure 12.16 which shows that under all conditions up to about 80°C this material is brittle in impact. This same comment applies even more cogently to the lower molecular weight materials required for processing by injection moulding or extrusion: data to support this conclusion are given in Figure 12.17 where the lower molecular weight grades are seen to have lower impact strengths than those given in Figure 12.16 for the cast sheet. The presentation in Figure 12.17 requires some comment: all the data refer to specimens with a notch depth of 0·110 in and notch tip radii as given on the $x$-axis. This scale is logarithmic and has been so chosen that the data over a wide range of notch tip radii can be summarized in a single presentation. The identification of a notch radius of about $2\frac{1}{2}$ in as equivalent to 'un-notched' has been adopted merely because the data thereby fit a smooth curve and as a practical convenience.

The influence of injection moulding cylinder temperature on impact strength for a general purpose moulding grade is given in Figure 12.18 where it can be seen that the influence of processing temperatures is not marked.

Impact strength vs. temperature curves for a modified craze-resistant grade are given in Figure 12.19 where a considerable improvement over the impact behaviour even of the cast acrylic sheet can be observed: other grades developed for improved impact behaviour show even greater improvement.

*Electrical Properties*

The electrical properties of acrylic plastics depend on frequency and temperature and to some extent on the amount of absorbed water in the material. The dependence of permittivity and loss tangent on frequency and temperature is shown in Figures 12.20 and 12.21 for dry acrylic cast sheet, and the variation of these quantities with frequency at 20°C over a wider frequency range is shown in Figures 12.22 and 12.23. These last two figures also indicate the by no means negligible effect of absorbed water on these two properties.

The dependence of apparent volume resistivity at 20°C on the time of electrification is shown in Figure 12.24 for dry cast sheet.

The electric strength of cast sheet measured on $\frac{1}{8}$ in thick samples at 20°C in transformer oil is 390 V/0·001 in (153 kV/cm) at 75% rh falling to 380 V/0·001 in (149 kV/cm) for samples immersed in water for 24 hours. These results cannot, however, be used as general design data under conditions different to those in the test. Under other conditions, acrylic sheet has been found to have an intrinsic electric strength approaching 25,000 V/0·001 in (10 Mv/cm).

*Optical Properties*

Acrylic plastics are the best-known plastics for transparency and find application in a wide range of optical instruments, in advertising and in building. Sheet acrylic plastics have been the principal materials used in canopies for subsonic aircraft. It is thus appropriate to describe the optical properties in considerable detail: these are for cast acrylic sheet of density 1·18 g/cm$^3$.

*Refraction.* Refractive index data are tabulated in Table 12.2 and plotted as an optical dispersion curve in Figure 12.25.

Reciprocal dispersive power $V_d = 59{\cdot}3$.
Critical angle ($\lambda = 5893$ Å) $= 42°\ 15'$.
Temperature coefficient of refractive index $= -0{\cdot}0001$/°C.
(The behaviour from 20–70°C is shown in Figure 12.26.)

TABLE 12.2

| Wave Length (Å) | Refractive Index at 20°C |
|---|---|
| 4358 | 1·50062 |
| 4861 | 1·49790 |
| 5461 | 1·49464 |
| 5893 | 1·49257 |
| 6563 | 1·48960 |

*Transparency*. Absorption for spectral wavelengths in the visible region is small and uniform, being less than 0·5% per inch in the body of the material. Some light is reflected, however, at each surface—an inevitable consequence of the change in refractive index at the interface. The amount of light reflection depends on the angle of incidence and its angular distribution on the nature of the surface in the manner shown in Figure 12.27, for light of wavelength 5461 Å. Transparency data are given in Table 12.3 for acrylic sheet of 3·08 mm (nominally $\frac{1}{8}$ in) for mercury light ($\lambda = 5461$ Å).

TABLE 12.3

| | |
|---|---|
| Direct transmission factor | |
| air-acrylic-air | 93% |
| corrected for surface reflection and defects | >99% |
| corrected to 1 mm thickness | >99% |
| Scattering coefficient | <0·01 $cm^{-1}$ |
| Forward scattering fraction uncorrected | 0·1% |
| corrected | <0·1% |

Because of these optical characteristics, particularly that of extremely low absorption, acrylic plastics can be used in applications exploiting total internal reflection. The critical angle, ~40°, means that quite a wide beam can be accepted and efficiently transmitted through long lengths of material. Light can be 'piped' in this way around curves, where the radius of curvature should be not less than three times the diameter to minimize the loss. It is important, too, that the plastic should be highly polished and free from scratches or surface defects because these will cause scattering of light and reduce the efficiency of the system.

*Light Transfer*. The low inherent absorption of acrylic plastics makes them eminently suitable for a variety of applications under this heading; glazing, lighting fittings and advertising are common examples. In the last two, a degree of translucency is sometimes desirable and compositions ranging from those of low diffusivity to those with values of this quantity approaching 0·9 are available. This translucency is coupled with light output ratios for spheres of over 90%; that is, the light 'lost' is less than 10% when a light source is totally enclosed in a sphere of the plastic.

*Chemical Properties*

Acrylics have very good resistance to water, alkalis, aqueous inorganic salt solutions and most dilute acids. Some dilute acids, e.g. hydrocyanic and hydrofluoric, however, attack acrylics, as do concentrated sulphuric, nitric and chromic acids. On the whole, the resistance to organic materials is not good: although some have no effect, some cause swelling and crazing and some dissolve the material completely.

TABLE 12.4

| | | |
|---|---|---|
| Equilibrium water content at 20°C | | |
| Immersed | 2·1% | |
| 60% rh | 0·9% | |
| Gas permeabilities ($cm^3$ (NTP) $cm/cm^2$ s cm Hg) $\times 10^9$ | | |
| Carbon dioxide | <0·01 | |
| Oxygen | <0·01 | |
| Nitrogen | <0·01 | |
| Resistance to: | | |
| Mineral acids (dilute) | Excellent | |
| Mineral acids (conc.) | Fair | |
| Alkalis | Excellent | |
| Solvents: | | |
| alcohols | Poor | promote crazing |
| ketones | Poor | promote crazing |
| aromatic hydrocarbons | Poor | promote crazing |
| chlorinated hydrocarbons | Poor | promote crazing |
| Detergent solutions | Excellent | |
| Greases and oils | Excellent | |

## APPLICATIONS

Acrylics often represent a very satisfactory replacement for glass with their advantages of light weight, ease of shaping and freedom from shattering, but their usefulness is not limited to this narrow field as their success in sanitary ware applications testifies.

*Typical Examples*

Signs, nameplates, lighting fittings, domelights, windscreens, canopies, reflex reflectors, car light housings and reflectors, lenses.

Sanitary ware (baths, basins, shower cabinets, bathroom screens, lavatory cisterns).

Machine guards, implosion guards, arc shutes, trays.

Gramophone pick-up arms, meter cases.

## LIST OF FIGURES

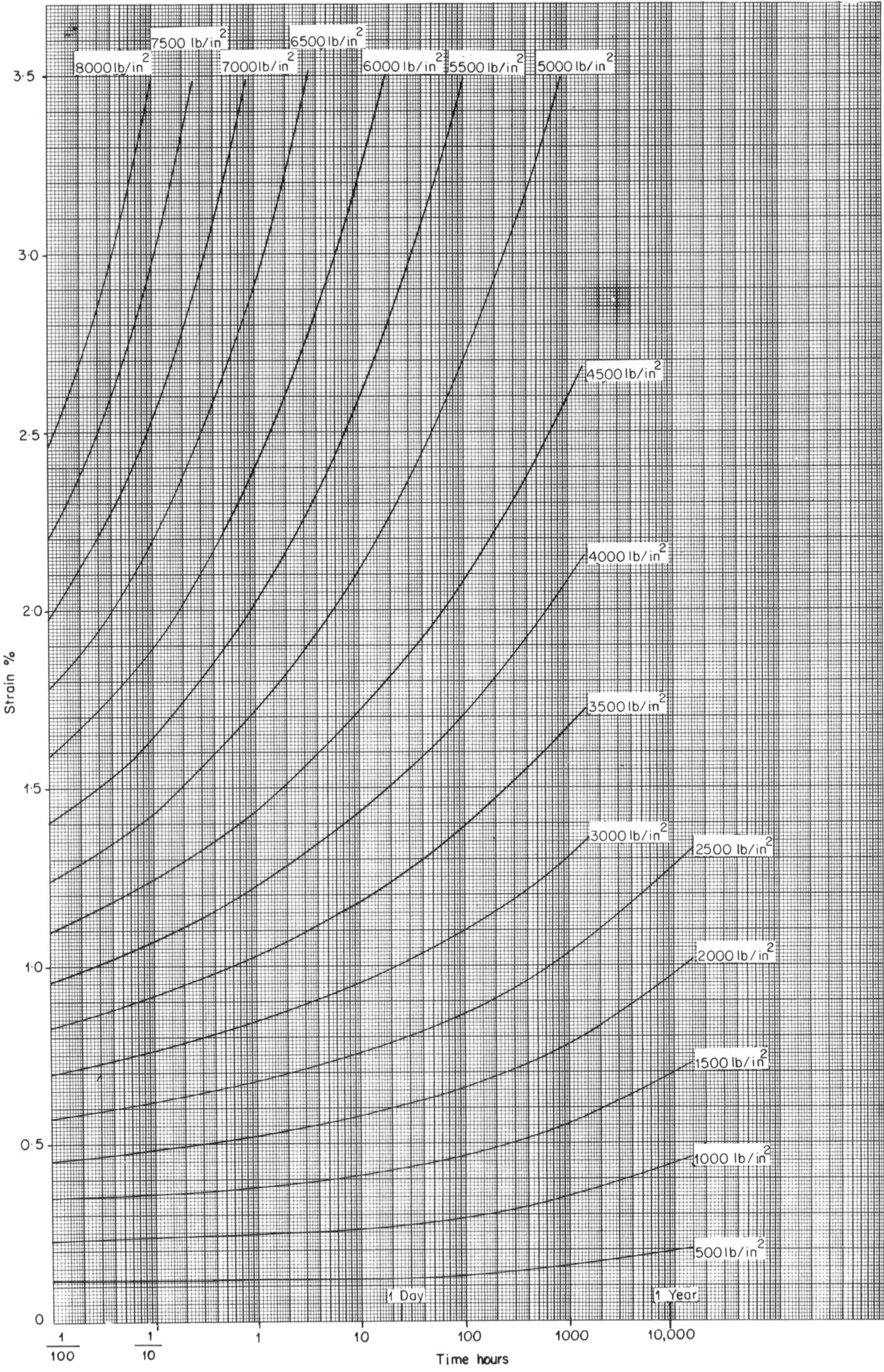


**Figure 12.1.** Creep curves in tension: 20°C, 65RH. Acrylic cast sheet ('Perspex')

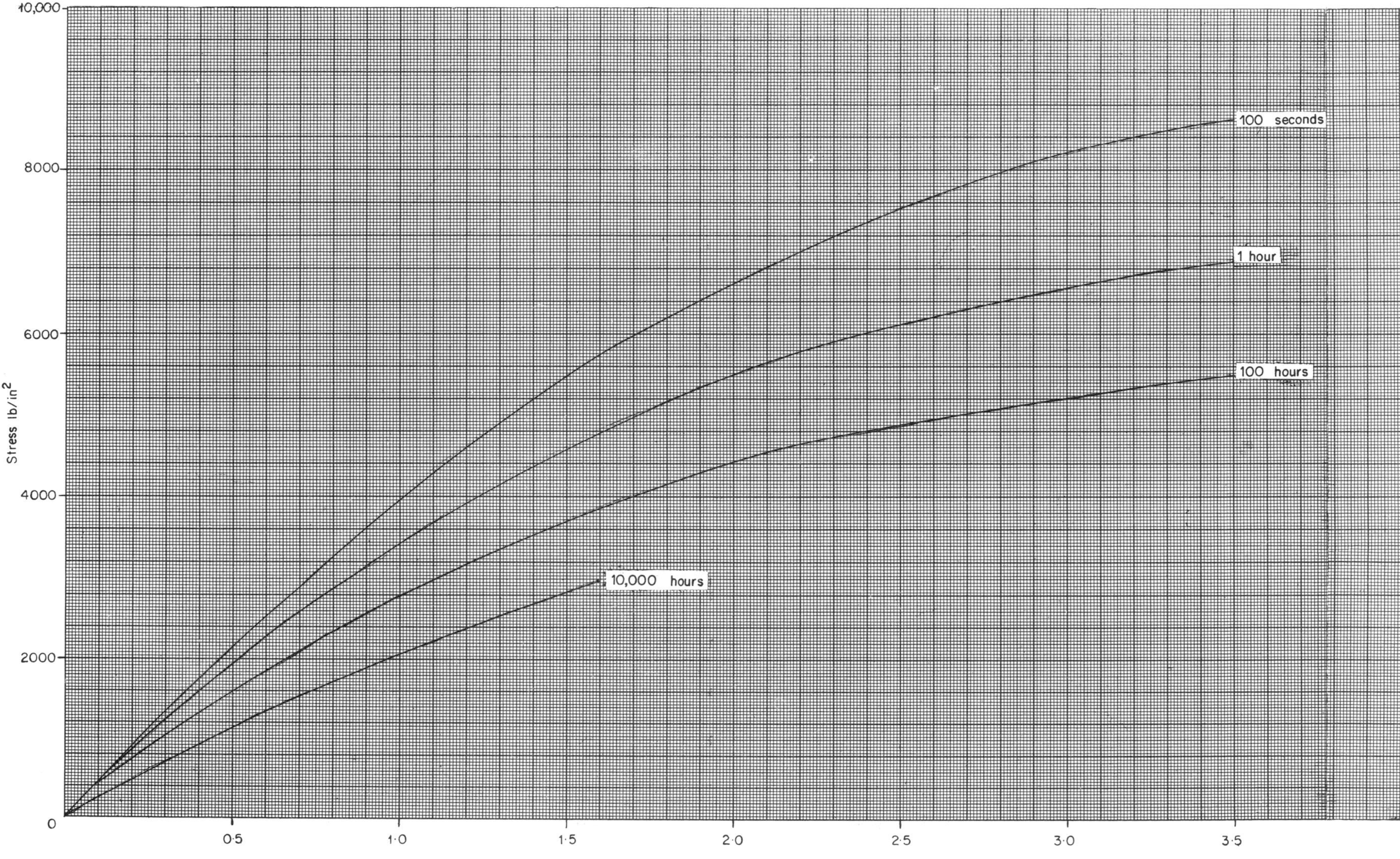


**Figure 12.2.** Isochronous stress vs strain curves: 20°C, 65RH. Acrylic cast sheet ('Perspex')

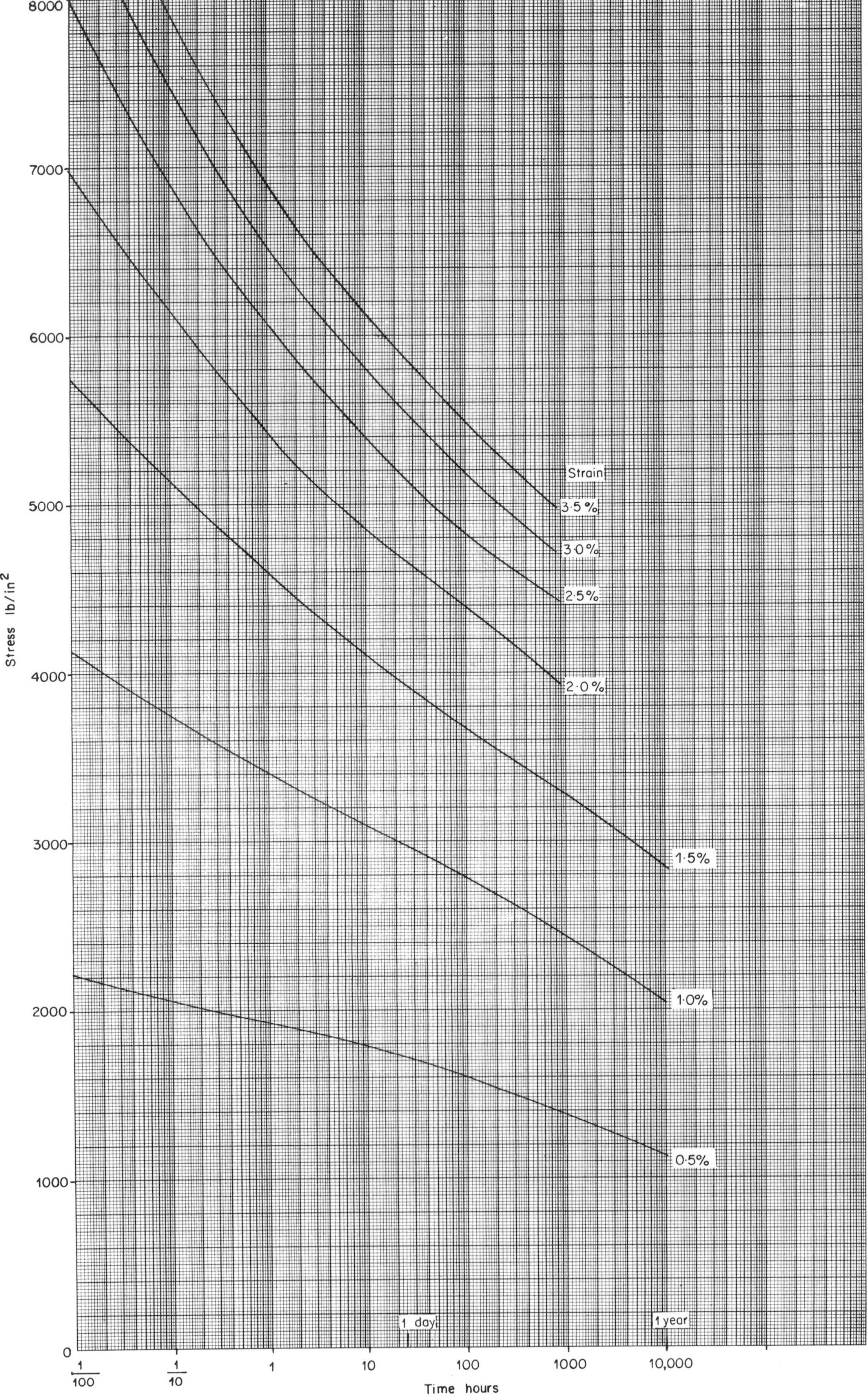


**Figure 12.3.** Isometric stress vs time curves: 20°C, 65RH. Acrylic cast sheet ('Perspex')

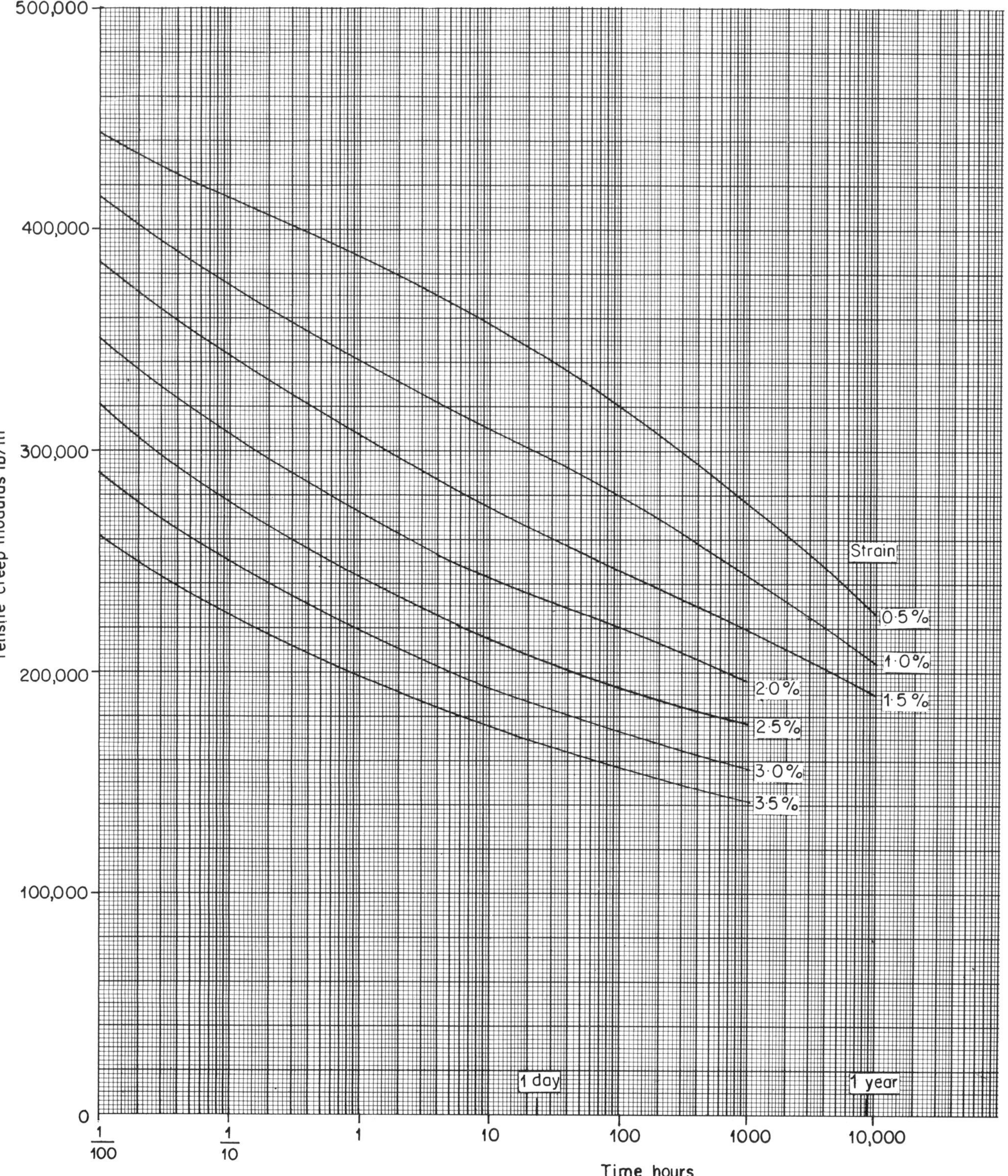


**Figure 12.4.** Tensile creep modulus vs time curves: 20°C, 65RH. Acrylic cast sheet ('Perspex')

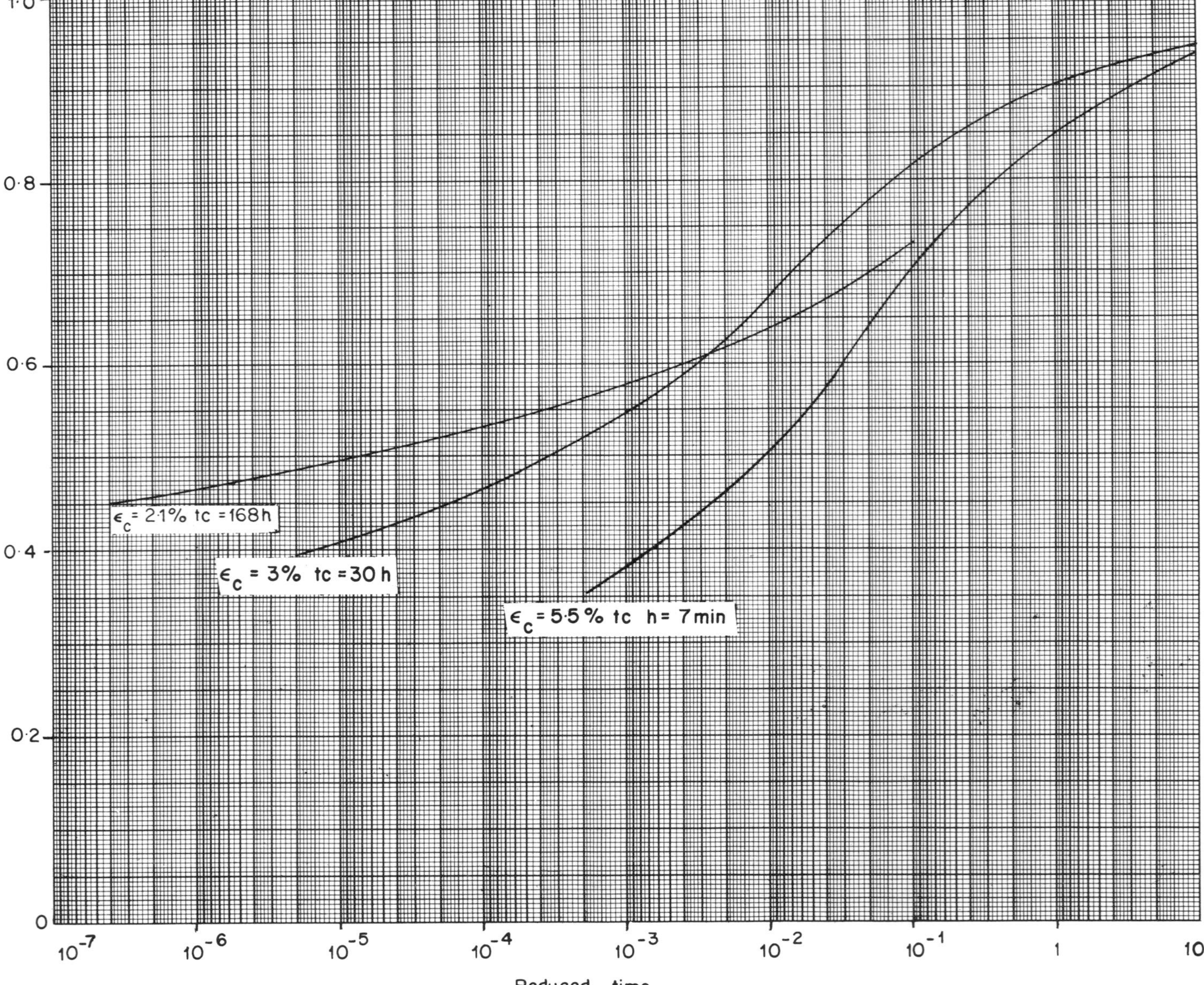


**Figure 12.5.** Recovery from creep in tension: 20°C, 65RH. Acrylic cast sheet ('Perspex')

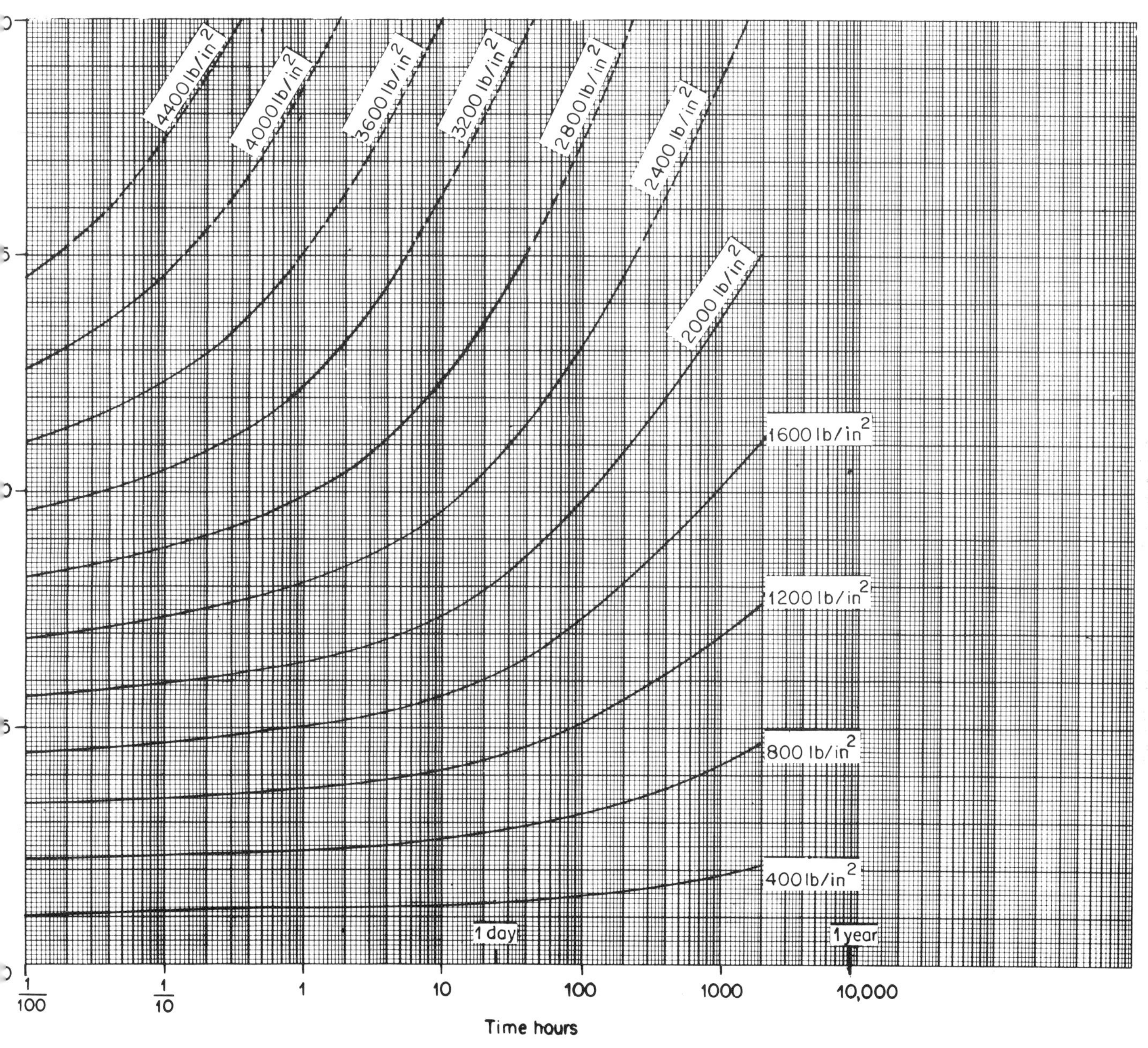


**Figure 12.6.** Creep curves in tension: 60°C. Acrylic cast sheet ('Perspex')

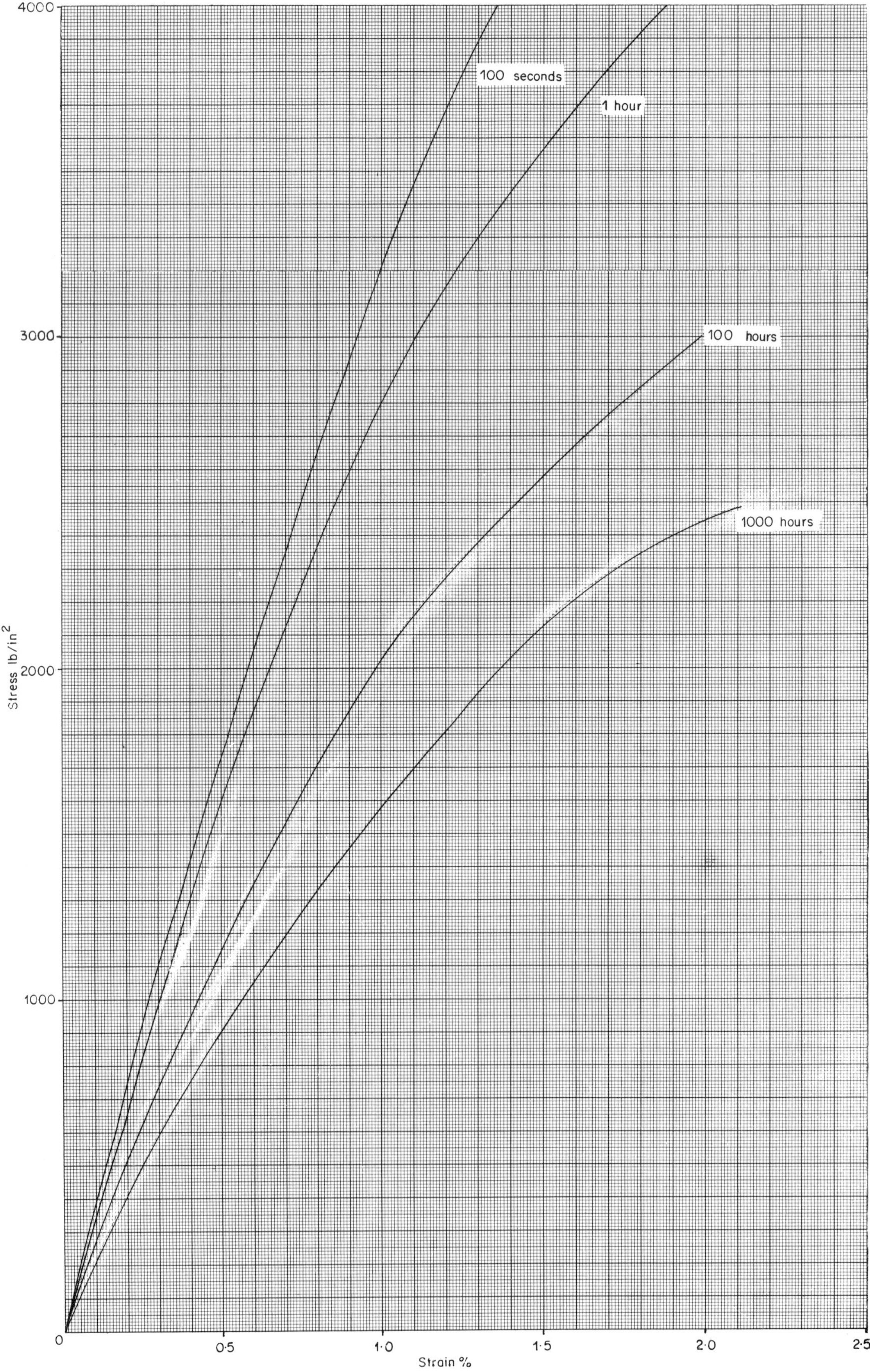


**Figure 12.7.** Isochronous stress vs strain curves: 60°C. Acrylic cast sheet ('Perspex')

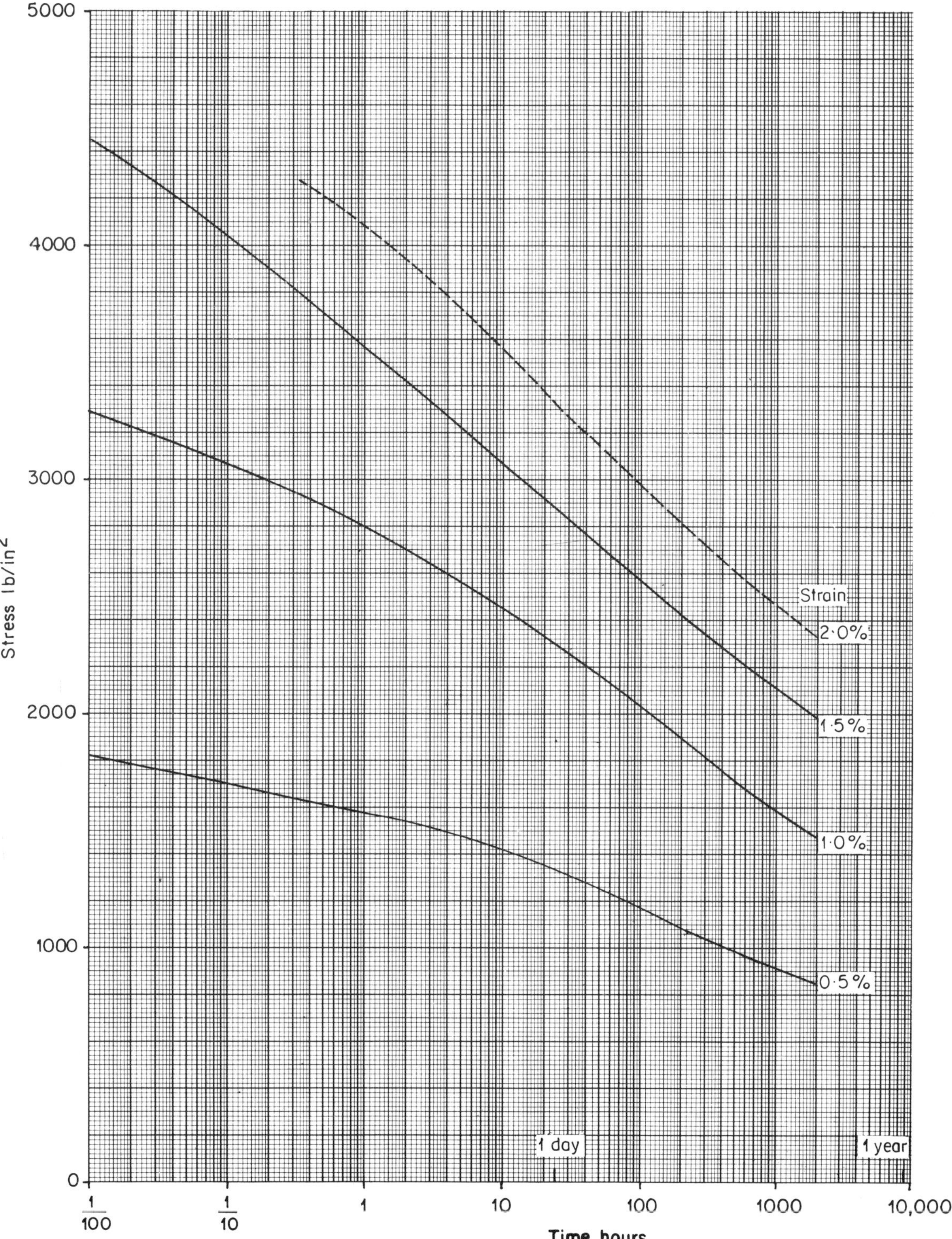


**Figure 12.8.** Isometric stress vs time curves: 60°C. Acrylic cast sheet ('Perspex')

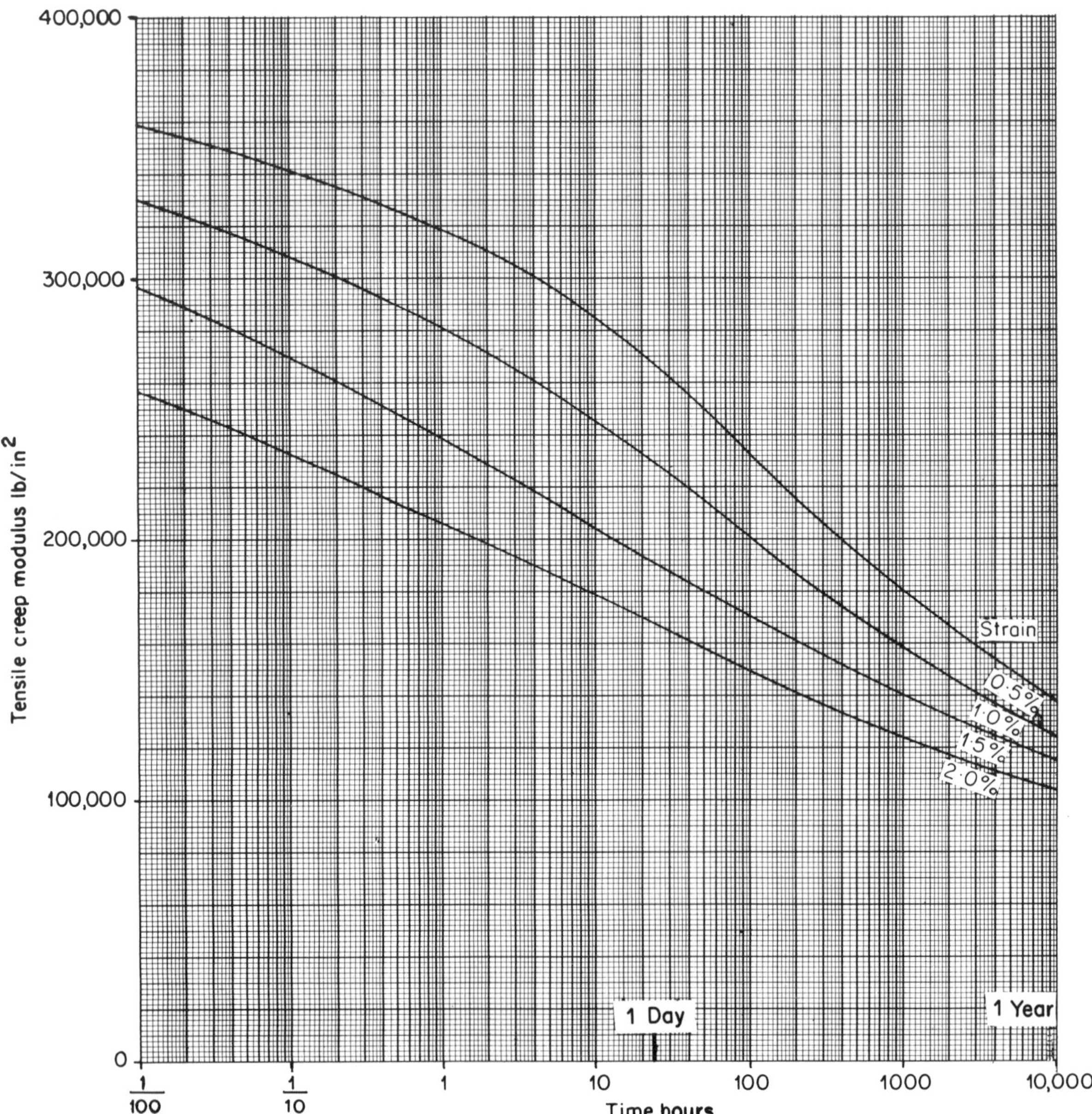


**Figure 12.9.** Tensile creep modulus vs time curves: 60°C. Acrylic cast sheet ('Perspex')

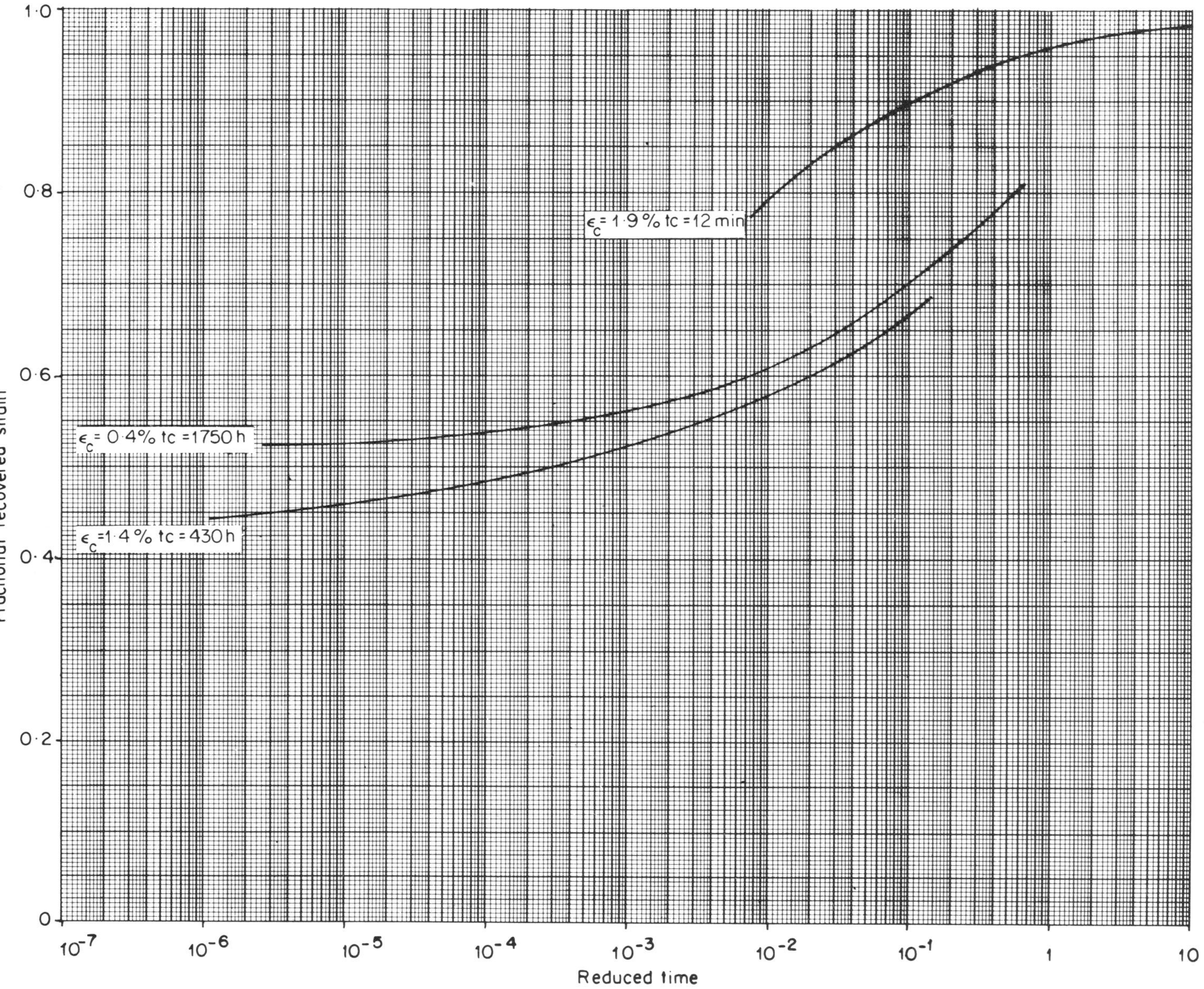


**Figure 12.10.** Recovery from creep in tension: 60°C. Acrylic cast sheet ('Perspex')

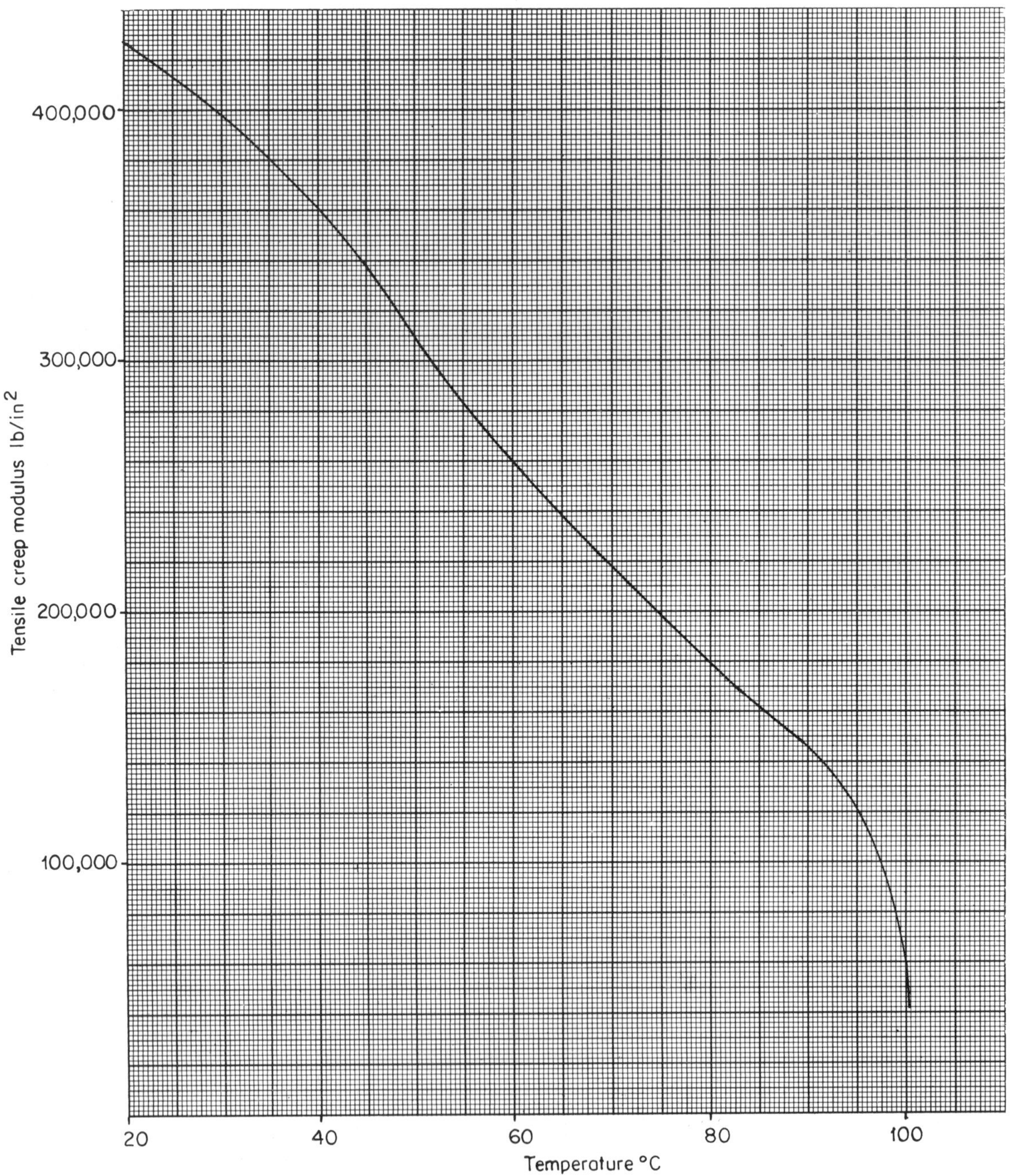


**Figure 12.11.** Tensile creep modulus (100 sec, 0·2 % strain) vs temperature. Acrylic cast sheet ('Perspex')

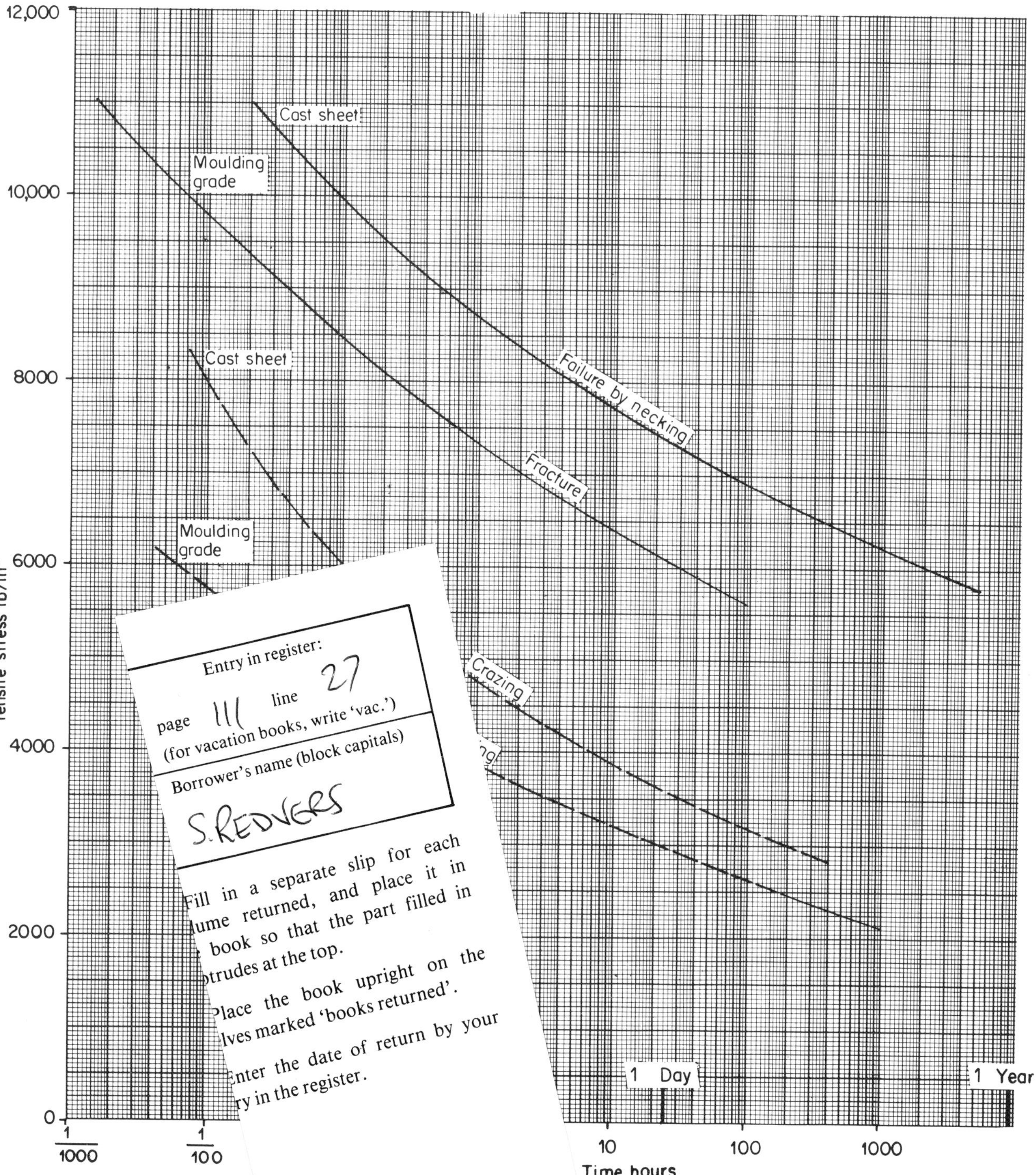


**Figure 12.12.** Creep rı … 0°C, 65RH. Acrylic cast sheet and moulding grade …on' MG)

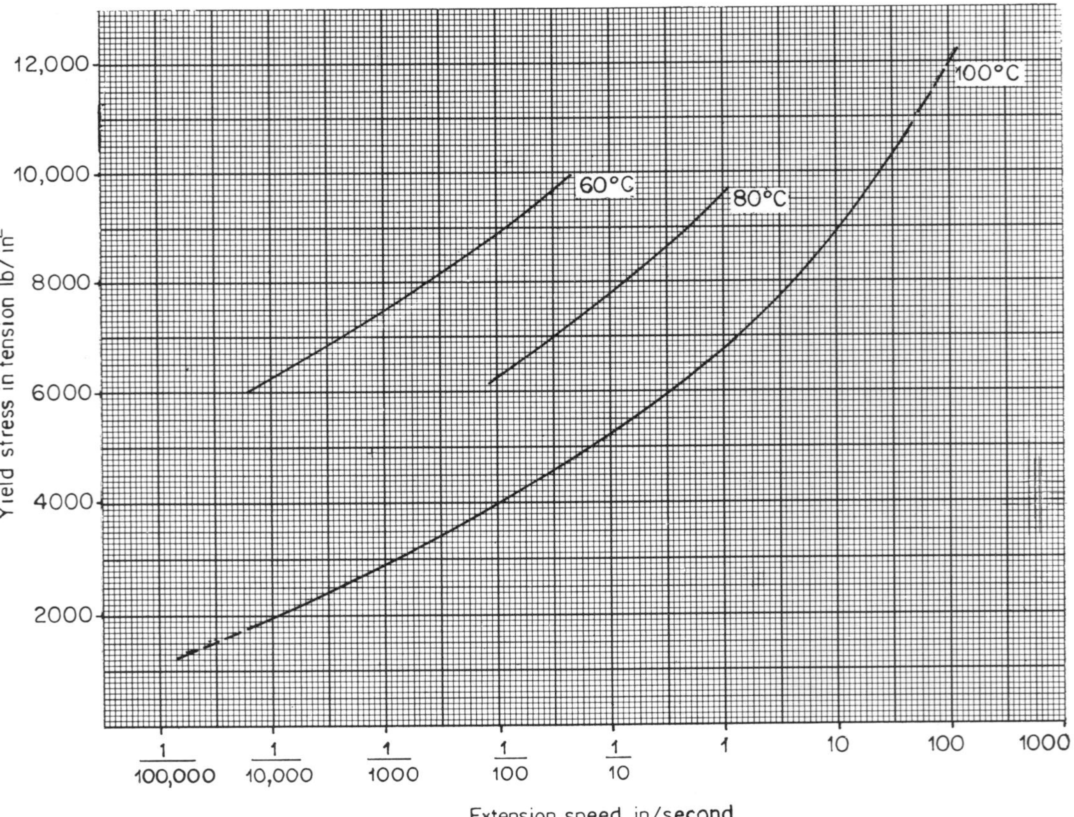


**Figure 12.13.** Yield stress in tension vs extension speed: 60°C, 80°C, 100°C. Acrylic cast sheet ('Perspex')

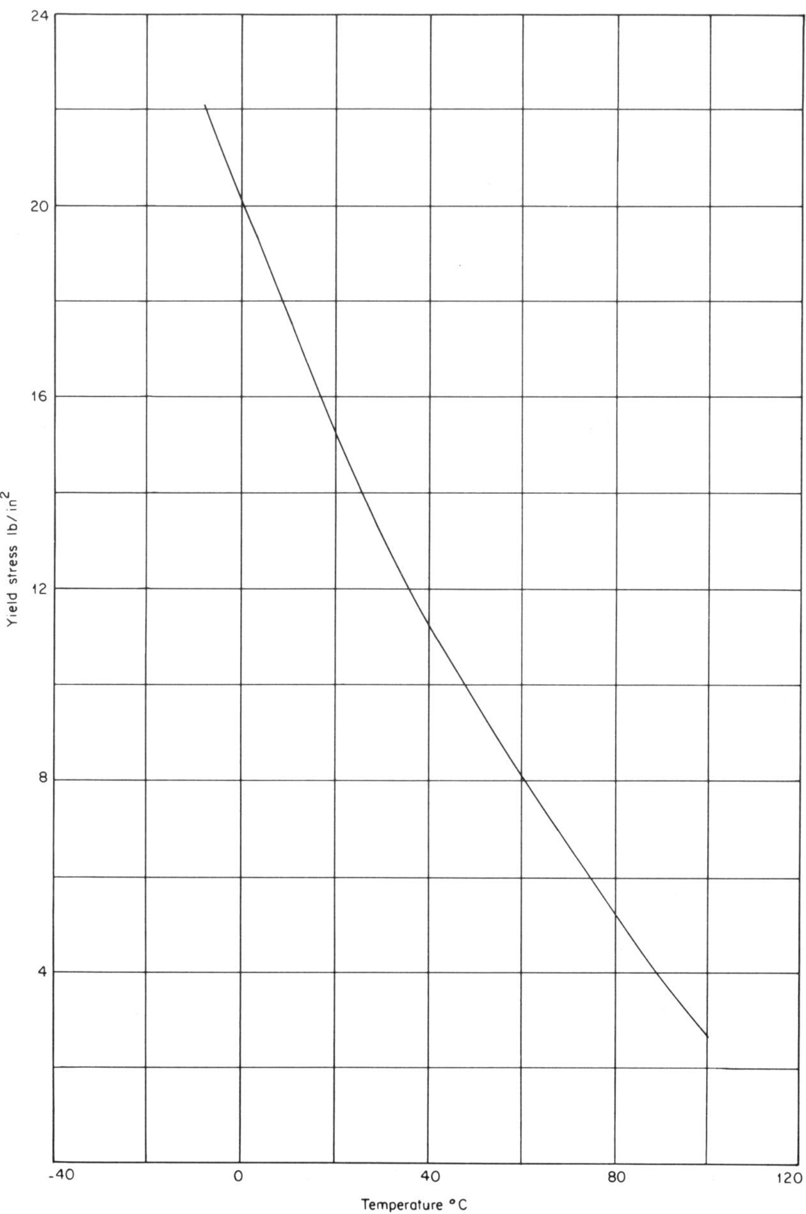


**Figure 12.14.** Yield stress in tension vs temperature: 50% per min straining rate. Acrylic cast sheet ('Perspex')

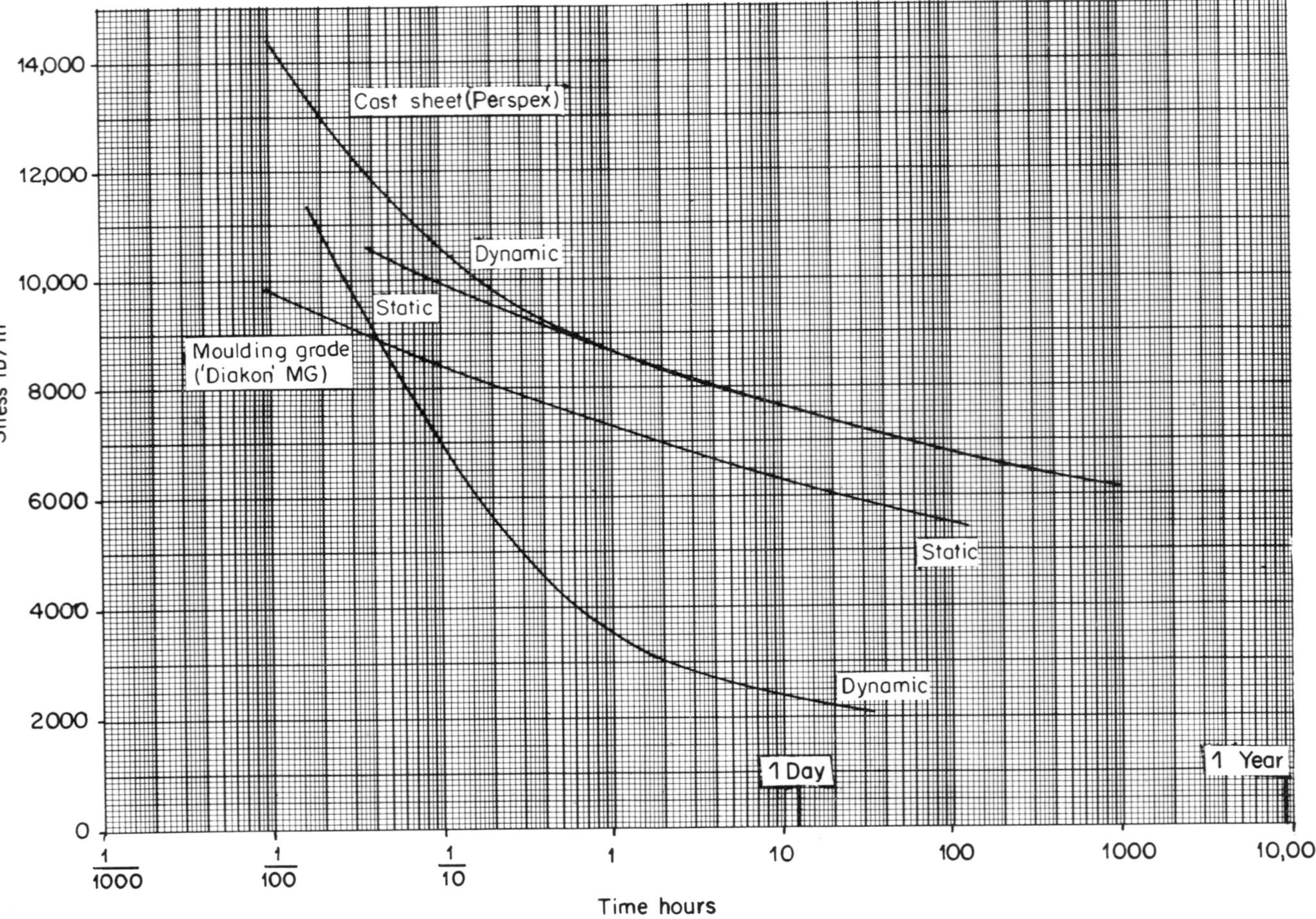


**Figure 12.15.** Stress vs time to failure: 20°C. 65RH. Dynamic fatigue in flexure. Acrylic cast sheet and moulding grade. ('Perspex', 'Diakon' MG)

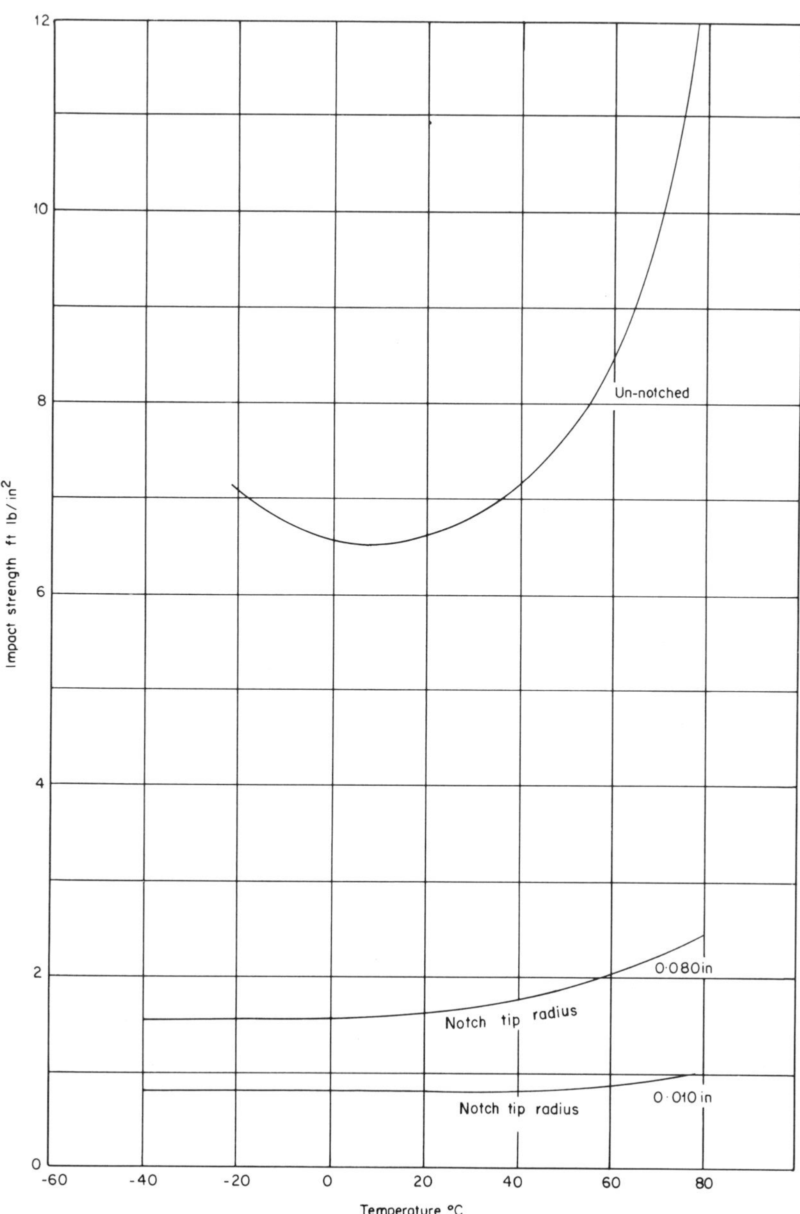


**Figure 12.16.** Impact strength vs temperature. Acrylic cast sheet ('Perspex')

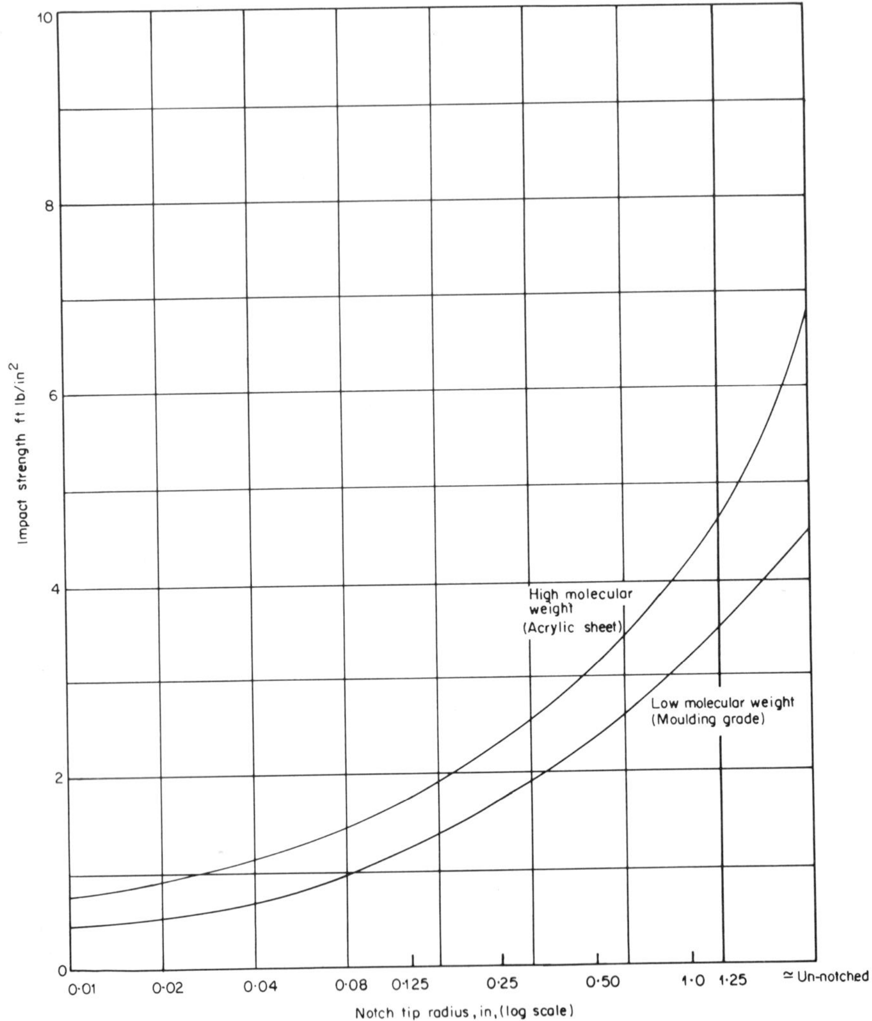


**Figure 12.17.** Impact strength vs notch tip radius: 20°C. Effect of molecular weight ('Perspex' acrylic sheet, various 'Diakon' grades)

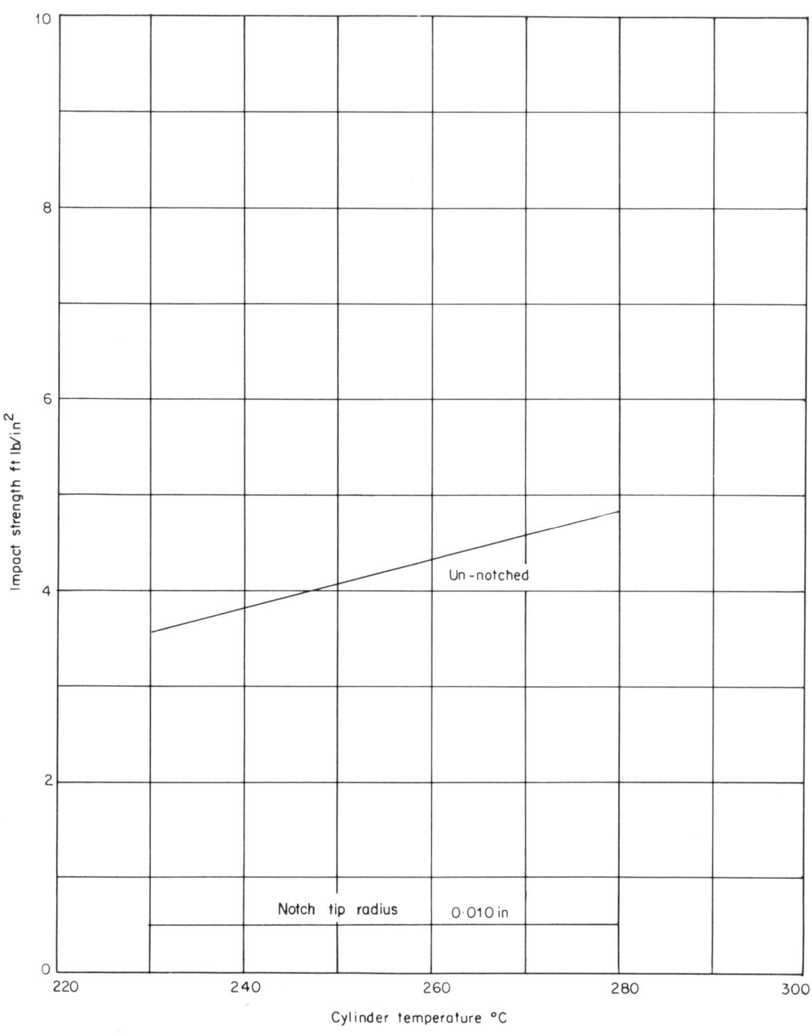


**Figure 12.18.** Impact strength vs injection moulding cylinder temperature: 20°C, 0·010 in notch tip radius, un-notched. Acrylic moulding grade ('Diakon' MG)

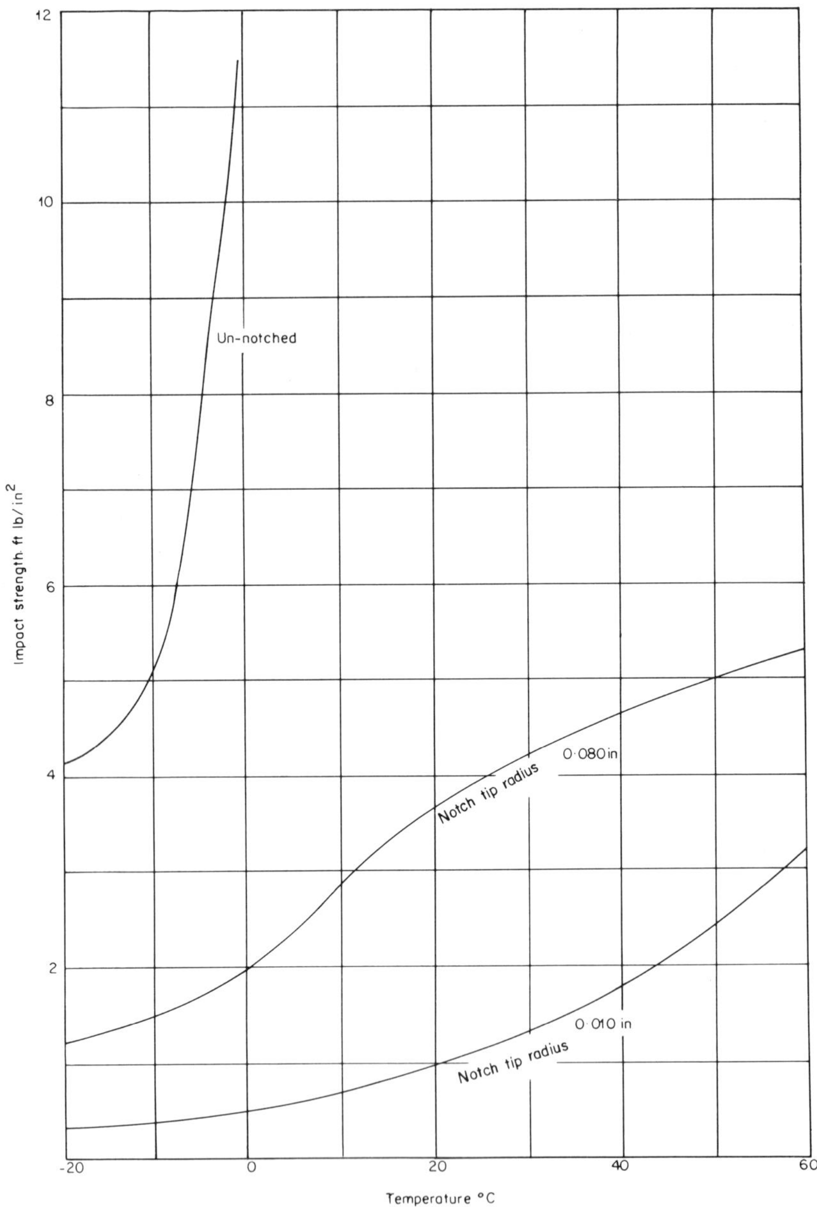


**Figure 12.19.** Impact strength vs temperature. Acrylic craze-resistant moulding grade ('Diakon' MC9450)

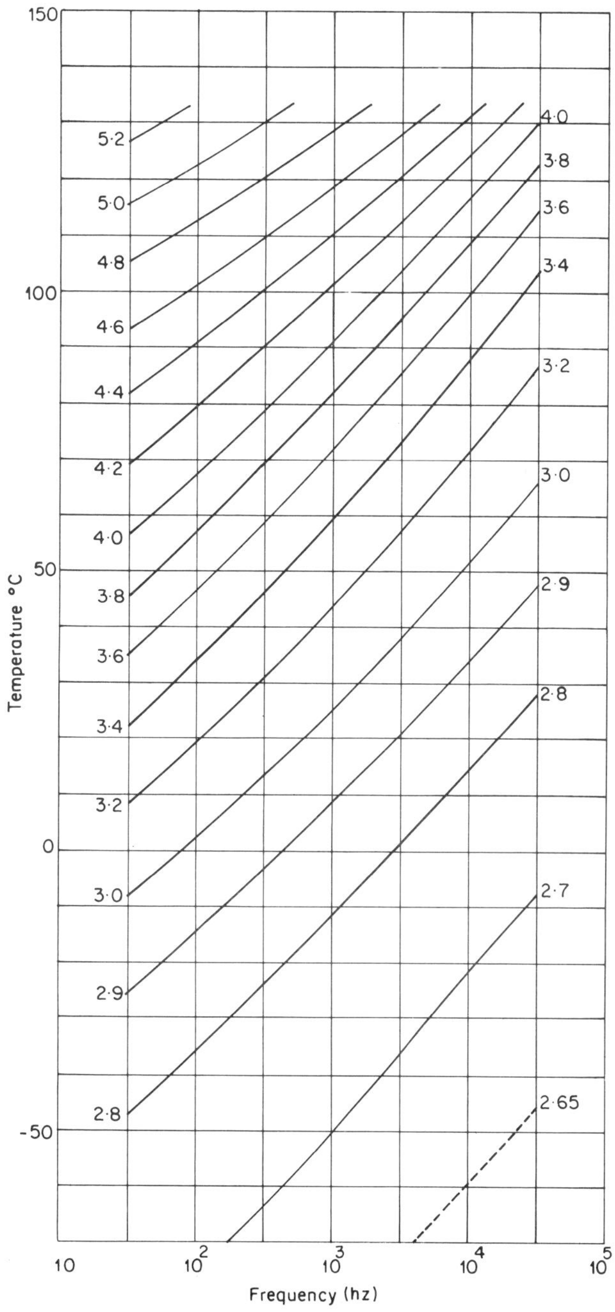


**Figure 12.20.** Permittivity vs frequency and temperature. Acrylic cast sheet, dry ('Perspex')

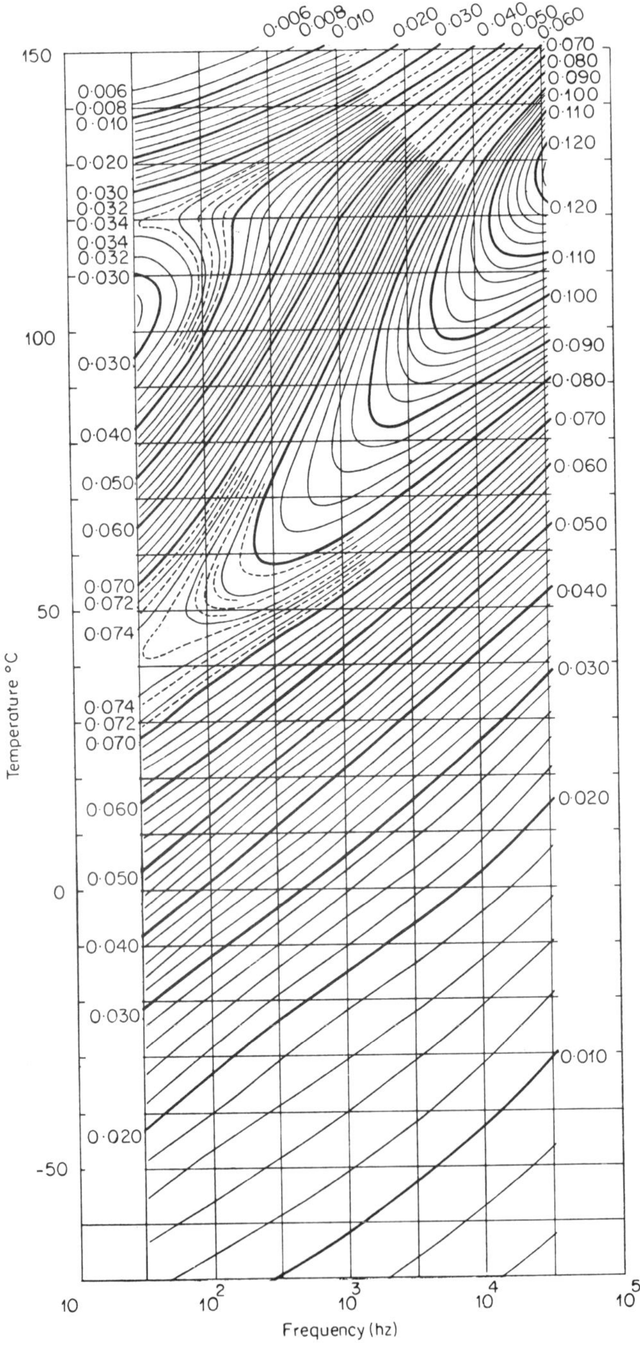


**Figure 12.21.** Loss tangent vs frequency and temperature. Acrylic cast sheet, dry ('Perspex')

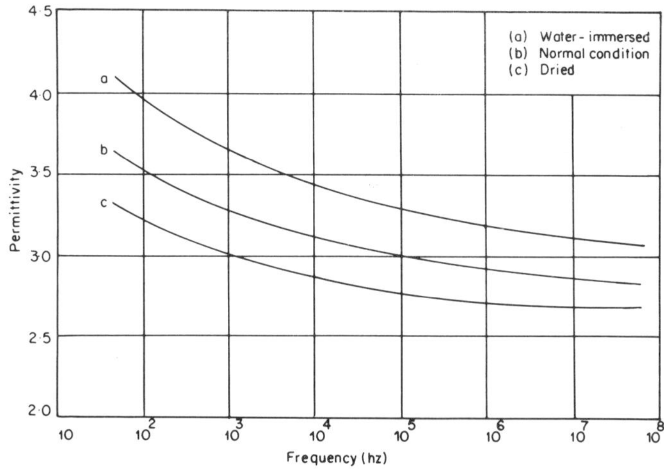


**Figure 12.22.** Permittivity vs frequency: 20°C. Effect of water content. Acrylic cast sheet ('Perspex')

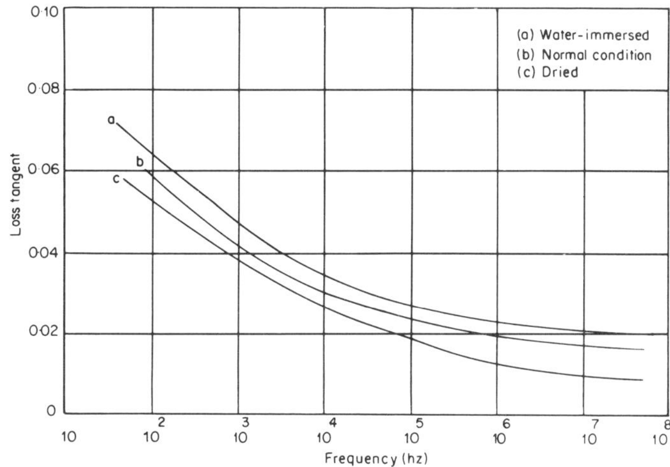


**Figure 12.23.** Loss tangent vs frequency: 20°C. Effect of water content. Acrylic cast sheet ('Perspex')

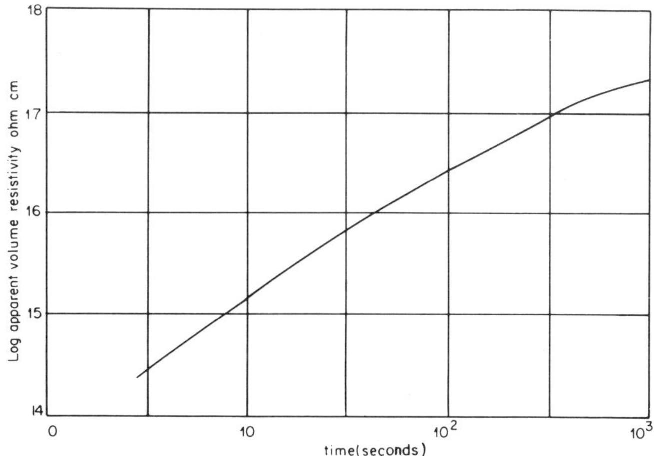


**Figure 12.24.** Apparent volume resistivity vs time of electrification: 20°C. Acrylic cast sheet, dry ('Perspex')

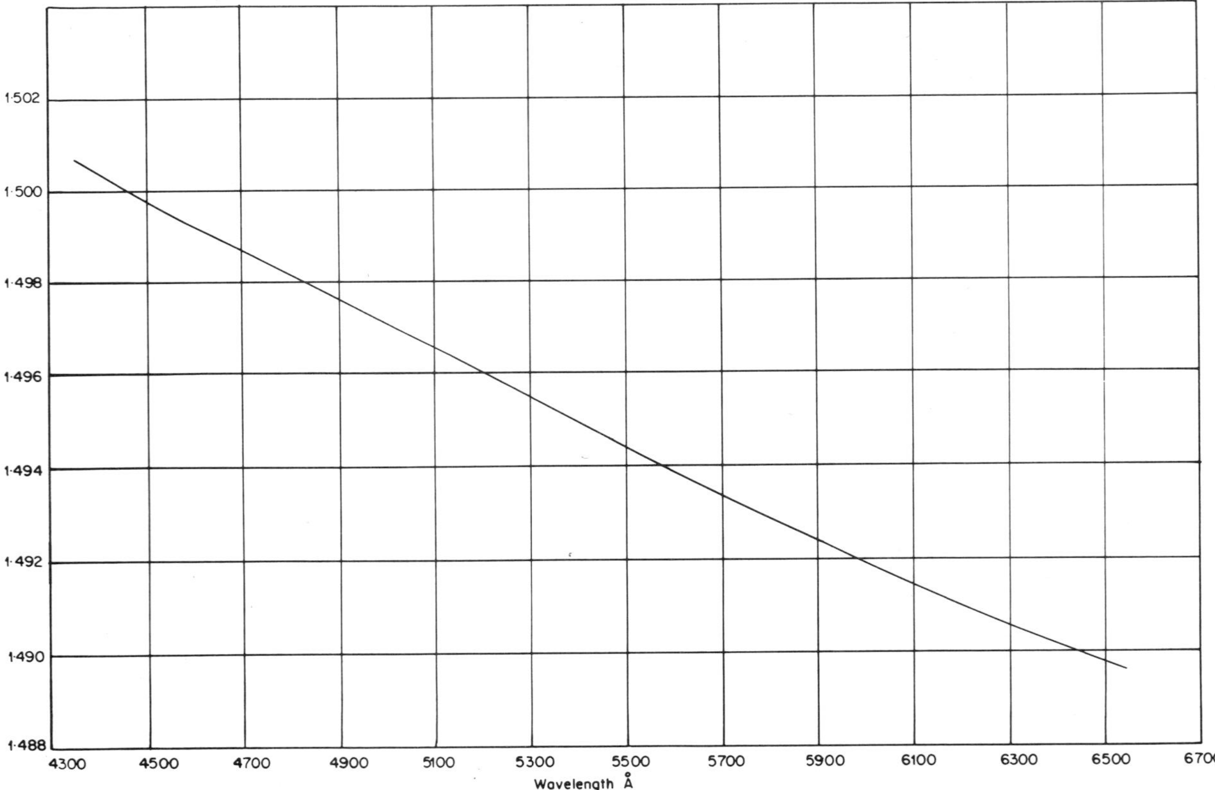


**Figure 12.25.** Optical dispersion curve: 20°C. Acrylic cast sheet, dry ('Perspex')

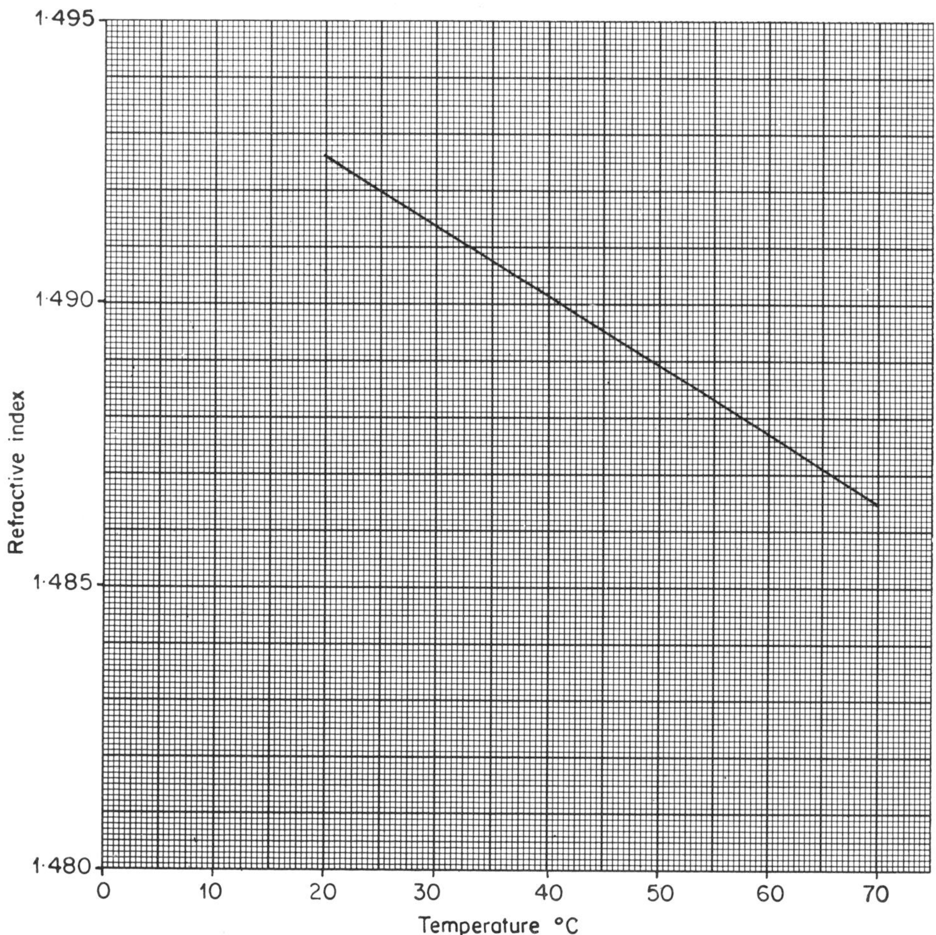


**Figure 12.26.** Refractive index vs temperature: 5893 Å. Acrylic cast sheet, dry ('Perspex')

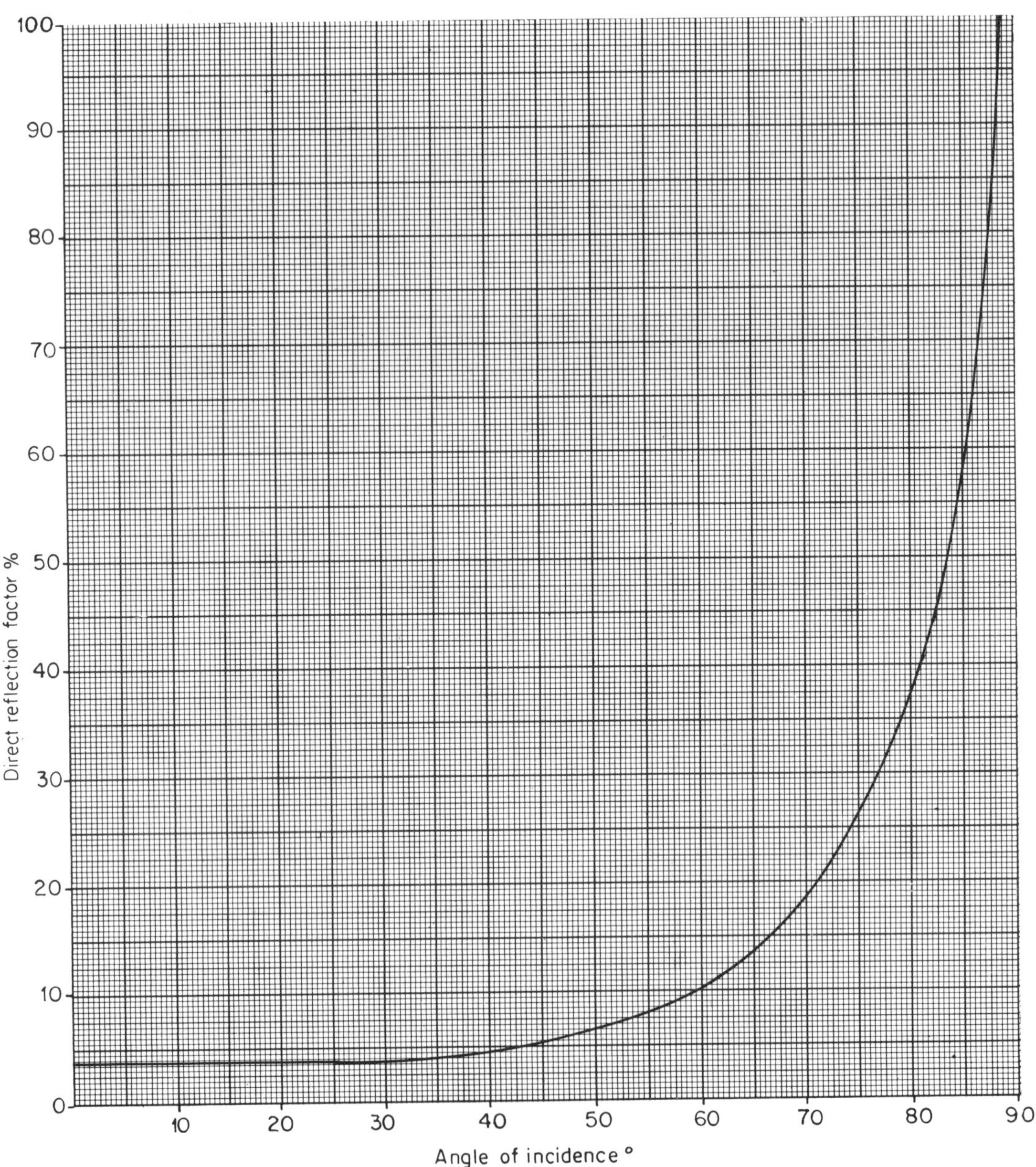


**Figure 12.27.** Direct reflection factor vs angle of incidence: 20°C, 5893 Å, air-acrylic interface. Acrylic cast sheet, dry ('Perspex')

# 13

# POLYVINYL CHLORIDE (PVC)

## INTRODUCTION

PVC is a term somewhat loosely applied to a wide range of materials varying considerably in formulation, form and properties, yet having in common the polymer polyvinyl chloride as a major constituent (see Chapter 2).

The monomer, vinyl chloride, was first reported in 1835 but it was not polymerized until 1872. A further period elapsed before, in 1912, Klatte in Germany and Ostromislensky in Russia, working independently, described in patents methods of producing polyvinyl chloride. However, for various reasons, technical and commercial, PVC was not successfully exploited until about 1930, but from then on progress became more rapid and by 1939, Germany and the U.S.A. were producing PVC, both in the plasticized and unplasticized forms. Manufacture began in Great Britain in 1942, mainly to provide a replacement for rubber in cable insulation and sheathing. Today PVC is produced in many countries throughout the world and ranks with polythene as one of the two most important plastics materials.

Early in the development of PVC, it had become apparent that its high melt viscosity, coupled with the risk of thermal degradation, would necessitate intensive research into polymers and processing methods if PVC plastics were to become of major commercial importance. As part of the work, polymerization techniques were varied to produce very pure polymers, suitable for the manufacture of compounds to meet the exacting requirements of cable insulation and sheathing; and materials requiring lower processing temperatures, particularly suited for calendering into rigid sheet. Investigations were carried out also on the addition of plasticizers and the co-polymerization of vinyl chloride with other monomers, e.g. vinyl acetate to produce softer, more easily processed materials. New branches of PVC technology sprang from the availability of paste-polymers suspended in plasticizer (plastisols) for coating and rotational moulding, and of suspensions of polymers and plasticizer in water (latices) for impregnation and coating.

Today, PVC, by virtue of its low cost and great variation in properties, is perhaps the most versatile of all plastics.

## NATURE OF PVC

Polyvinyl chloride is structurally similar to polythene, but differing from it in having one hydrogen atom on alternate carbon atoms replaced by a chlorine atom. As normally made, the structure is largely atactic, and because of the large size of the chlorine atoms, close packing of the chains is impossible and the polymer is almost wholly amorphous.

## GRADES AND FORMS

Vinyl chloride polymers and copolymers are made in many forms, differing in molecular weight[1], particle size, size distribution, shape and characteristics, e.g. some have hard, smooth particles and some soft, crumbly ones.

The polymers are mixed with additives to give a very large number of compounds, which may be in granular or powder form, pastes, and latices. Because of the many different additives which are compatible with PVC, compounds can be formulated to give properties within a very wide range. Among the characteristics obtainable are: clarity, colour, freedom from toxic hazards, excellent low temperature flexibility, and good high temperature properties. Pastes too are available in various grades to suit specific requirements: all contain a heat stabilizer. The processing temperatures for PVC compounds and the mechanical properties of the resultant products are determined in part by the $K$-value of the polymer used.

Semifabricated forms of PVC are widely used, the main ones being rigid sheet and foil, plasticized calendered sheeting, paste-coated fabric ('leathercloth'), film, pipe (rigid and flexible) and foams.

## PROCESSING

PVC can be handled by all conventional techniques, although some are more widely used than others. PVC is extruded into sheet, foil, film, pipe and wire and cable covering, and calendered into sheet and foil. It can be injection moulded on machines equipped with screw plasticization, and blow moulded on equipment incorporating suitably designed crosshead dies. Powders can be used for coating by the fluidized bed technique. Pastes can be applied by spreading, spraying or dipping; they can be converted into solid articles by injection moulding, rotational casting and dip forming; and, by incorporating a blowing agent, they can be made into foams of either open cell (sponge) or closed cell (expanded) structure.

PVC sheet can be shaped (copolymer sheet is particularly well suited to vacuum forming) and is easily machined, welded and cemented. Pipe can be joined by solvent cementing and by means of moulded PVC fittings.

## PROPERTIES

Although PVC has no sharp softening point, and begins to soften at fairly low temperatures, the effect of the chlorine atom is to increase the interchain attraction and thus the stiffness of the material, making it resistant to flow under pressure at temperatures much below 160°C; at such temperatures it is difficult to prevent decomposition, discolouration and loss of strength. The chlorine atom is responsible also for the dipolar nature of the PVC molecule. This limits the use of PVC, compared with polythene for example, as an insulator in applications involving high frequencies, where its loss factor is very high. On the other hand, unlike polythene, it can be readily welded by radio frequency techniques. Finally, the chlorine atom renders PVC self-extinguishing, a property rare among thermoplastics.

As already stated, the properties of PVC compounds depend so much on the nature and amount of the additives present that it is not possible to detail them comprehensively. Table 13.1 gives some typical figures for an unplasticized PVC. It is with such rigid compounds that this chapter is concerned.

[1] Polymers are usually characterized not by their molecular weight but in terms of their $K$-value. This is a figure related to the average molecular weight and is obtained from measurements of the viscosity of a solution of the polymer at a given concentration and temperature. Its value depends on the nature and concentration of the solvent and therefore different $K$-values may be quoted for equivalent polymers from different sources.

*Deformation*

The deformational behaviour at 20°C is given by the tensile creep curves in Figure 13.1. It can be seen that, compared with the data for the other plastics in this book, there is a pronounced upswing of the curves at long times, indicating the onset of runaway creep. The curves are appropriate to specimens tested soon after moulding, and it has been found that storage of such specimens at 60°C delays the runaway creep process markedly as illustrated in Figure 13.1, for

TABLE 13.1

| Property | Units | Typical Value |
|---|---|---|
| Density | g/cm³ | 1·39 |
| Softening region | °C | 75–85 |
| Coefficient of linear thermal expansion, | | |
| −20 to 60°C | per $°C^{-1}$ | $7 \times 10^{-5}$ |
| | per $°F^{-1}$ | $3{\cdot}8 \times 10^{-5}$ |
| Thermal conductivity, 20 to 40°C | cal/cm s °C | $4{\cdot}5 \times 10^{-4}$ |
| | B ThU in/ft² h °F | 1·3 |
| Specific heat, 20 to 80°C | cal/g °C | 0·26 |
| | B ThU/lb °F | 0·26 |
| Flammability | | Self-extinguishing |

the 20°C tensile creep curves and in Figure 13.5 for the 60°C tensile creep curves. It can be seen that this storage effect is of less importance for the 20°C than for the 60°C tensile creep data and, in any case, provides an additional safety factor in design. The 100 second isochronous stress vs. strain curve, corresponding to the tensile creep curves of Figure 13.1, is given in Figure 13.2.

Recovery data at 20°C are given in Figure 13.3. The usual pattern of behaviour of almost immediate and complete recovery for small strains and short times, and less rapid recovery from large strains and long times is followed.

The deformational behaviour of unplasticized PVC is very dependent on temperature, particularly for temperatures above 40°C. For this reason, of all the plastics discussed in this book, unplasticized PVC has the lowest maximum temperature of usefulness, although at room temperature it is one of the most rigid of thermoplastics in the short term.

The deformational behaviour at 60°C is given by the tensile creep curves in Figure 13.4. The corresponding 100 second isochronous stress vs. strain curve is given in Figure 13.6. The effect of storage at 60°C of the specimens before testing, already mentioned in connection with the 20°C data, is illustrated by the tensile creep curves in Figure 13.5. The considerable effect of temperature on the deformational behaviour can be assessed either by comparing the 20°C and 60°C tensile creep data or from the 100 second tensile creep modulus (at 0·2% strain) vs. temperature curve shown in Figure 13.7.

*Limiting Stress*

PVC is a ductile material in the absence of stress concentrations so that failure at long times under load is essentially by necking and the limiting behaviour is defined by the onset of this process. Two other phenomena precede necking, however, and might in certain applications limit the usefulness of a component: they are voiding and surface cracking. All the relevant data for 20°C are given in Figure 13.8. Similar data derived from tensile load-elongation curves at different straining rates are given in Figure 13.9.

The temperature-dependence of failure stress has not yet been obtained for long times of loading and the only data available at temperatures other than 20°C are in terms of the yield stress in a conventional tensile test. Such data are presented in Figure 13.10 for unplasticized PVC.

The effect of additives on the yield stress in a conventional test varies with the nature of the additive. Ingredients insoluble in PVC tend to decrease the yield stress; lubricants, pigments and fillers act in this way. It is noteworthy that although they reduce the yield stress, fillers increase the rigidity at very low strains. On the other hand, small quantities of soluble additives, plasticizers etc., increase the yield stress and make the product more brittle. Indications of behaviour are given in Figure 13.11.

### *Impact Behaviour*

The basic impact behaviour of unplasticized PVC is given in Figure 13.12 which shows curves of impact strength vs. temperature for specimens of various notch tip radii and for unnotched specimens. The pronounced effect even of blunt notches in reducing the impact strength is an important feature. These results were obtained on sheet material but basically similar data are obtained for extrusion and moulding compounds.

The impact behaviour of PVC is affected very considerably by additives which, as has been mentioned previously, are very much involved in the technology of the material. The impact strength is affected adversely by small amounts of additives soluble in PVC; specimens both with sharp notches and blunt notches are similarly affected. The deterioration is considerable, corresponding to a change of some 15°C for an additive concentration of some 5% by weight. This is shown in Figures 13.13 and 13.14 for sharp and blunt notches respectively.

The impact strength of PVC is also affected by insoluble additives—mineral fillers such as titanium dioxide are typical examples. There is an improvement in behaviour, particularly in the impact strength of sharply notched specimens: Figure 13.15 gives data appropriate to a titanium dioxide content of approximately 10% by weight and shows the pronounced improvement observed for sharply notched specimens. Another common insoluble additive is rubber and certain synthetic rubbers can be blended with PVC to give rigid compounds of good impact behaviour.

### *Electrical Properties*

The molecules of PVC are strongly dipolar because of the chlorine atoms in the structure. Consequently, the dielectric properties are strongly dependent on temperature and frequency. The dielectric behaviour of unplasticized PVC as a function of temperature and frequency is given in Figures 13.16 and 13.17, in terms of permittivity and loss tangent, respectively. Curves of apparent volume resistivity with temperature and time of electrification are given in Figure 13.18.

Additives and plasticizers have a significant effect on the dielectric properties of PVC plastics. In fact, they give improvement in dielectric properties by reducing the power loss at high frequencies. A full description of this phenomenon is, however, outside the scope of this work.

Compared with those of other plastics, the electrical properties of PVC plastics are not outstanding. Nevertheless, PVC is useful in non-exacting applications, for which, by the choice of suitable additives, a range of insulating materials has been developed. The high frequency dielectric behaviour of PVC is of practical importance in one respect: the power losses are such that radio frequency dielectric heating may be used to weld it.

*Optical Properties*

PVC plastics can have a wide range of optical properties, depending on the types of additives present and fabrication conditions. For instance, the transparency can vary so that some grades are opaque while others are reasonably transparent with useful optical properties. The data given below are for a grade developed for good optical properties. The specimen is pressed sheet of 3·36 mm thickness illuminated at 5461 Å and the results are corrected for surface defects and reflection.

| | |
|---|---|
| Direct transmission factor | 82% |
| Direct transmission factor corrected to 1 mm thickness | 94% |
| Scattering coefficient | 0·6 $cm^{-1}$ |

*Refraction*

Refractive index data are given in Table 13.2 for a specimen of density 1·384 $g/cm^3$.

TABLE 13.2

| Wavelength (Å) | Refractive Index |
|---|---|
| 4861 | 1·54806 |
| 5893 | 1·54151 |
| 6563 | 1·53843 |

Reciprocal dispersive power $V_d = 59{\cdot}3$
Critical angle ($\lambda = 5893$ Å) $= 56{\cdot}23$
Temperature coefficient of refractive index $= -0{\cdot}0001142°/°C$.

*Chemical Properties*

As with other properties it is essential to distinguish between the behaviour of plasticized and unplasticized PVC. Unplasticized homopolymers have the best chemical resistance; the effect of copolymerization and plasticization is to impair the chemical resistance: the addition of other additives, e.g. stabilizers, lubricants and pigments, while lowering the resistance, is of lesser significance.

Unplasticized PVC has excellent chemical resistance to acids, alkalis, aqueous solutions, and all but the most severe oxidizing agents, e.g. sulphuric acid below 90% concentration has no effect at temperatures up to 60°C, whereas stronger acid should not be used above 50°C. Oils, fats, alcohols and petrol can all be used in contact with unplasticized PVC, but solvents such as aromatic and chlorinated hydrocarbons, ketones and esters cause swelling. The suitability of unplasticized PVC for use with town gas depends on the level of aromatic constituents in the gas and each case must be carefully considered.

The chemical resistance of plasticized PVC will depend on the type of plasticizer used and also on the amount—the more plasticizer present, the poorer the resistance. The comments below apply to a compound containing dioctyl phthalate (DOP), a commonly used, fairly resistant plasticizer.

TABLE 13.3

| Property | UPVC | Plasticized PVC |
|---|---|---|
| Gas permeabilities (cm$^3$ (NTP) cm/cm$^2$ s cm Hg) $\times 10^9$ | | |
| Carbon dioxide | 0·06 | 0·2–2 |
| Oxygen | 0·01 | 0·04–0·4 |
| Nitrogen | 0·002 | 0·01–0·1 |
| Water vapour permeability (g/m$^2$ 24 h) at 38°C and 90% rh for 0·001 in thickness | 25 | 50–180 |
| Resistance to: | | |
| Mineral acids (dilute) | Excellent | Excellent |
| Mineral acids (conc.) | Excellent | Good |
| Alkalis | Excellent | Good |
| Solvents: | | |
| alcohols | Excellent | Fair to poor |
| ketones | Fair | Fair to poor |
| aromatic hydrocarbons | Fair | Fair to poor |
| chlorinated hydrocarbons | Fair | Poor |
| Detergent solutions | Excellent | Good to fair |
| Greases and oils | Good | Good to poor |

## APPLICATIONS

Because of its versatility, coupled with low cost, PVC is used extensively in a whole variety of applications where toughness, abrasion and chemical resistance, and self-extinguishing properties are required. The list below is divided according to the form of the material.

*Compounds*

Extrusions—unplasticized:

Pipes for potable and non-potable water supplies, chemical plant, waste and soil systems. Rainwater goods, curtain rails.

Extrusions—plasticized:

Cable sheathing and domestic wiring insulation (low frequency).

Decorative trim, hosepipe, sachets for shampoos.

Injection mouldings—unplasticized:

Pipe fittings, textile bobbins, electrical conduit fittings, junction boxes, door handles, equipment housings, telephone handsets, chair backs and seats.

Injection mouldings—plasticized:

Toys, watch straps, cable grommets, gaskets, handle coverings, beach shoes, sandals.

Powder coatings:

Metal protection.

*Pastes*

Spreading:

Leathercloth for upholstery in transport, furniture; handbags, clothing.

Rotational casting and slush moulding:

Toys, balls, arms rests.

Dip forming and dip coating:

Gloves, handle-bar grips, spouts, wire basket covering.

Spraying:
Tank linings.

*Latices*
Paper and textile impregnation.

*Rigid Sheet*
Internal wall-cladding for chemical plant, food stores and tunnels; lighting fittings, reflectors and signs, fume ducting, tank linings, food processing equipment.

*Foil*
Decorative signs and displays, refrigerator components, liners, trays, packaging, liners for concrete shuttering.

*Plasticized Sheeting*
Protective coverings, flooring, clothing, (such as raincoats), hats, galoshes, inflammable toys, air-beds, fancy goods. Transparent 'windows' in flexible doors, etc. Lining of swimming pools.

*Film*
In-store packaging; cooked meats, cheese, produce; packing of toys, games, stationery, cosmetics, confectionery, pharmaceuticals.

*Foam*
Cushioning: mattresses, pillows, etc.; insulation.

## LIST OF FIGURES

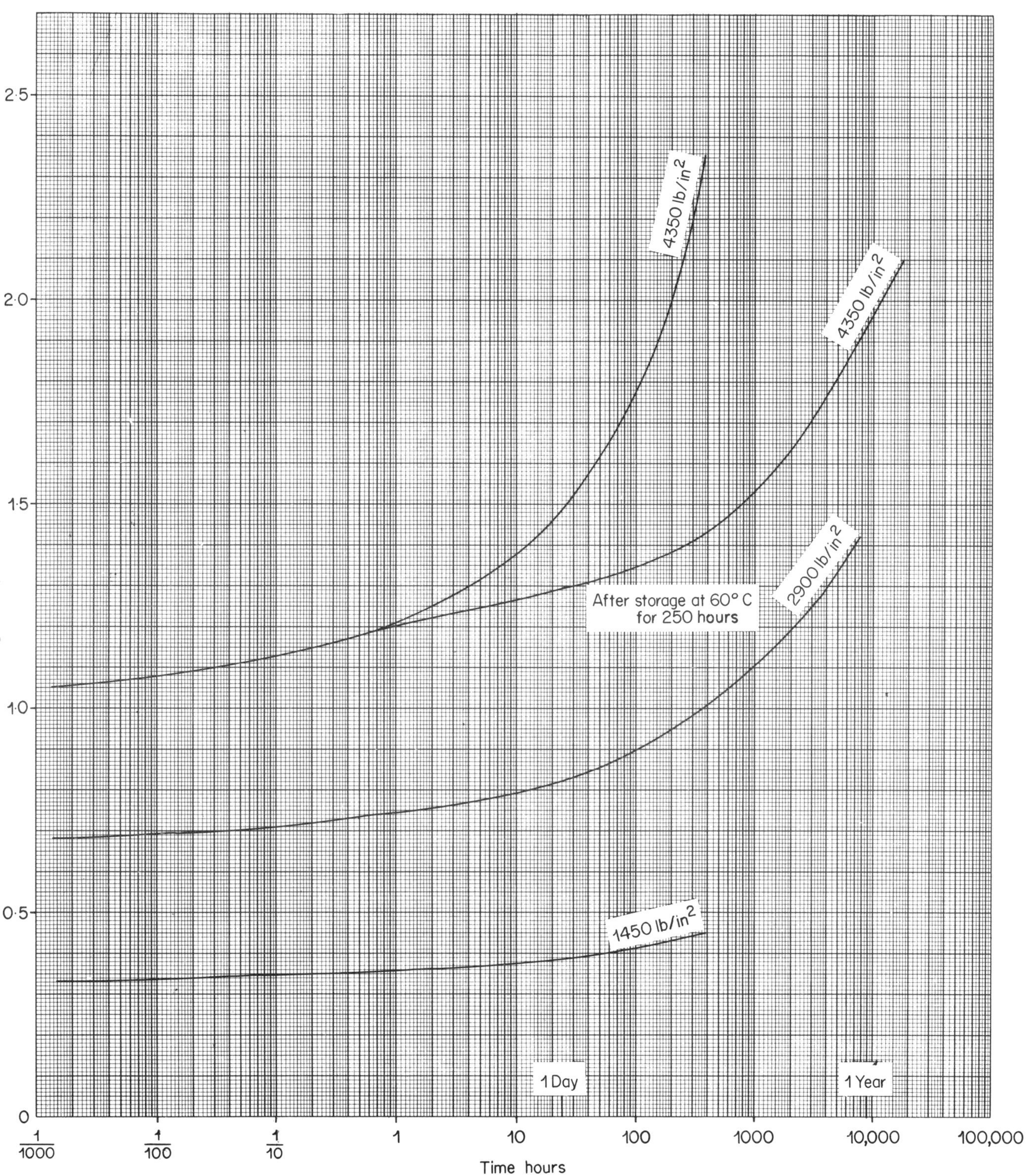


**Figure 13.1.** Creep curves in tension: 20°C. Also shows effect of storage at 60°C for 250 hours prior to test. Unplasticized PVC ('Darvic' 110)

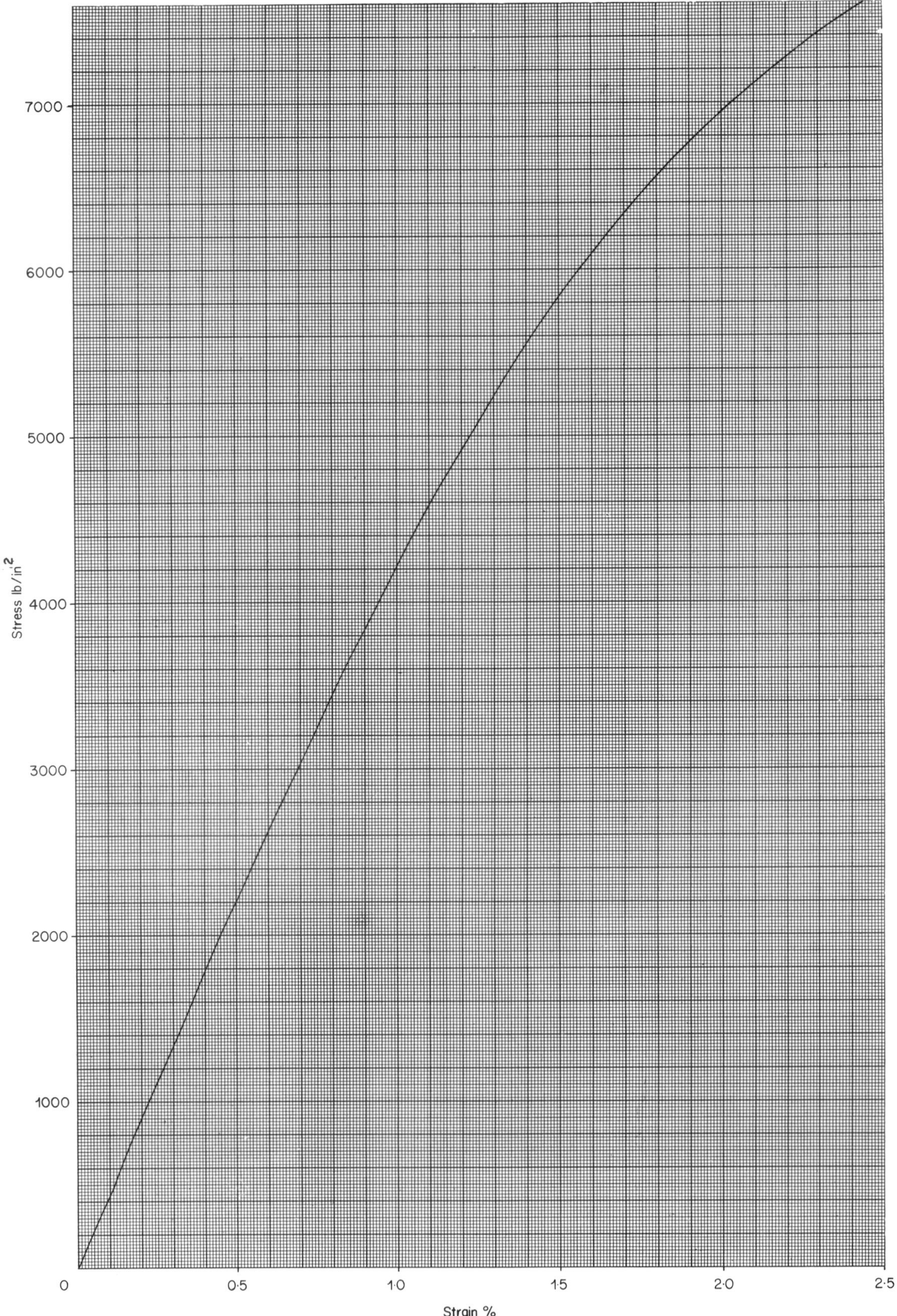


**Figure 13.2.** Isochronous stress vs strain curve: 20°C, 100 sec. Unplasticized PVC ('Darvic' 110)

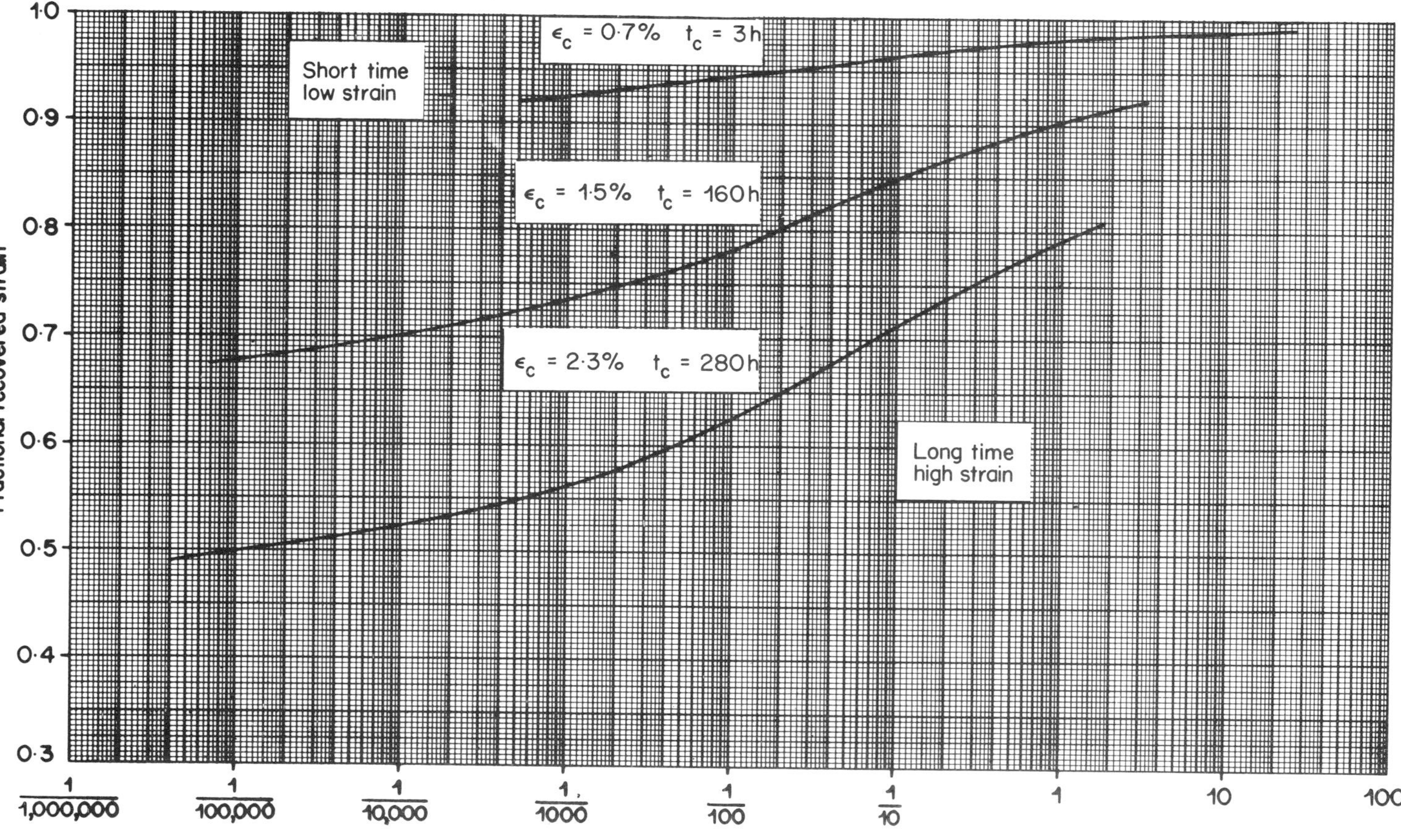


**Figure 13.3.** Recovery from creep in tension: 20°C. Unplasticized PVC ('Darvic' 110)

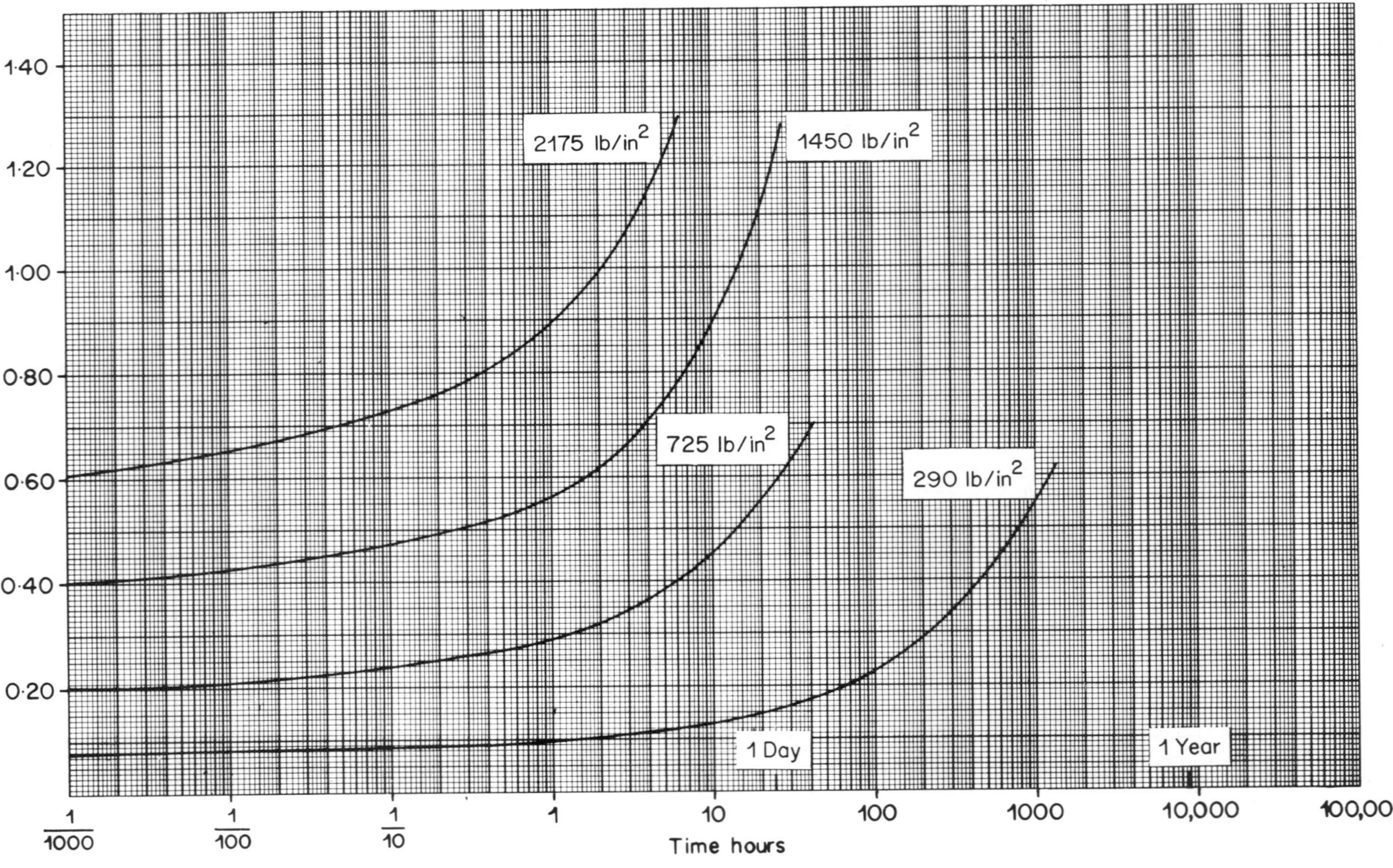


**Figure 13.4.** Creep curves in tension: 60°C. Unplasticized PVC ('Darvic' 110)

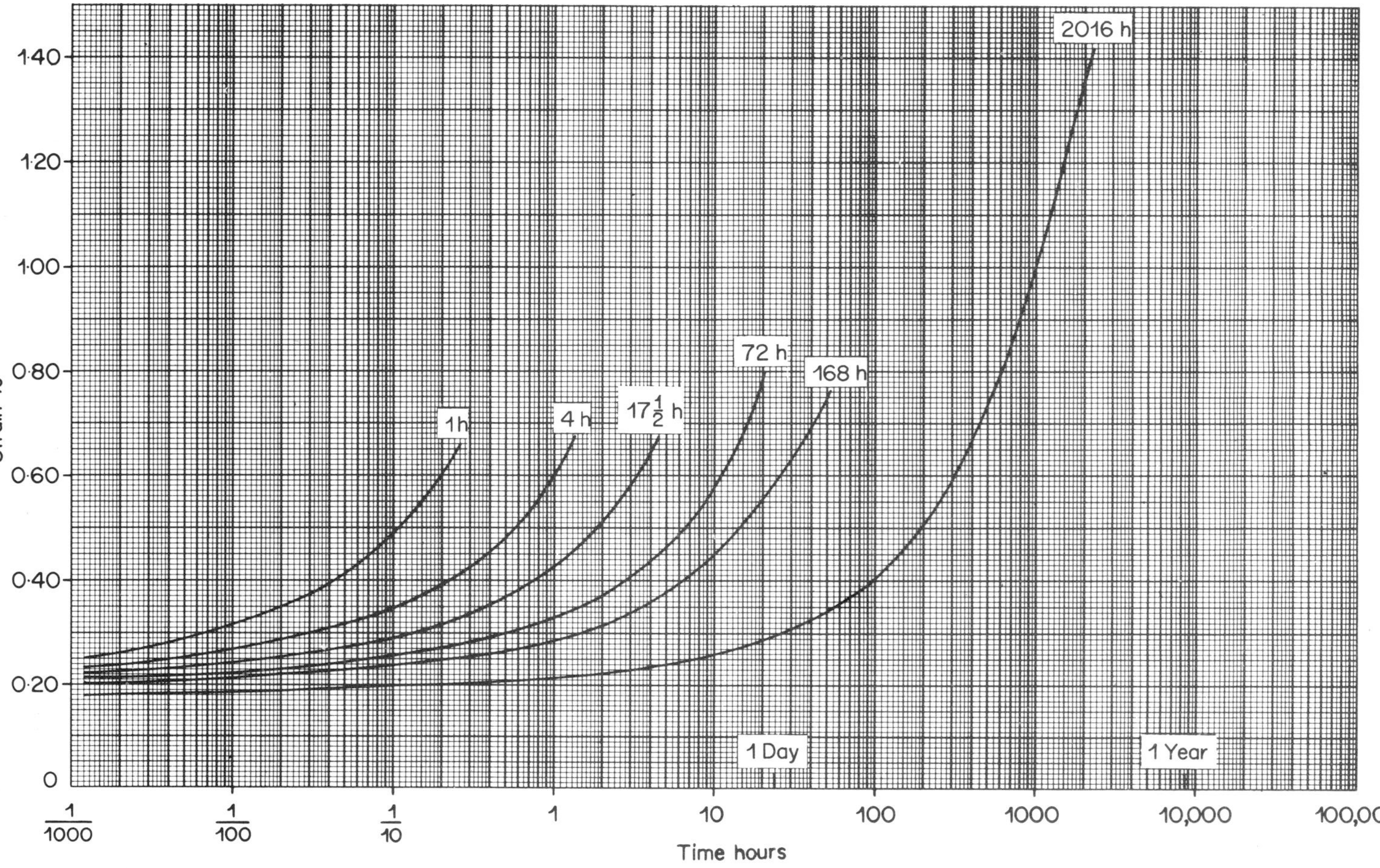


**Figure 13.5.** Creep curves in tension: 60°C, 725 lb/in$^2$. Effect of storage at 60°C prior to test. Unplasticized PVC ('Darvic' 110)

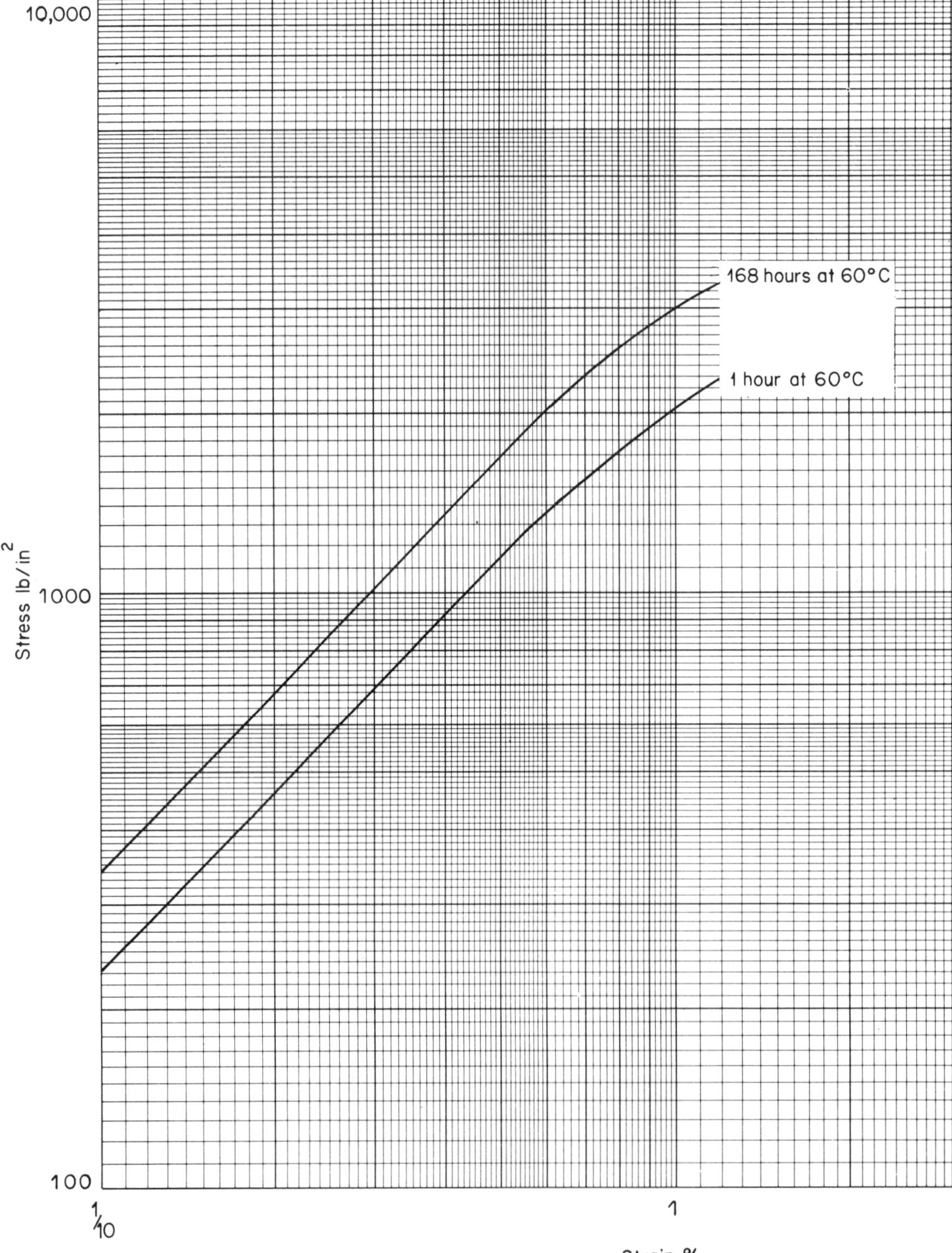


**Figure 13.6.** Isochronous stress vs strain curves: 60°C, 100 sec, logarithmic axes. Effect of storage at 60°C prior to test. Unplasticized PVC ('Darvic' 110)

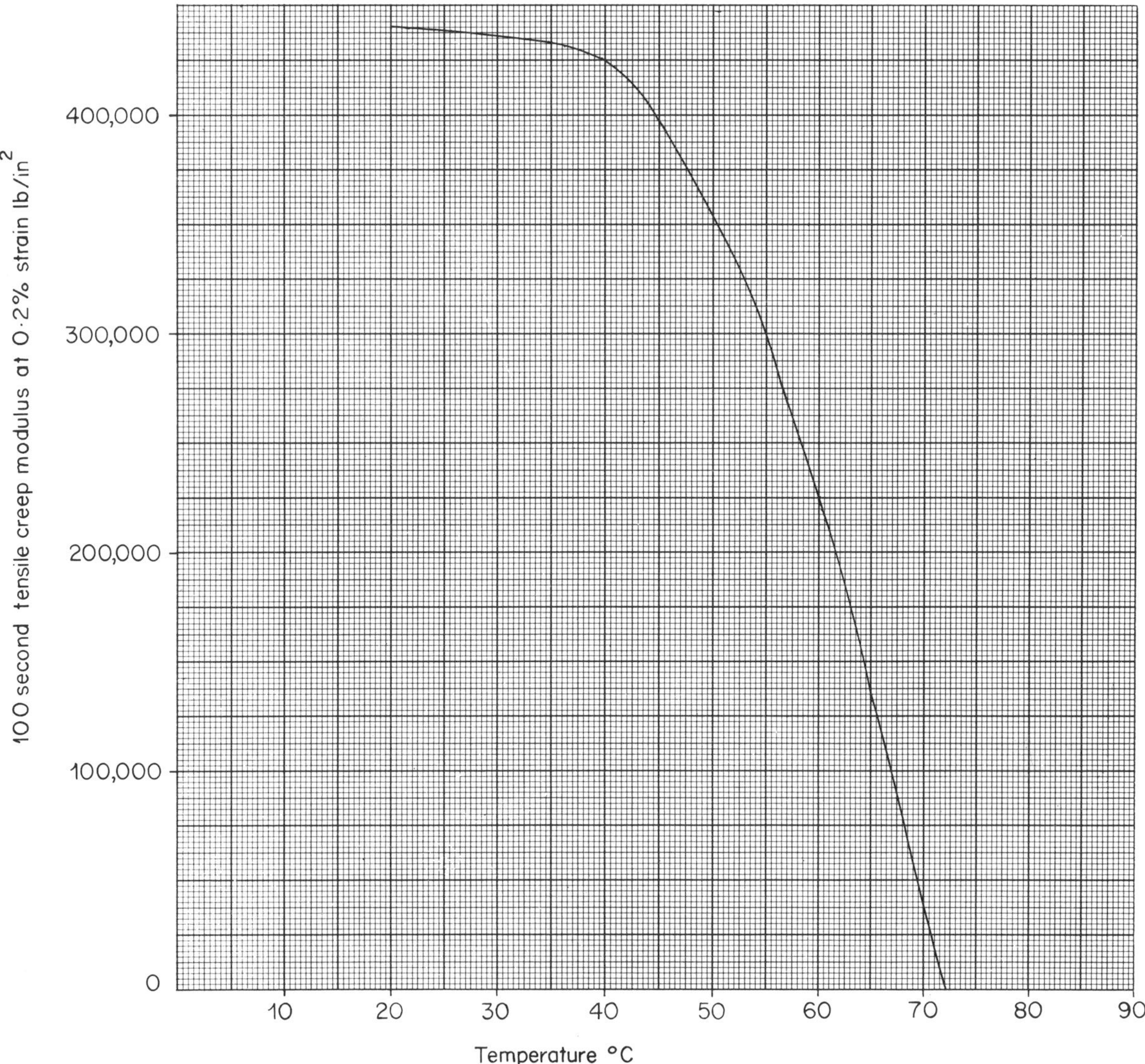


**Figure 13.7.** Tensile creep modulus (100 sec, 0·2% strain) vs temperature. Unplasticized PVC ('Darvic' 110)

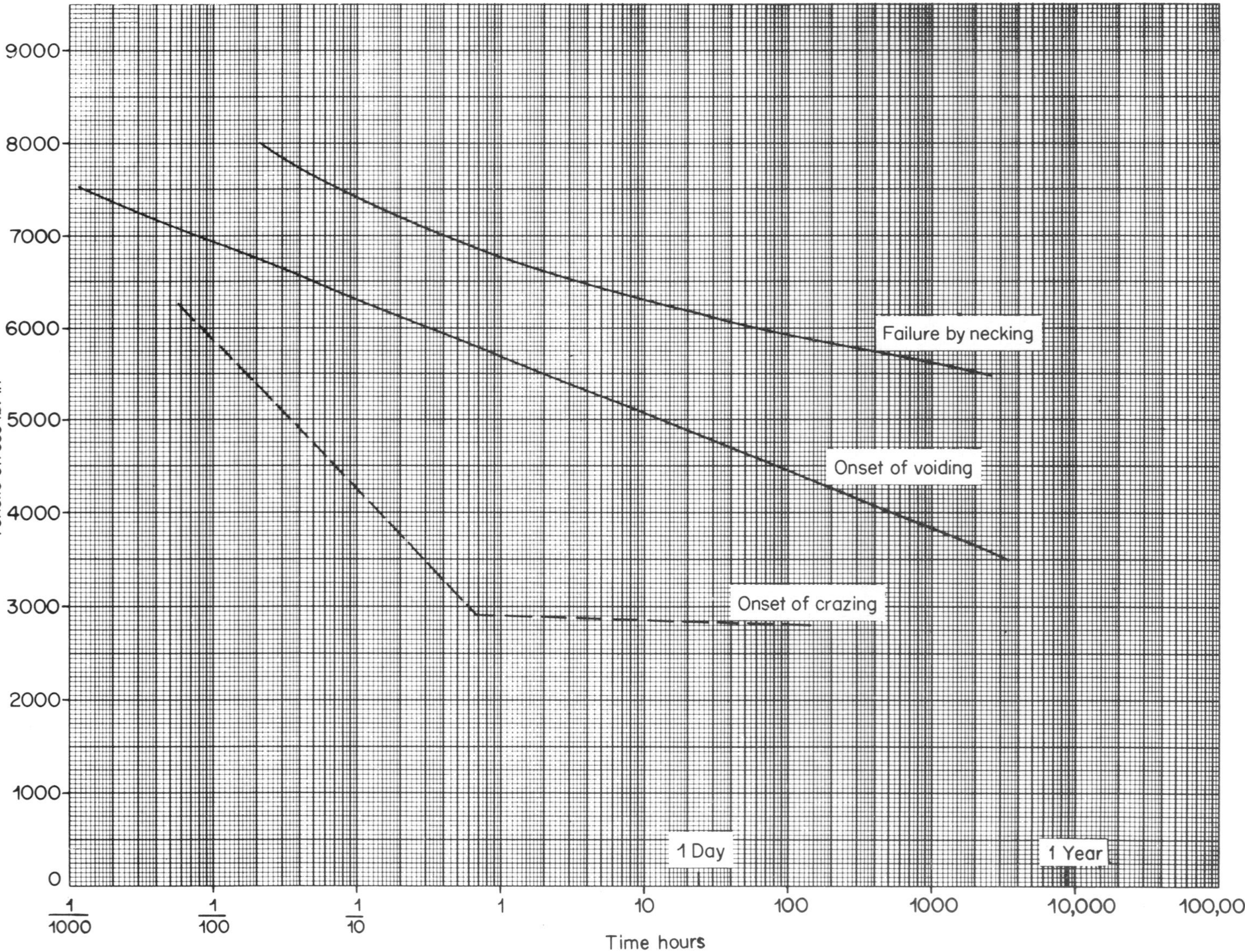


**Figure 13.8.** Creep rupture stress in tension vs time to failure: 20°C. Unplasticized PVC ('Darvic' 110)

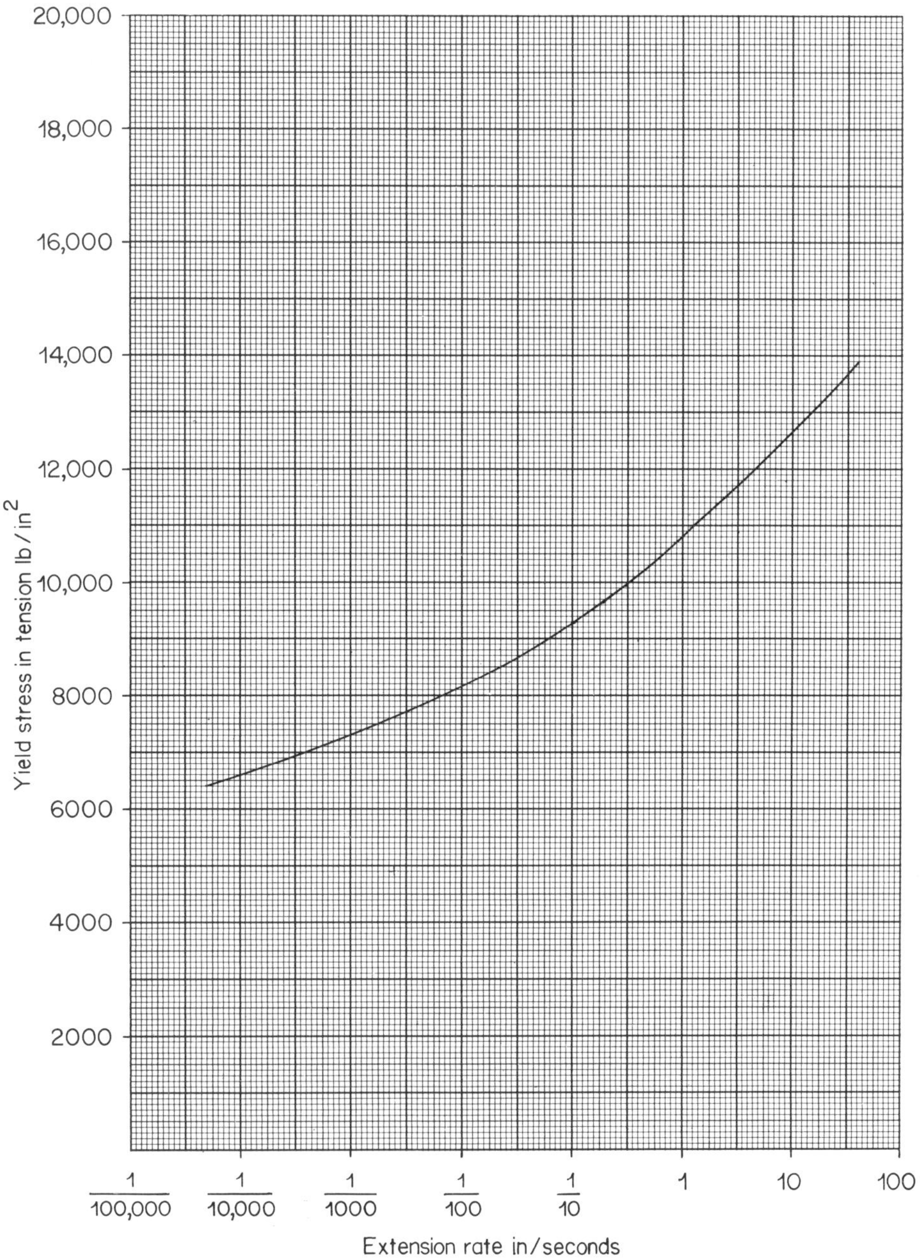


**Figure 13.9.** Yield stress in tension vs straining rate: 20°C. Unplasticized PVC, calendered foil

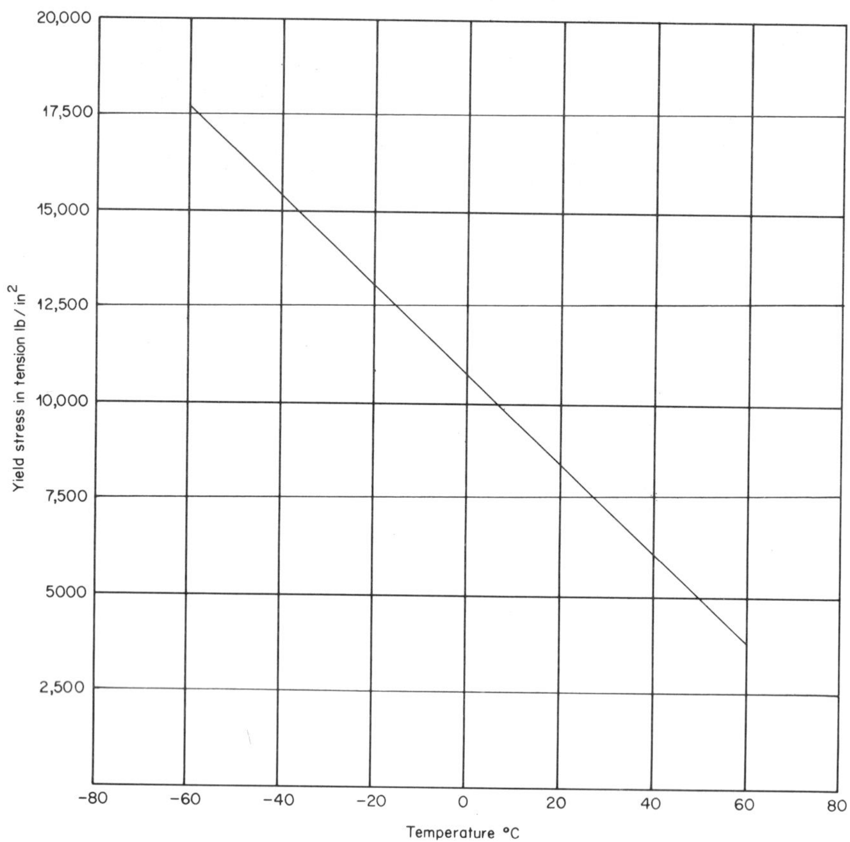


**Figure 13.10.** Yield stress in tension vs temperature: 50 % per min straining rate. Unplasticized PVC ('Darvic' 110)

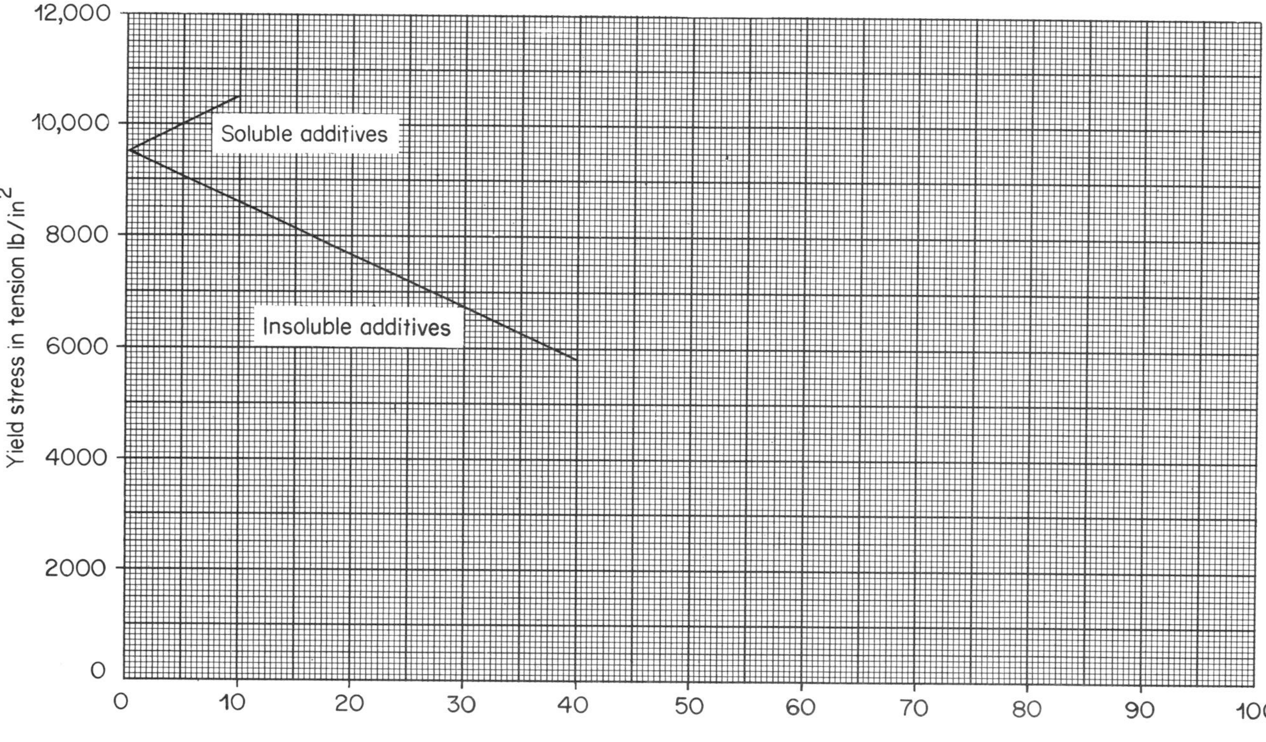


**Figure 13.11.** Yield stress in tension: 20°C. Effect of additives. PVC rigid compositions

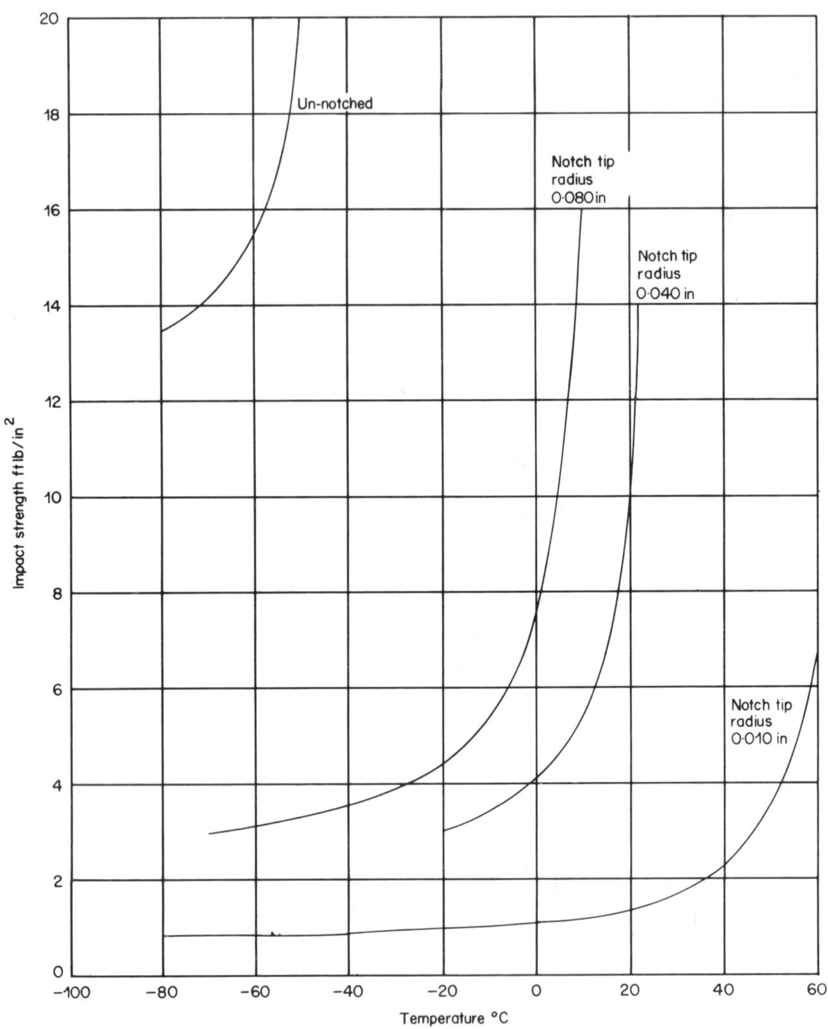


**Figure 13.12.** Impact strength vs temperature. Unplasticized PVC ('Darvic' 110).

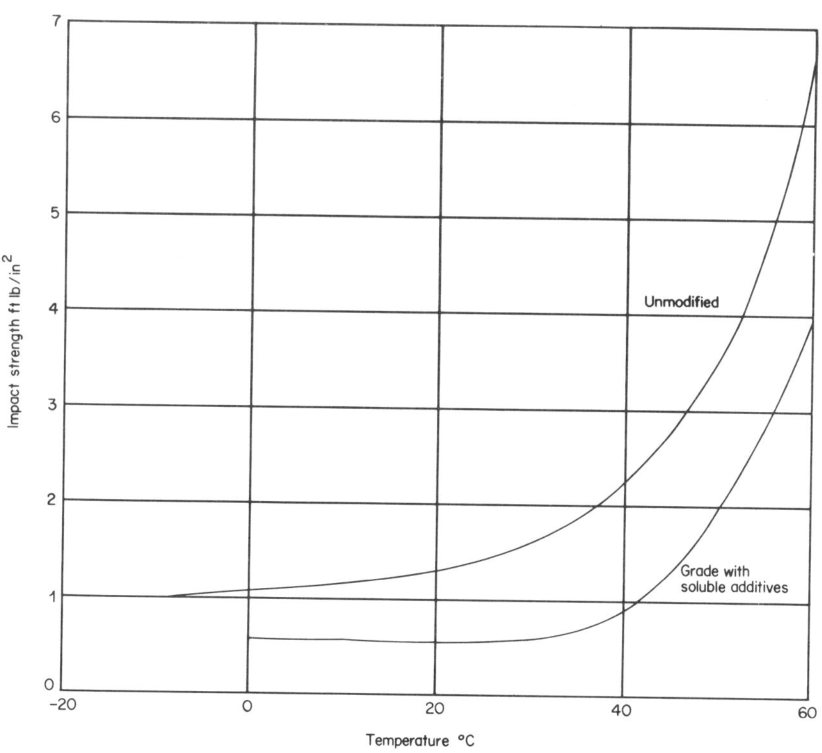


**Figure 13.13.** Impact strength vs temperature: 0·010 in notch tip radius. Effect of soluble additives. PVC rigid compositions

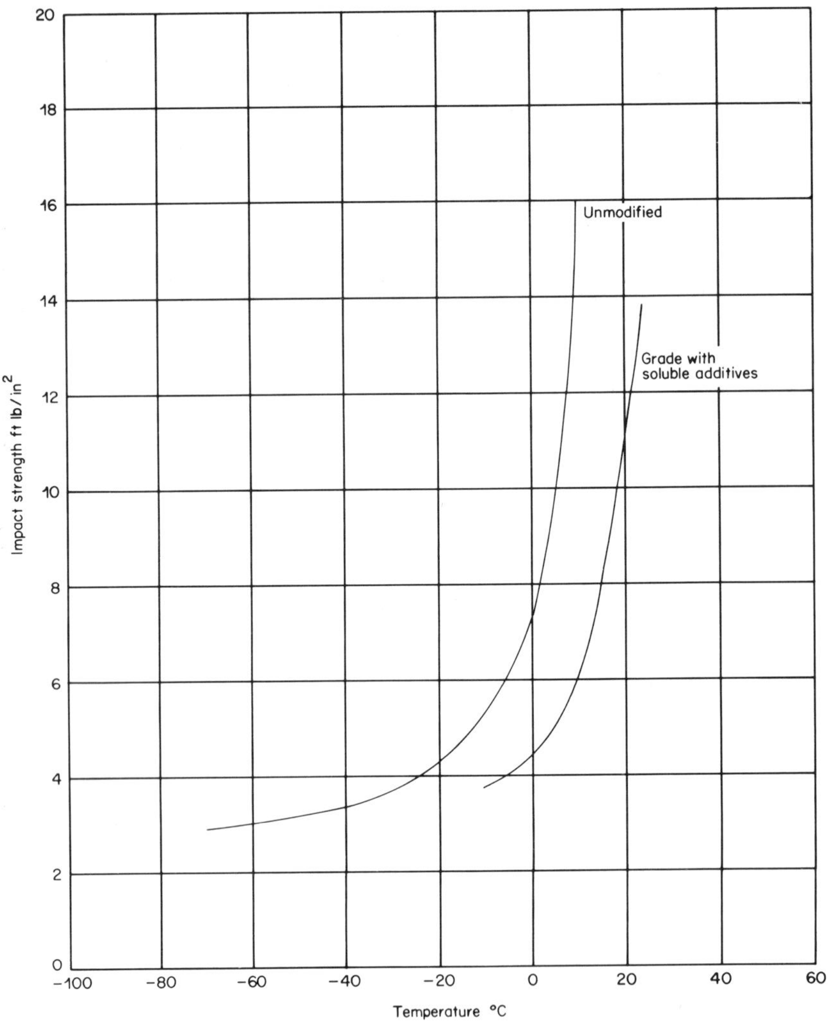


**Figure 13.14.** Impact strength vs temperature: 0·080 in notch tip radius. Effect of soluble additives. PVC rigid compositions

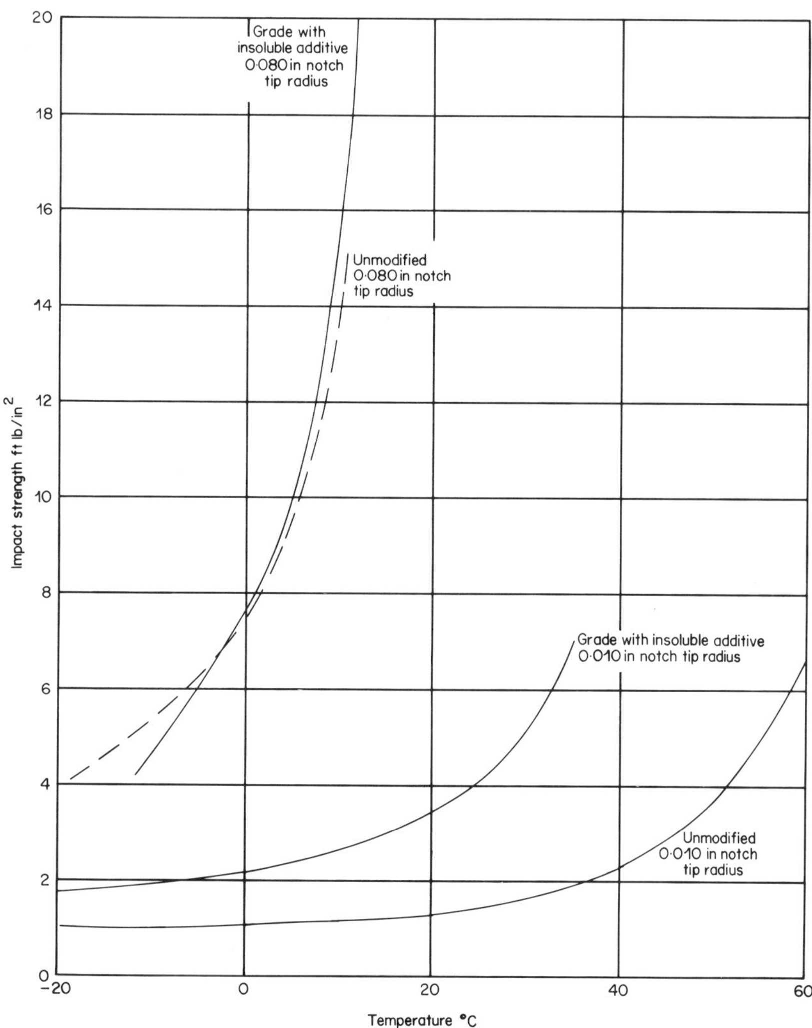


**Figure 13.15.** Impact strength vs temperature: 0·010 in, 0·080 in notch tip radius. Effect of insoluble additives. PVC rigid compositions

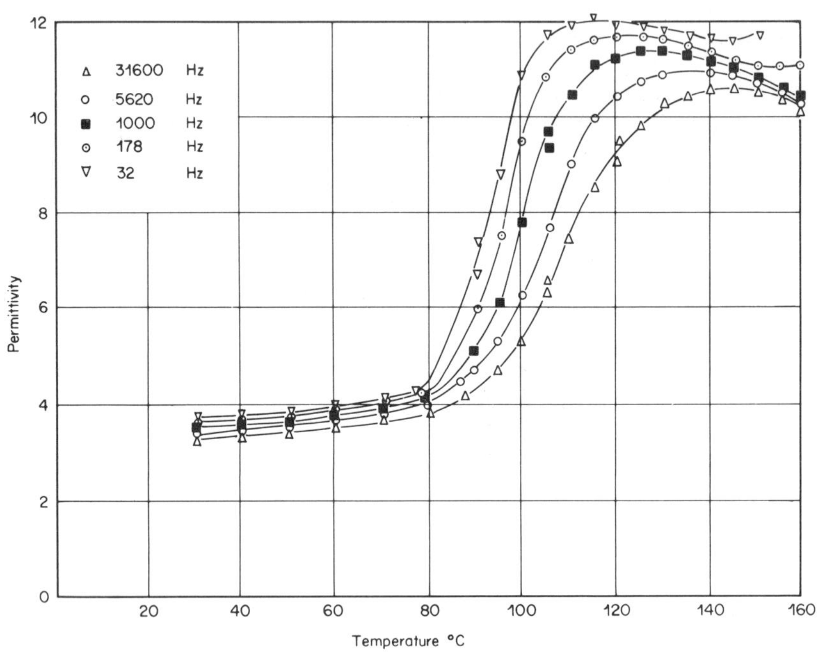


**Figure 13.16.** Permittivity vs frequency and temperature. Unplasticized PVC ('Darvic' 110)

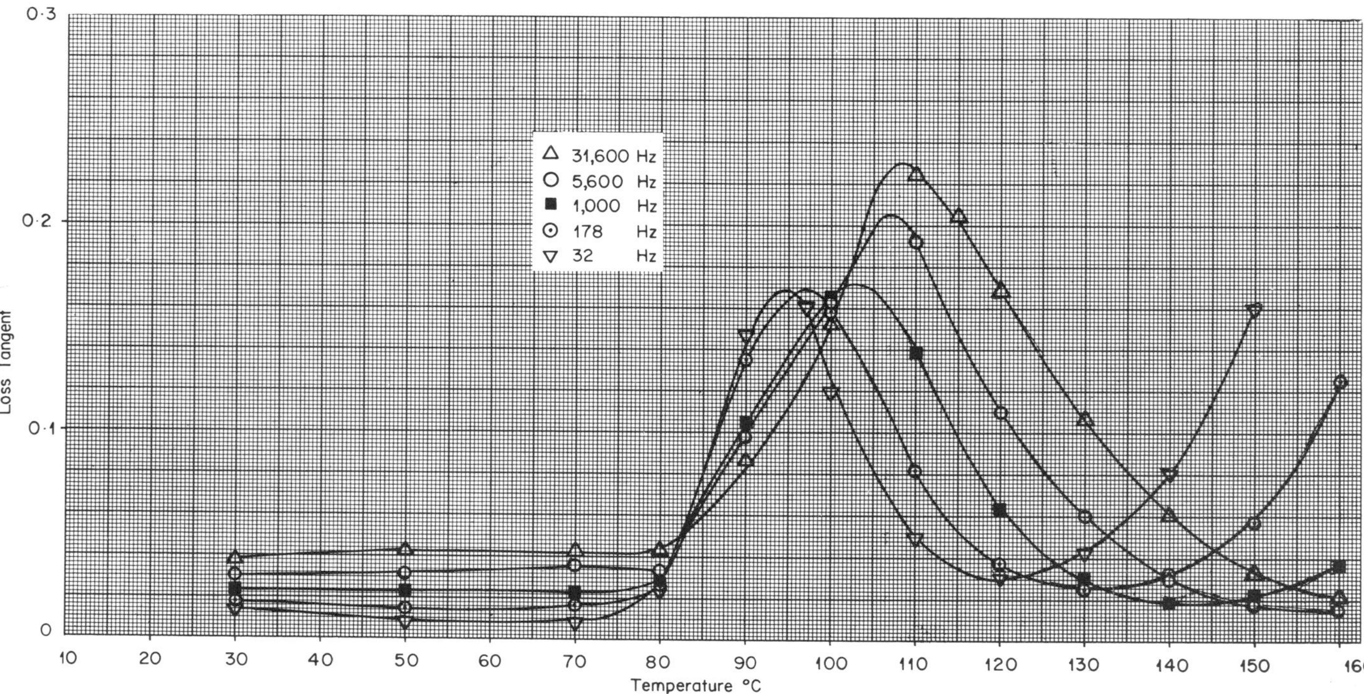


**Figure 13.17.** Loss tangent vs frequency and temperature. Unplasticized PVC ('Darvic' 110)

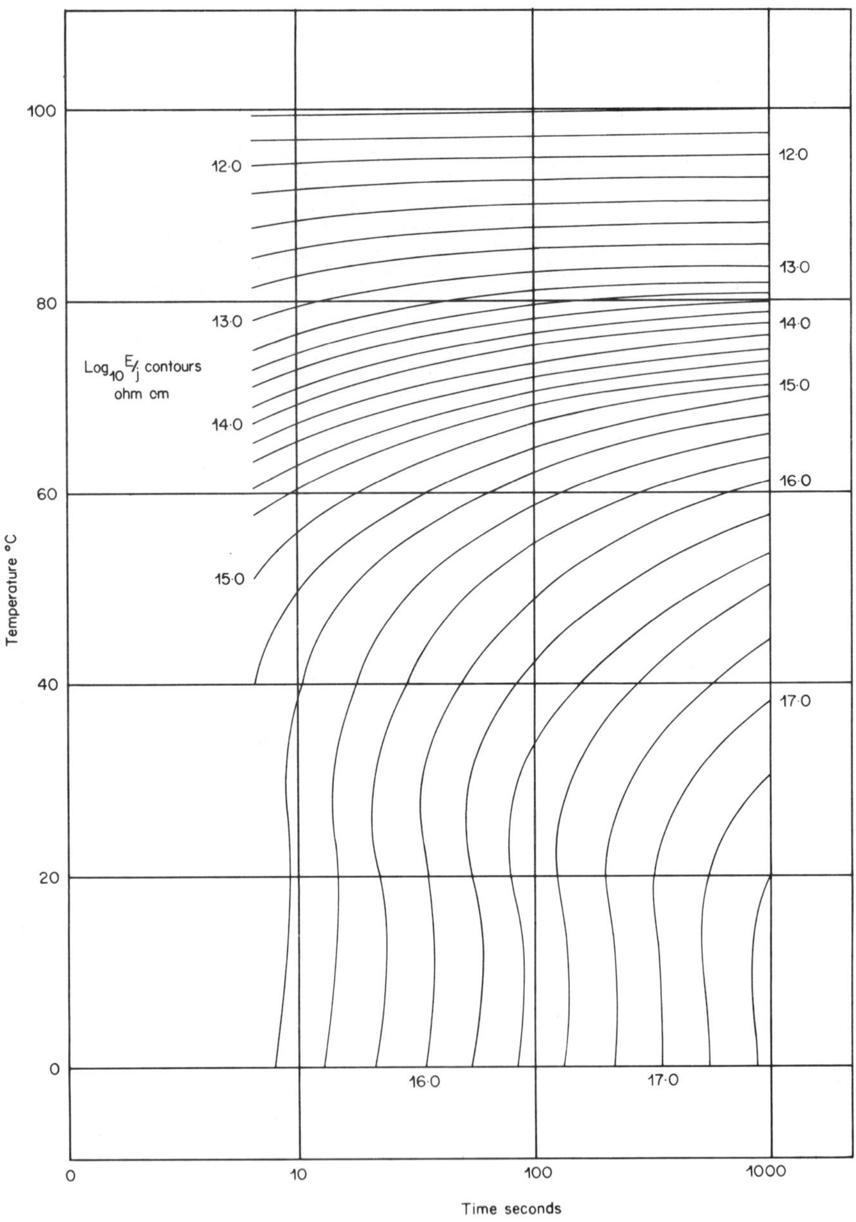


**Figure 13.18.** Apparent volume resistivity ($\log_{10}$ values) vs time of electrification and temperature. Unplasticized PVC ('Darvic' 110)

# 14

# POLYTETRAFLUOROETHYLENE (PTFE)

## INTRODUCTION

On 6th April 1938, Dr. R. J. Plunkett, engaged in the U.S.A. in research on refrigerant gases, found that a cylinder believed to contain a kilogram of tetrafluoroethylene displayed no gas pressure, though the weight of the cylinder showed that material was still inside.

On cutting open the cylinder a white solid was found which proved to be polytetrafluoroethylene. This discovery was rapidly exploited, so much so that PTFE played a significant role during the Second World War. Its first application was, in fact, in the production of the original atomic bomb where the corrosive gases involved demanded, for their safe handling, chemically inert gaskets and PTFE was used successfully for making them.

## NATURE OF PTFE

The PTFE molecule consists of a carbon chain wholly saturated with fluorine atoms. Its conformation is dictated by the necessity to accommodate the large fluorine atoms and this is achieved by the molecule twisting itself into a helix. PTFE has a very high molecular weight—over a million—and as manufactured is 90–95% crystalline, suggesting that the molecule is linear and unbranched. This high crystallinity is never completely recovered after processing; slowly cooled PTFE will be about 75% crystalline and quench-cooled PTFE possibly no more than 50% crystalline.

## GRADES AND FORMS

PTFE is available in three basic forms: granular powders, coagulated dispersion powders (commonly known as paste polymers), and aqueous dispersions. Granular powders can be filled with inorganic materials such as glass, asbestos and powdered metals to give materials with improved wear and deformation resistance and better dimensional stability.

The natural polymer is white, but coloured products can be obtained by tumbling the powders with pigment before fabrication and by stirring a pigment paste into the aqueous dispersions.

Many semifabricated forms are commercially available, including film, sheet, tape, tube and rod.

## PROCESSING

PTFE cannot be processed by the methods normally used with other thermoplastics, such as injection moulding and melt extrusion, because its high molecular weight and rigid conformation prevent it from becoming a low-viscosity liquid.

The method common to all PTFE fabrication is to bring the material into the desired shape in the cold, and then to fuse the polymer by the application of heat so that adjacent particles coalesce in a process known as 'sintering' to give a homogeneous product.

The processing of each of the three basic forms is, in outline, as follows.

*Granular Polymers*

Granular PTFE can be moulded into billets, sheets and rings, the operation consisting essentially of three steps; preforming, sintering and cooling. Rods and thick-walled tubes are made by ram extrusion, a process in which charges of loose polymer are compacted in a long straight tube by a reciprocating ram. The compacted polymer moves along the tube to a heated zone in which it is sintered and becomes fused into a homogeneous rod or tube. This process is suitable for the production of rod in diameters from $\frac{1}{8}$–3 in and for tubing of about $\frac{1}{8}$ in thickness and above.

The direct processing of granular PTFE is suitable only for making objects of simple and regular shape; complex and very accurate shapes are made by machining stock produced by moulding or extrusion. Fortunately, PTFE can be machined fairly easily.

*Coagulated Dispersion Polymers*

Coagulated dispersion polymers are used for the production of thin sections such as thin-walled tube or insulated wire covering and they are always extruded. To do this a lubricant or extrusion aid is first incorporated in the polymer and the mixture is then lightly compacted into a cake before being transferred to the extrusion chamber of the extruder. A constant-rate ram forces the lubricated cake through a die and a thin-walled tube or thin rod of unsintered PTFE results. The lubricant is then extracted and finally the extrudate is sintered and cooled.

As a result of the cold-working of the polymer during extrusion, molecules become highly oriented so that a fibrous structure is built up. This serves to support the extrudate during sintering and results in the final products having excellent flex life and high strength.

*Aqueous Dispersions*

Aqueous dispersions of PTFE are used for the impregnation of porous materials such as asbestos or glass fabric, for the coating of thermally stable substrates such as metal, ceramic or glass, and for a number of other less widespread applications. The dispersions are applied by dipping, spraying or flow-coating, and may be sintered or unsintered depending on the application. The most important products made from PTFE aqueous dispersions are impregnated asbestos, impregnated glasscloth and coated metals.

## PROPERTIES

PTFE within its working temperature range of 250°C (300°C for short periods) down to almost absolute zero will withstand almost any chemical environment, certainly all those met with in general engineering practice. This extreme inertness arises from the strength of the carbon–carbon and carbon–fluorine bonds and from the shielding action of the sheath of fluorine atoms.

The other great advantages of PTFE are its dielectric performance, which matches that of polythene but over a much wider temperature range, and its very low coefficient of friction. PTFE does not display 'stick-slip' movement and no known material will truly bond to it.

Some physical properties are shown in Table 14.1.

In some applications, the poor wear resistance and low deformation resistance are disadvantages. By careful selection of fillers, products can be obtained with better mechanical properties at the expense of a slight loss in chemical inertness and dielectric properties.

TABLE 14.1

| Property | Units | Typical Value |
|---|---|---|
| Density | g/cm³ | 2·15–2·24 |
| Crystalline melting point | °C | 327 |
| Coefficient of linear thermal expansion | | See Figure 14.7 |
| Thermal conductivity, 20–35°C | cal/cm s °C | $6 \times 10^{-4}$ |
| | B ThU in/ft² h °F | 1·7 |
| Specific heat, above 40°C | cal/g °C | 0·23 |
| | B ThU/lb °F | 0·23 |
| Flammability | | Does not burn |

*Deformation*

PTFE is well known for its low dielectric loss, low coefficients of friction and its extreme chemical inertness, but its deformational behaviour must be known also if it is to be used in load-bearing applications. Generally, components made of it will be subjected to compressive loads, and so tests have been confined to measuring the creep response to uniaxial stress in compression.

As for all other partially crystalline thermoplastics the mechanical properties of PTFE are dependent on time, temperature, and degree of crystallinity. There is, however, a further complication: both the particulate and crystalline structure and hence the deformational behaviour of PTFE are extremely sensitive to slight changes in the sintering procedure carried out during processing. The scatter that can occur through slight variations in the sintering process is indicated in Figure 14.1 where the 100 second isochronous stress vs. strain data for PTFE and 25% by weight glass-filled PTFE are represented as bands. Creep data are as yet very limited. Preliminary creep data at a stress of 1000 lb/in² are given in Figure 14.2, for PTFE and 25% by weight glass-filled PTFE.

Recovery data are shown in Figure 14.3.

*Limiting Stress*

The failure characteristics at long times are not yet known. Figure 14.4 shows the dependence of yield stress on temperature determined by the conventional constant straining rate tensile test, but these data are applicable to short times of loading only.

*Impact Behaviour*

Data on the resistance to impact of PTFE, determined by the Charpy impact test, show that even under severe conditions, e.g. with sharp notches, PTFE samples are not truly brittle in impact at temperatures above −20°C. Because of this toughness, impact resistance is not a serious problem with properly fabricated PTFE components. Tests at −20°C with a notch tip radius of 0·010 in gave impact strength values of approximately 4 ft-lb/in² over a range of grades. Addition of fillers appeared to have very little effect on this behaviour.

*Friction*

A number of workers have examined the frictional properties of PTFE, presenting a variety of data in the literature. However, there is general agreement that the coefficient of friction of this plastic in contact with a variety of other materials can be extremely low: values of 0·04–0·05 are frequently quoted. It also seems certain that coefficients of friction some 3–5 times higher than this can be obtained in other circumstances. For example, recent work at 20°C[1] has shown that for a 1 cm diameter steel indenter loaded with a 1 kg weight sliding over a plate of PTFE the coefficient of friction increases from 0·045±0·01 at 0·1 cm/s to 0·17±0·015 at 10 cm/s. At lower speeds (down to 0·01 cm/s) and higher speeds (up to 100 cm/s) there is little further change. This finding agrees with the work of McLaren and Tabor[2] and Ludema and Tabor[3].

*Electrical Properties*

PTFE is a non-polar material and the dielectric losses are sufficiently low for the permittivity to be independent of frequency. At a density of 2·174 g/cm$^3$ the permittivity is 2·05. From studies of the effect of changes of crystallinity by quenching and slow cooling it has been shown that any loss tangent peaks present are attributable to amorphous regions in the polymer: appropriate data are given in Figure 14.5. However, the presence of polar impurities can lead to much higher loss values.

Steady state conduction does not occur in PTFE except at temperatures above about 160°C. The apparent volume resistivity can be judged from Figure 14.6 of apparent volume resistivity vs. time of electrification for various temperatures; also on this figure are contours of constant loss angle.

PTFE is a non-tracking material but fails at high voltages, that is in the presence of surface discharges, by erosion. Because tracking does not occur PTFE has useful surface characteristics for exploitation in outdoor applications. When used as bulk insulation high quality fabrication is required because electrical failure will occur only by discharge through internal voids.

*Optical Properties*

PTFE is not at all transparent and is in no sense a material for the optical industry. This is indicated by the following data for a cast sheet specimen of PTFE of thickness 0·048 mm illuminated at 5461 Å. The results are corrected for surface defects and reflection.

| | |
|---|---|
| Direct transmission factor | 68% |
| Direct transmission factor corrected to 1 mm thickness | ~0 |
| Scattering coefficient | 80 |

*Refraction*

The following data are for PTFE of density 2·1516 g/cm$^3$.

| | | |
|---|---|---|
| Refractive index | $n_f$ (4861 Å) | 1·385 |
| | $n_D$ (5893 Å) | 1·325 |
| | $n_c$ (6563 Å) | 1·300 |
| Critical angle, | $a_D$, (Na, 5893 Å) | 52° 58′ |
| Constringence | | 3·8 |

*Chemical Properties*

PTFE suffers chemical attack only from the alkali metals, from elementary fluorine under certain conditions, and from a few fluorinated compounds at high temperatures. It is not dissolved

[1] E. C. Clark, I.C.I. Plastics Division, unpublished work.
[2] K. G. McLaren and D. Tabor, *Nature*, **197**, pp. 856–9, 1963.
[3] K. C. Ludema and D. Tabor, *Wear*, **9**, p. 329, 1066.

or swollen by any solvent within its normal working temperature range.

| | |
|---|---|
| Gas permeabilities: | ($cm^3$ (NTP) $cm/cm^2$ s cm Hg) $\times 10^9$ |
| Oxygen | 1·0 |
| Nitrogen | 0·35 |
| Water vapour permeability: | extremely low |

## APPLICATIONS

PTFE is a relatively expensive material; it has, for a plastic, a high density; and it has to be fabricated by complex and often slow techniques. There is therefore seldom any question of competition between PTFE and other plastics for a particular application. PTFE is used whenever its unique combination of properties offers the best performance at the lowest overall cost, i.e. a cost that takes into account the likely savings in maintenance and replacement charges. In using PTFE, the aim should be to use the minimum amount consistent with adequate functioning of the component.

Among the industries using PTFE are the chemical, electrical, electronic, aircraft, bakery and confectionery industries.

### *Typical examples*

Gaskets, diaphragms, valves, O-rings, seals, bellows and couplings.

PTFE-lined pipe and flexible hose, stopcocks.

Insulating tapes and sheets; sleevings and hermetic seals for condensers and transformers.

Dielectric and insulator in high temperature cables; spacers and connectors in high frequency cables.

Printed circuits.

Pipe-thread seals.

Dry and self-lubricating bearings.

Coverings for rollers handling sticky materials.

Linings for hoppers, chutes and guides.

Coatings for cooking utensils—frying pans, cake tins, etc.

LIST OF FIGURES

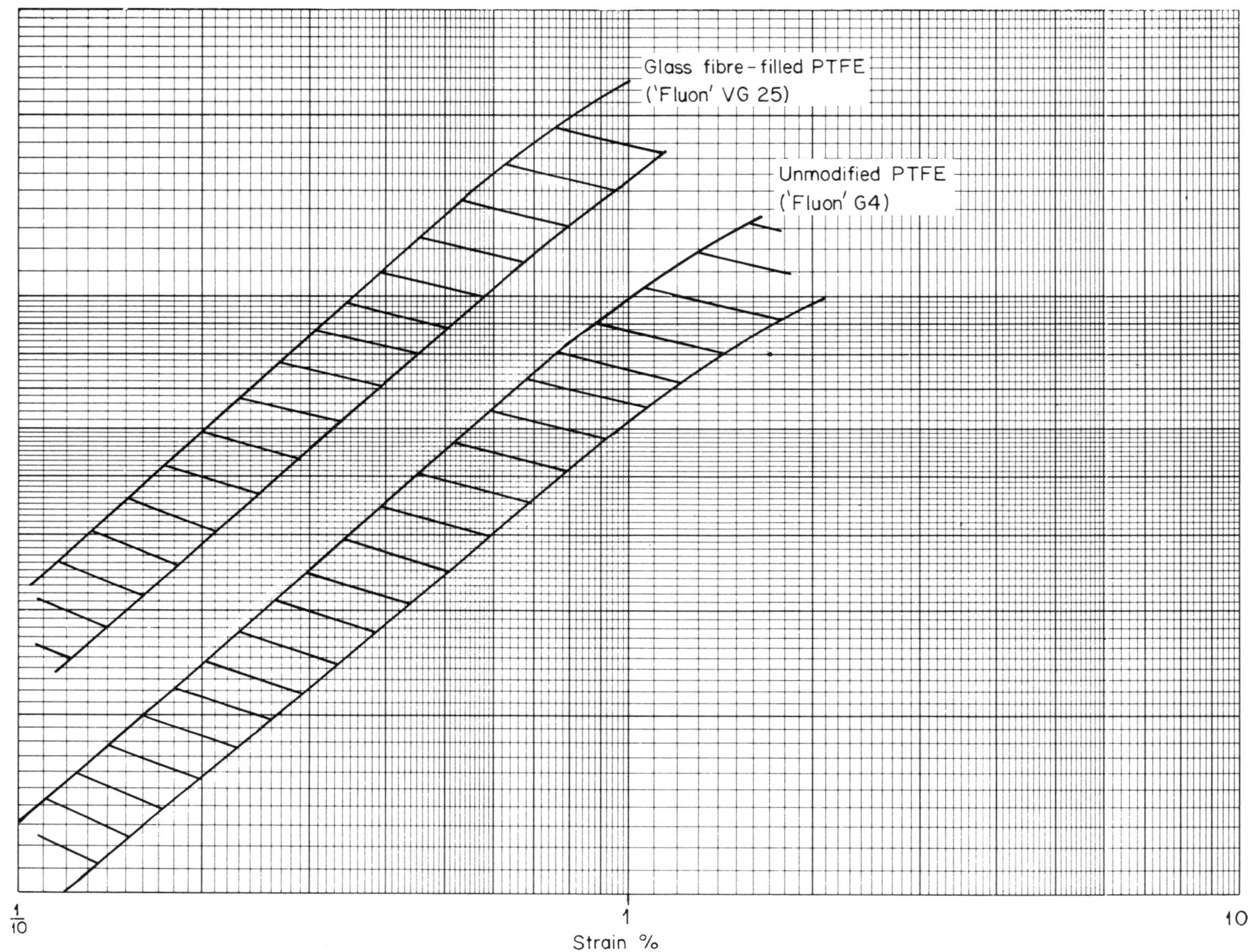


**Figure 14.1.** Isochronous stress vs strain curves in compression: 20°C. Effect of glass fibre filling (25% w/w) ('Fluon' G4, VG25)

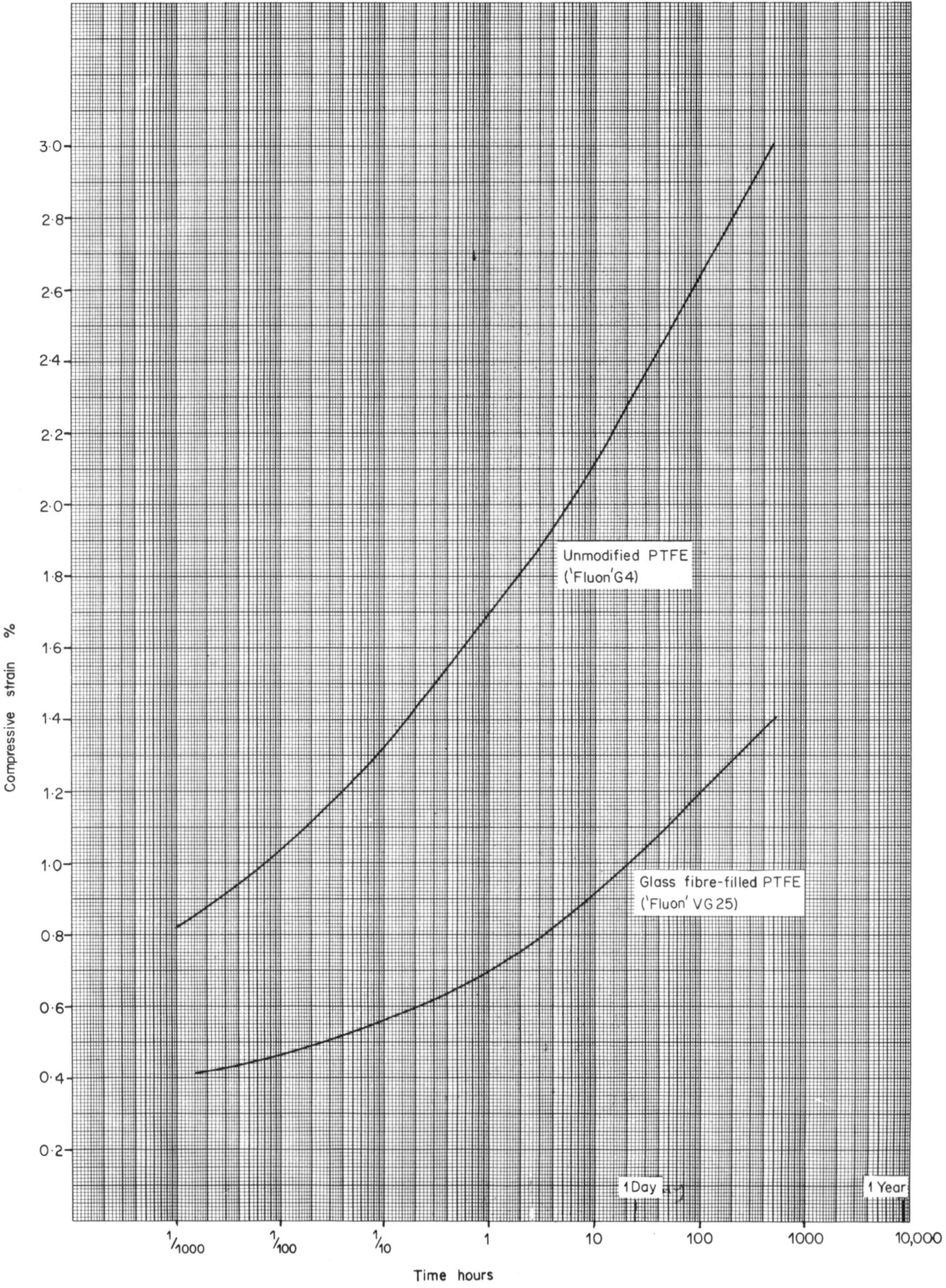


**Figure 14.2.** Creep curves in compression: 20°C, 1000 lb/in$^2$. Unmodified and 25% w/w glass fibre-filled PTFE ('Fluon' G4, VG25)

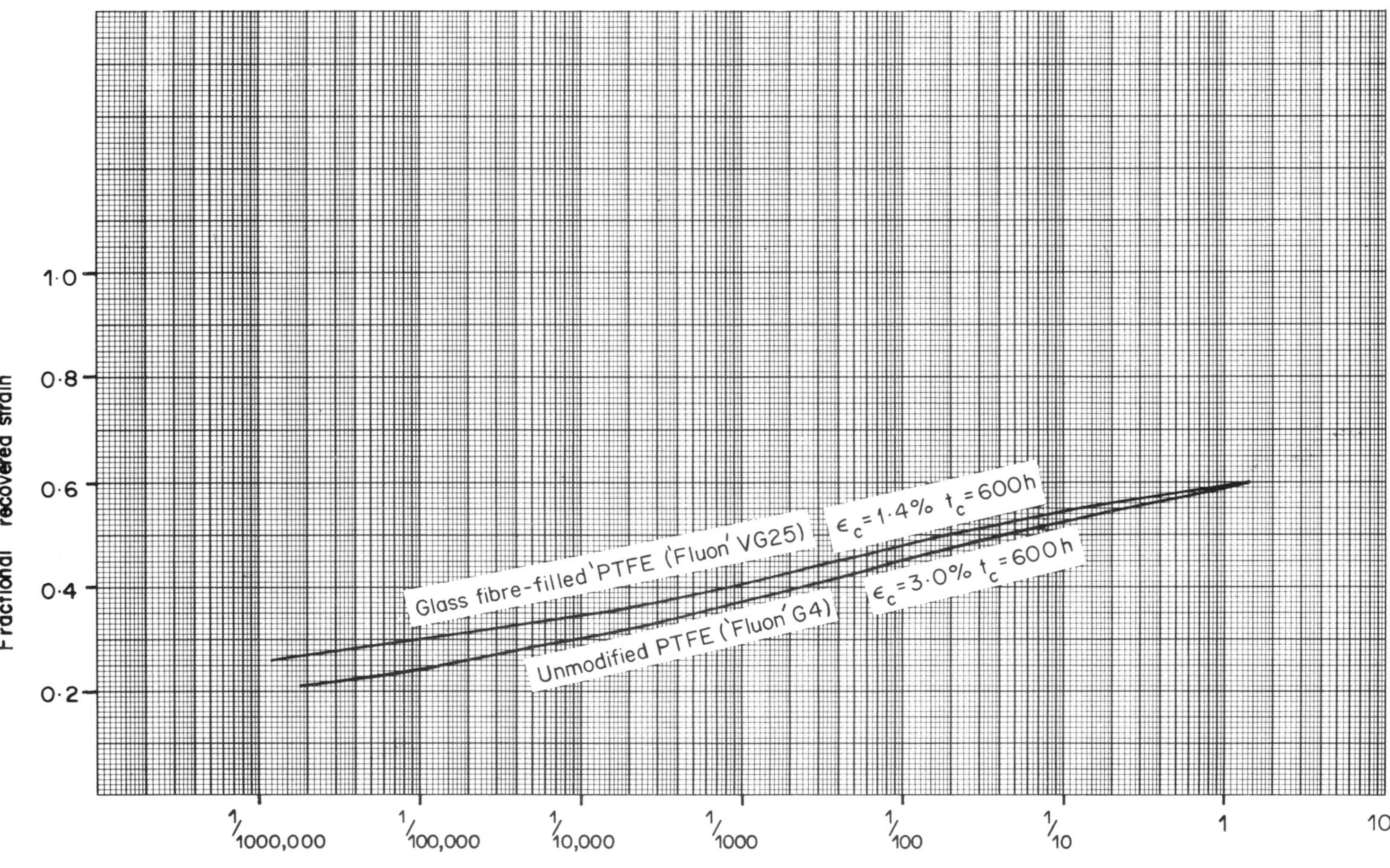


**Figure 14.3.** Recovery from creep in compression: 20°C. Unmodified and 25% w/w glass fibre-filled PTFE ('Fluon' G4, VG25)

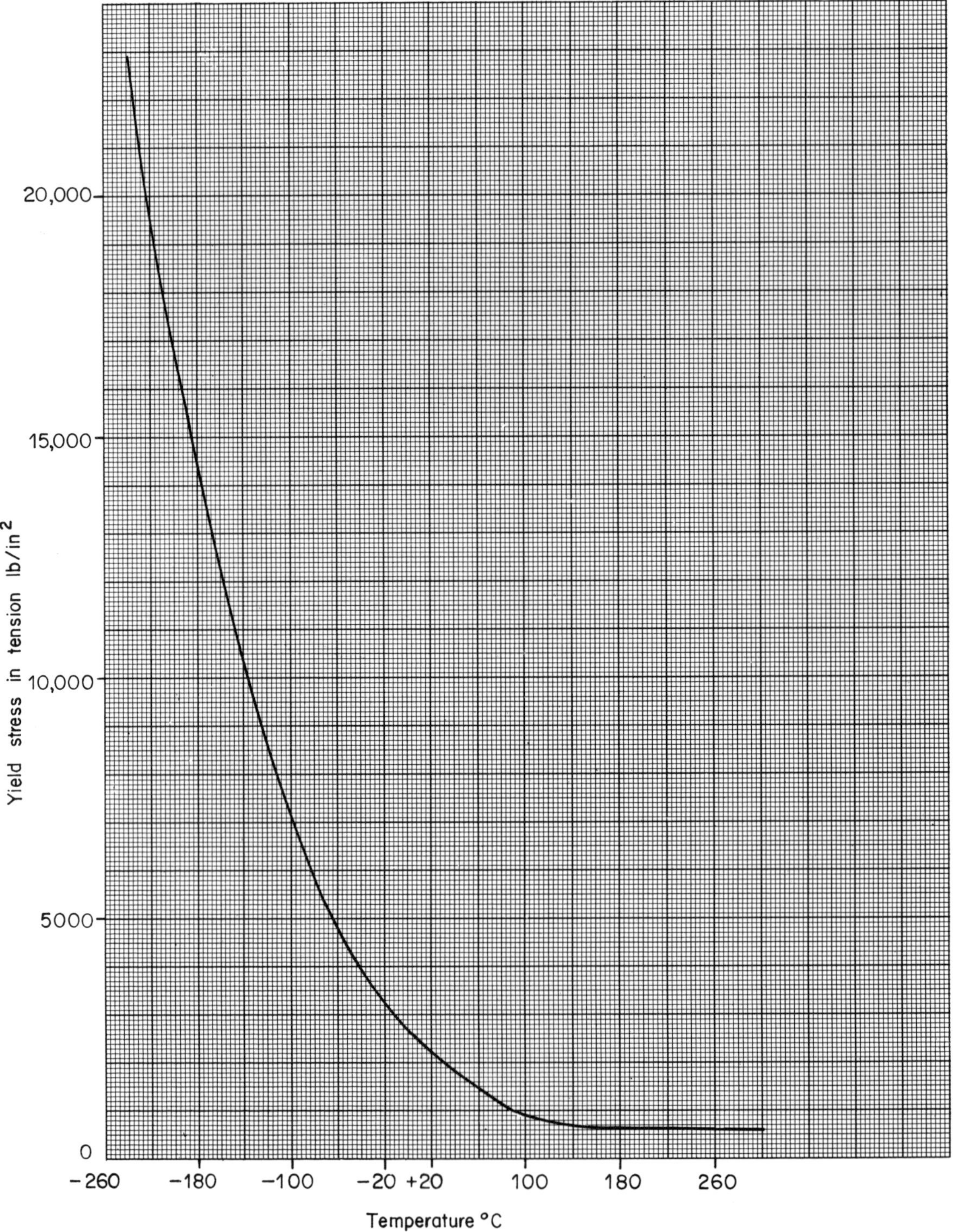


**Figure 14.4.** Yield stress in tension vs temperature: 50% per min straining rate. Unmodified PTFE ('Fluon' G4)

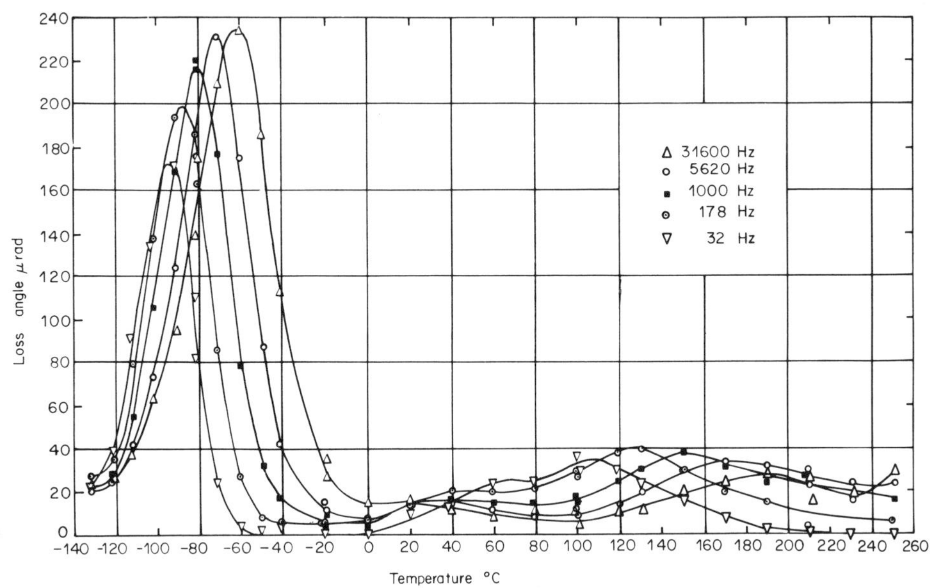


**Figure 14.5.** Loss angle vs temperature: 32 Hz to 31·6 kHz. Unmodified PTFE ('Fluon' G1)

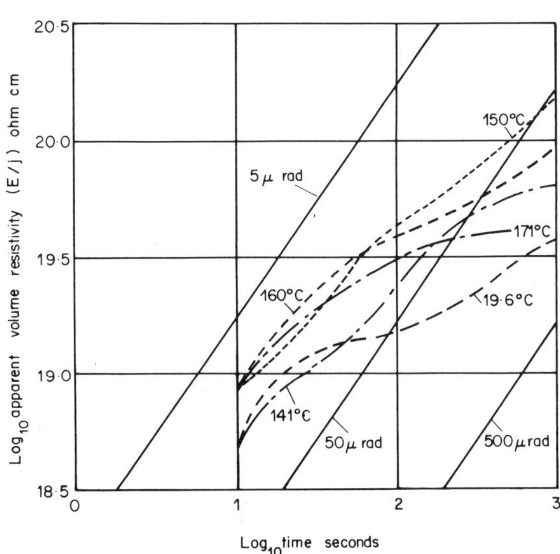


**Figure 14.6.** Apparent volume resistivity vs time of electrification. Effect of temperature; also showing contours of constant loss angle. Unmodified PTFE ('Fluon' G1)

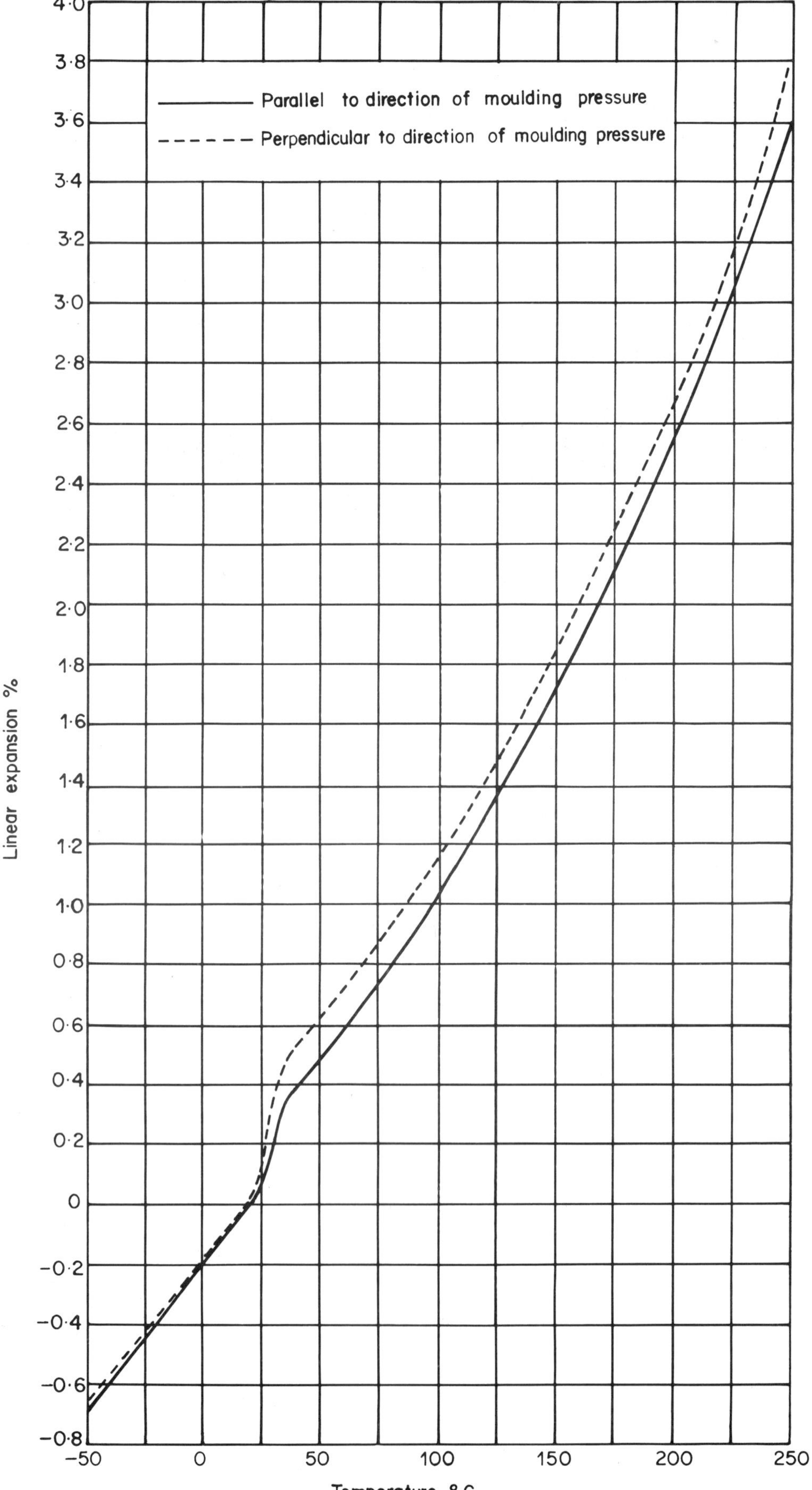


**Figure 14.7.** Linear thermal expansion vs temperature. Unmodified PTFE ('Fluon')

# Part IV

## DESIGNING OF PLASTICS COMPONENTS

# 15

# ECONOMICS AND DESIGN

The task confronting every designer is to design a component which will meet the functional requirements and which can be produced at an acceptable cost.

The emphasis on cost does not mean, of course, that the performance of products should be sacrificed for the sake of reducing costs, although, unfortunately, this has happened on occasions. Obvious examples were some of the first cheap plastics toys, which were poorly designed, often made in an entirely unsuitable plastics material, and which lasted only a few hours. Nowadays, however, considerably more is known about plastics, a wider range of plastics is available, and it is possible to select plastics materials to do certain jobs more efficiently than any other material, at the same time achieving considerable cost savings.

There are four principal ways in which plastics can contribute to lower costs. The first of these is lower raw material cost. Because the prices of the various plastics materials are different, the choice of raw materials can affect considerably the cost of a component. Therefore a valid comparison of material costs is essential and this is best made on the basis of volume costs. For example, if nylon costs 66d per lb and 7 gear wheels can be moulded from one pound, the material cost of each gear is about 9½d. From one pound of brass, one gear wheel of the same size as the nylon ones can be made and, assuming the price of brass to be 13d per lb, the material cost of this gear is 13d, i.e., about one and a half times that of a nylon gear. In other words, the lower density of nylon more than compensates for the higher selling price on a weight basis.

The lower volume costs of certain plastics compared with those of some metals is often a telling factor in the selection of materials. Thus, plastics such as nylon and acetals have replaced brass in many applications, particularly those requiring good corrosion resistance: the solenoid valve shown in Plate I is an excellent example; the cost saving in the manufacture of this component was about 15%.

In comparing raw material costs, it is important to bear in mind also the variation in properties other than density among plastics materials. For example, if high rigidity is required in a component, the plastics material which has the highest modulus is not necessarily the best choice. The modulus at room temperature of polypropylene is only about half that of dry nylon 66, and therefore, to obtain equal apparent stiffness, the thickness of a polypropylene section would have to be about 26% greater than that of the corresponding section in nylon. When correction is made for the different densities, the weights of polypropylene and nylon required are, for all practical purposes, identical. At current prices, polypropylene would be markedly less expensive in effective material cost than nylon. The 26% increase in section thickness would result in slower mould cycling, but the piece cost would be likely to remain lower in polypropylene.

The second area of cost saving in which plastics can offer considerable advantages is that of fabrication, and it is here that very often the biggest savings can be made. As was seen in Chapter 3, the processes by which thermoplastics can be shaped are extremely versatile, and many features

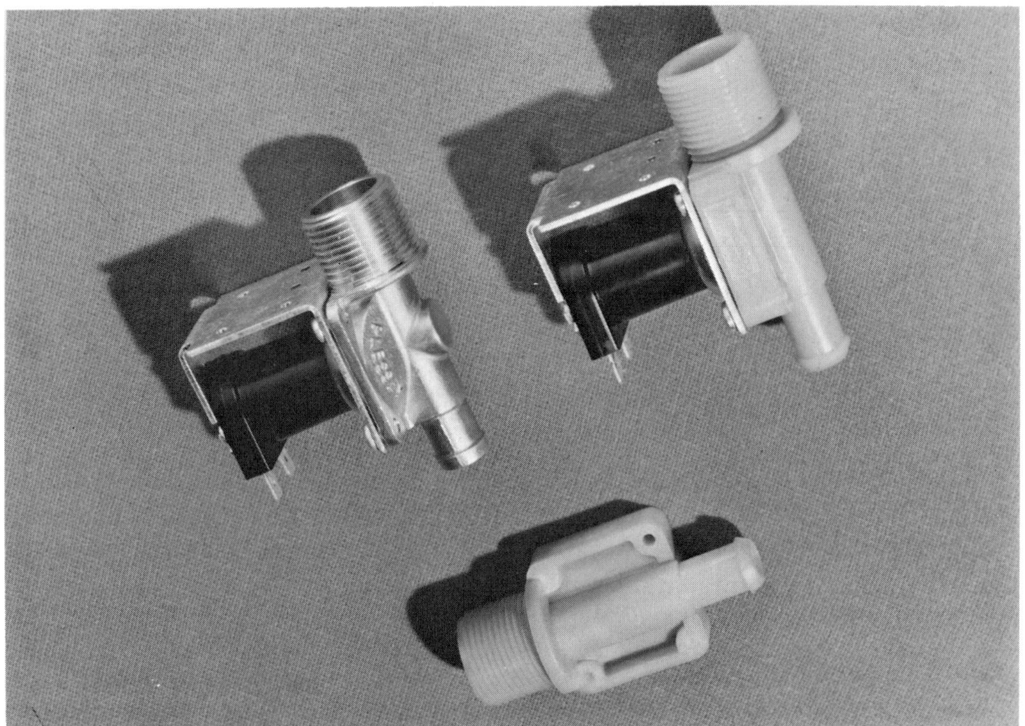

**Plate 1.** *Right*: body of solenoid-operated water valve for washing machine, moulded in glass-filled nylon. *Left*: superseded machined brass component. (Moulded by: Fraser & Glass Limited. Valve by: The Plessey Co. Ltd., Wound Components Division. Machine: Frigidaire Jetamatic)

can be 'built in' to the mould or die so that components can be simplified, and assembly and finishing operations eliminated. Often several metal parts can be replaced by one plastics part. The thirty-hour clock movement shown in Plate 2 used to contain 66 metal components. When a completely new design in plastics was undertaken, the number of components was reduced to 43. The polypropylene accelerator pedal (Plate 3) is an excellent example of the way in which the integrally moulded hinge described in Chapter 7 can be used to simplify and effect economies in fabrication techniques. Similarly, the acetal ball-valve components in Plate 4 show how functional features such as threads can be moulded in; when made in brass, these threads have to be machined.

Decorating is another finishing operation, often necessary with metals, which can be eliminated. Most plastics are available in a wide range of colours, and surface effects and textures can readily be obtained as an integral part of the processing operation. The polypropylene blow moulding which forms the console for the Vauxhall Ventora has a grained appearance which is a faithful reproduction of the mould surface (Plate 5).

Another great processing advantage of plastics is that by their very nature they can offer a wider choice of conversion techniques than is usually available for traditional materials. For example, a dustbin may be made from polythene by injection moulding, blow moulding or rotational moulding; a trough shaped cover for a lighting fitting may be shaped from sheet, extruded or injection moulded; and a car trim panel may be vacuum formed or injection moulded.

**Plate 2.** Thirty-hour mechanical clock movement containing many train components such as gears and shafts in acetal copolymer. (Smiths Industries, Clock and Watch Division)

Frequently the choice of process will be determined by the scale and rate of production required and by the size of the component; only rarely need it be determined by the design of the product because design can usually be suited to a process without any significant effect upon the performance of the part in service.

When a process has been selected, variations in procedure are often possible. If a large number of injection moulded parts are required, there may be a choice between using one large moulding machine with a mould having many cavities or a number of much smaller machines with moulds

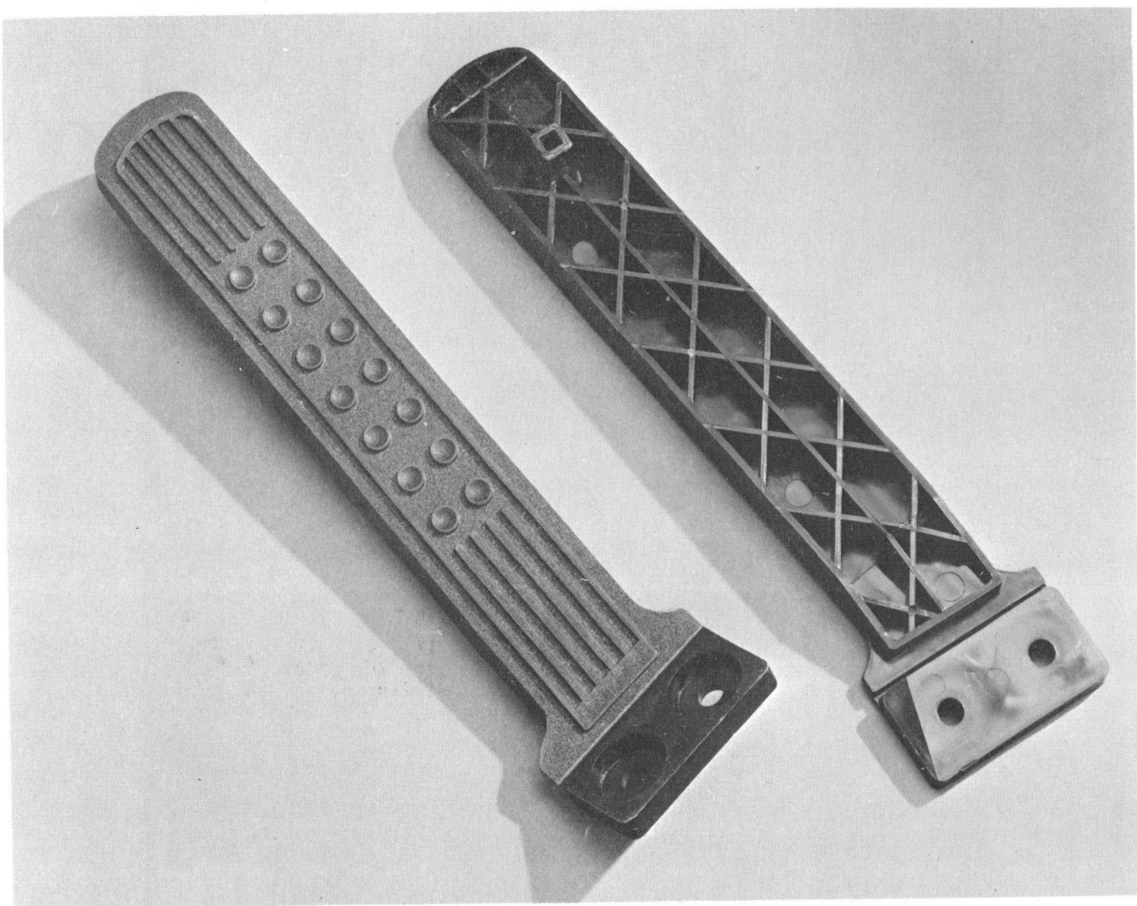

**Plate 3.** Polypropylene accelerator pedal with integrally moulded hinge. (Moulded by Fraser and Glass Limited and Industrial Mouldings (Warwick) Ltd.)

having a few cavities or even only a single cavity. The failure of a single large machine may cause some subsequent assembly operation to be closed down because of lack of parts; the failure of one or two machines out of a large number of small units, would usually permit the 'customer line' to continue, although at a reduced output. The likely higher capital cost of the number of smaller machines may be offset by the lower risk of lost production in the later stages of a manufacturing process. With vacuum forming there is a choice between a fully integrated line consisting of extruder, vacuum forming machine and trimming plant with automatic refeed of trimmings back to the extruder, and a more simple system of an extruder and a separate vacuum forming machine with a higher labour requirement. These are only two examples of possible diversifications.

Large injection moulding machines are very costly but the relationship between shot capacity and cost is not linear. Pro rata, the cost per unit volume of capacity decreases as the total capacity increases; a line of ten 2 oz machines will cost substantially more than one 20 oz machine, and ten single-cavity moulds for the small machines will be more expensive than one 10-cavity mould for the large one. However, usually production will be more quickly established with the small

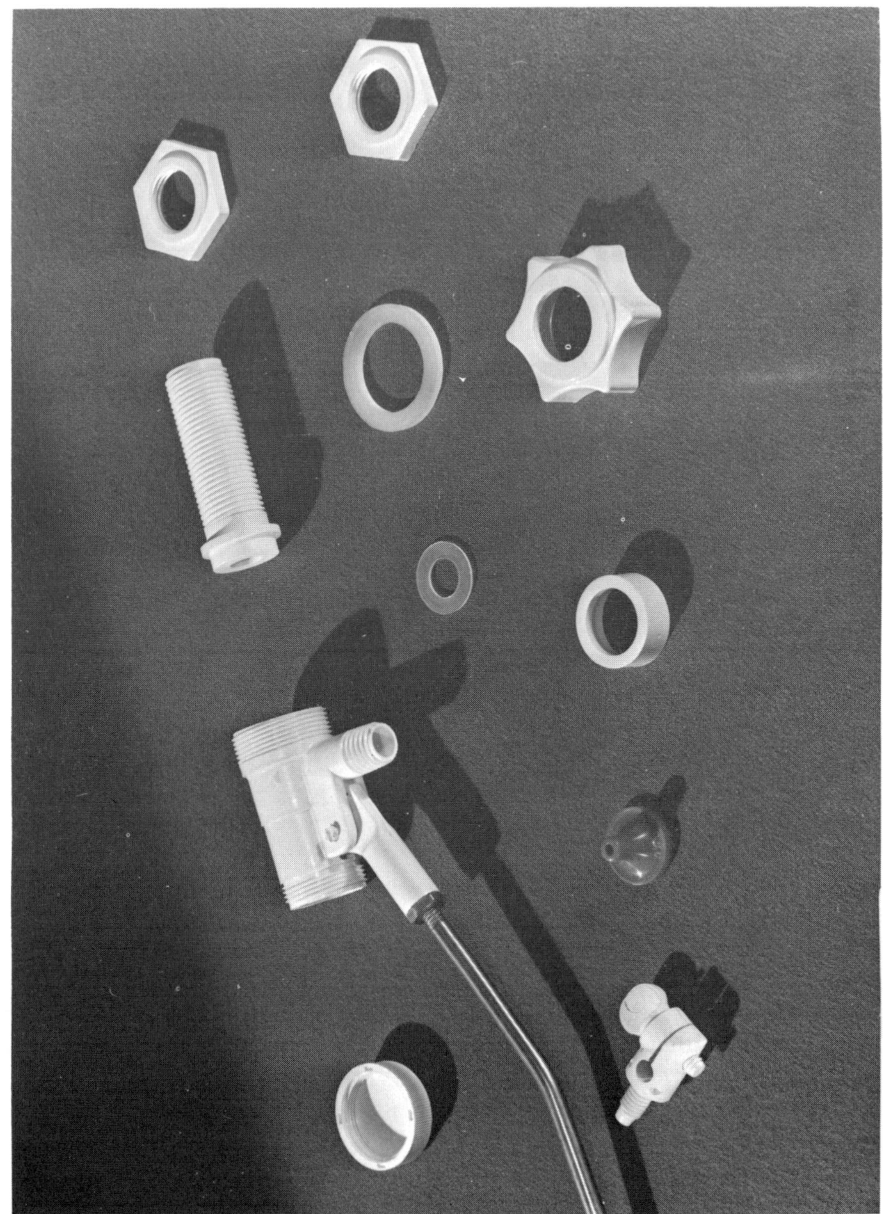

**Plate 4.** Ball-valve components moulded from acetal copolymer and marketed by Valor Cisterns

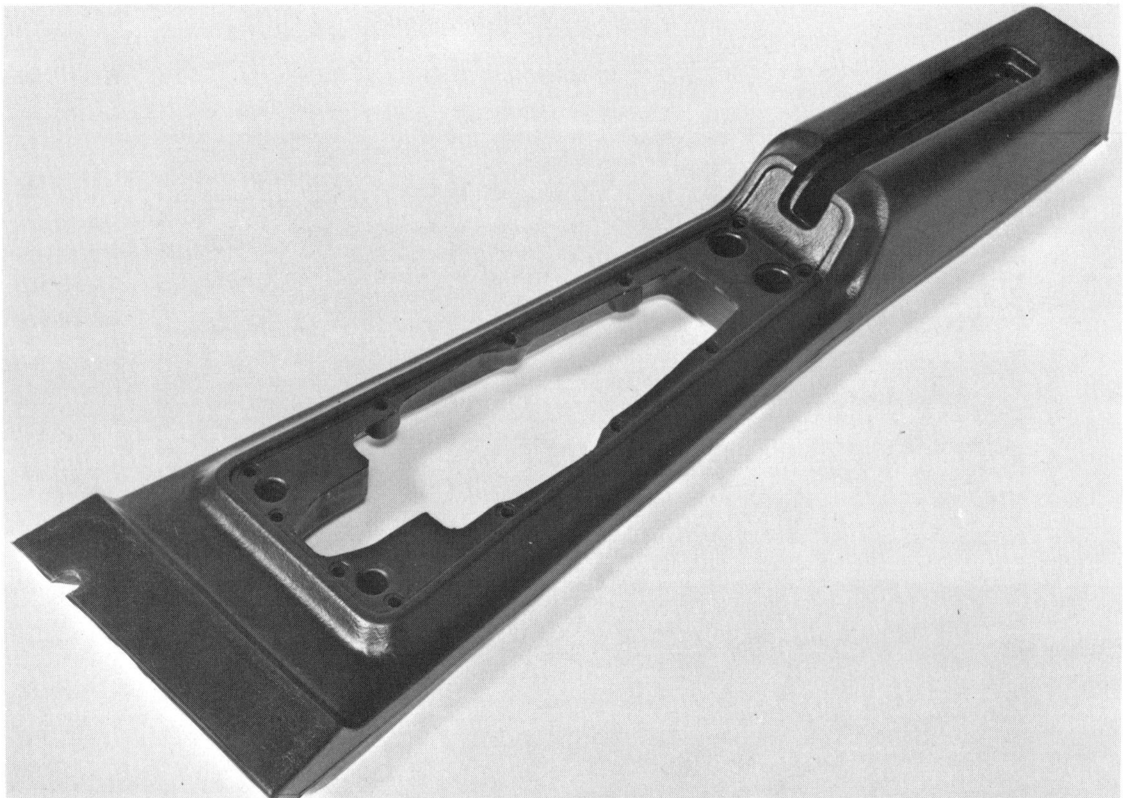

**Plate 5.** Polypropylene console for the Vauxhall Ventora (Blow moulded by Willamot Industrial Mouldings Ltd.)

moulds. With the multicavity mould, much more complicated constructions will be necessary and delays may be caused by the need to balance flow of the melt to all the cavities; additionally, maintenance costs will be higher. Labour costs may not be greatly different because small machines can often be operated fully automatically at very fast cycles with a minimum of supervision.

Injection moulding tools are expensive, particularly those required for long runs of precision mouldings, when high quality steels have to be used. Tools for blow moulding are much cheaper; because of the low pressures used in blow moulding, tools cast in zinc alloy are usually adequate. Tools for rotational moulding are usually cheaper still. For sheet shaping, wooden or phenolic laminate tools are satisfactory for the production of small quantities of items; for longer runs, aluminium castings may be used.

From considerations such as those outlined above it is not too difficult to form an idea as to the relative costs of fabricating an item in various ways. A more accurate and complete picture can be obtained only after detailed study of all aspects of the chosen material and conversion process, a study which should preferably be done in conjunction with the raw material supplier and converter.

Handling, transport and installation is the third area where the use of plastics can lead to considerable cost savings, largely as a result of their light weight. An interesting example of this is

**Plate 6.** Ten-inch PVC mains water pipe. (Made by Yorkshire Imperial Plastics Limited for Manchester Corporation Waterworks)

provided by a scheme carried out by Manchester Corporation Waterworks, in which 6 in cast iron water mains were replaced by 3000 yards of 10 in rigid PVC pipe. The pipes carry water from Bleawater, a small mountain lake, to the Manchester main. Because the pack-horse track which the corporation had had specially made to drag up the 6 in iron pipe in 1938 could not have been used to carry up 10 in iron pipe, it was decided to use PVC pipe, which is much more easily handled (Plate 6).

Fourthly, plastics can contribute to low 'in-service' costs of components. This is a factor which is quite often forgotten or ignored, but nevertheless it can be significant and may indeed be decisive in whether a plastics component will be cheaper in the long run. Probably the most important property of plastics here is their corrosion resistance. This is of particular value in chemical plant and in any equipment which is exposed to the elements. Plate 7 shows an agricultural seed drill made in Finland in which no fewer than 203 components are made from nylon. Compared with the cast iron parts they replaced, the nylon parts are about 100 lb lighter in weight, more wear resistant, unaffected by corrosive chemicals, and need no maintenance of any kind.

The costs of maintenance, repair and replacement have, of course, to be balanced against the desired service life and the likely service life. An extreme example illustrates the importance of these factors. In a certain chemical installation, expansion joints in a long pipe system regularly

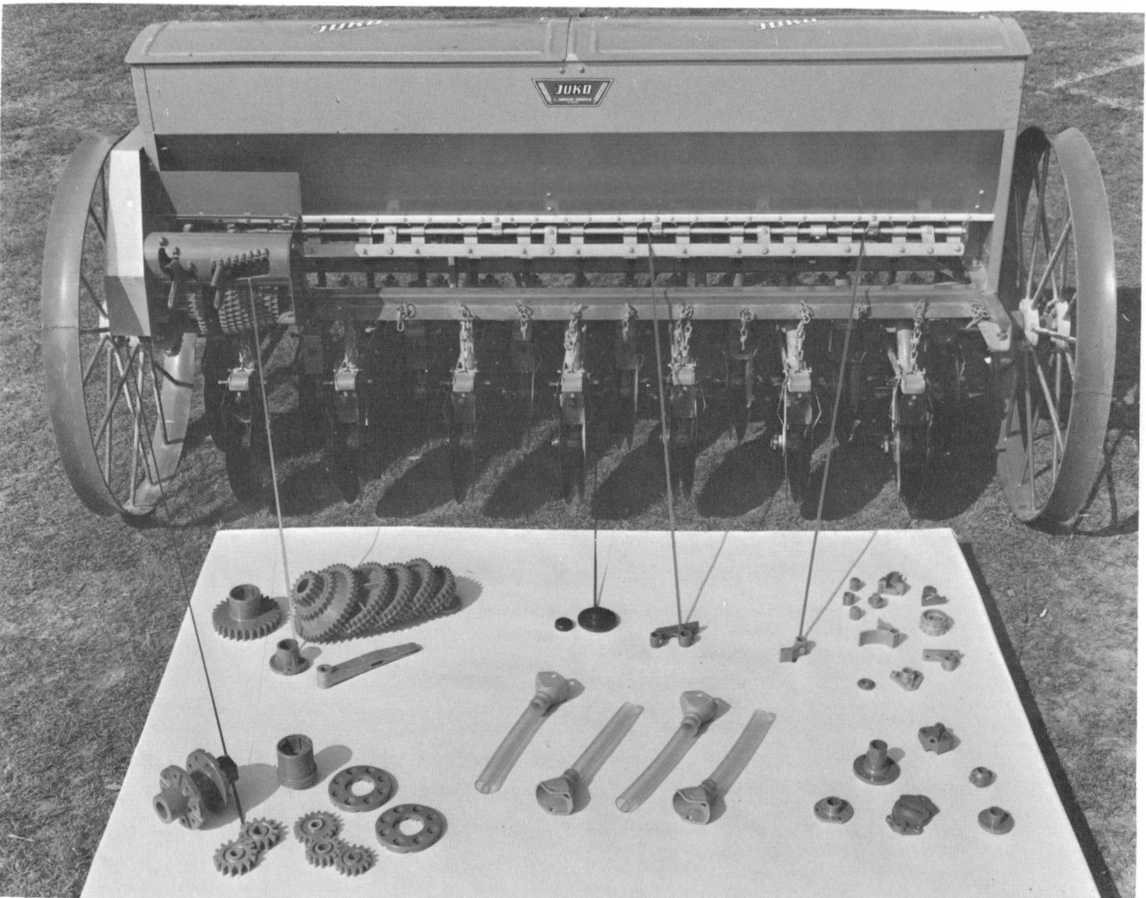


**Plate 7.** Two hundred and three components of this seed drill are made from nylon. The weight saving over cast iron amounted to about 100 lb. ('Juko' seed drill by Junnilan Konepaja, Finland)

failed after only a few weeks of service; it was then necessary to close down the plant while the joints were repaired. Not only were the continual labour and raw material costs excessive, but there was loss of production during the periods of closing down and of starting up, as well as during the actual shut-down period. PTFE was considered for the expansion joints (Plate 8) because it was clear that its technical merits were such that it would stand up for very much longer periods to the conditions of service and, although the PTFE joints cost 15 times as much as those already in use, they were put into service. When the plant had run for over six months without a single stoppage they had already paid for themselves.

It must not be thought, however, that the advantages of plastics are to be obtained solely from cost savings. Many examples can be given of plastics components which surpass in performance and appearance the components which they have replaced.

Light weight has already been mentioned as a factor which can lead to cost savings, but it can lead also to improved quality. For example, gravity conveyor rollers made of nylon have a low mass and hence a low inertia. The racks in which they are supported can therefore be tilted at a lower angle than that required for efficient operation of metal rollers, and thus less space is required. Plate 9 shows a 'live storage' unit consisting of several inclined racks. Packages loaded at the back of the racks move down towards the front where they are removed by operators

**Plate 8**. Expansion bellows made from PTFE (Makers: Crane Packing Limited)

**Plate 9.** Storage racking system incorporating nylon wheels and acetal snap bearing straps. (System designed by Atlantic Conveying Equipment Ltd. in association with Gallaher Limited)

making up mixed orders. The use of nylon rollers in this installation enabled small, very light packages to be handled satisfactorily.

Moving parts can also be considerably quieter in use when made from plastics, a particularly important consideration in the design of gears and bearings. Plate 10 shows a cage moulded in acetal copolymer for an angular contact ball bearing. Compared with the original brass cage, the new one affords a quieter running bearing, as well as greater efficiency in the distribution of lubricant to the rolling elements and the elimination of machining operations; friction is minimized and abrasion resistance is high.

All plastics have low coefficients of friction, but PTFE is outstanding in this respect. Excellent bearing materials can be made by combining PTFE and other materials in some way; an example of their use as expansion bearings is shown in Plate 11.

Corrosion resistance leading to maintenance-free operation is another important property of plastics materials. Working tops and floor coverings are established applications for plastics, but there is growing use of panelling and cladding. A civil engineering application is shown in Plate 12, the cladding of the walls of the Clyde tunnel with rigid PVC sheet to present an impervious, easily cleaned, continuous surface. Resistance to chemical attack was also a factor in the selection of polypropylene for the one-piece moulding forming the twin tubs of a particular type of washing machine (Plate 13).

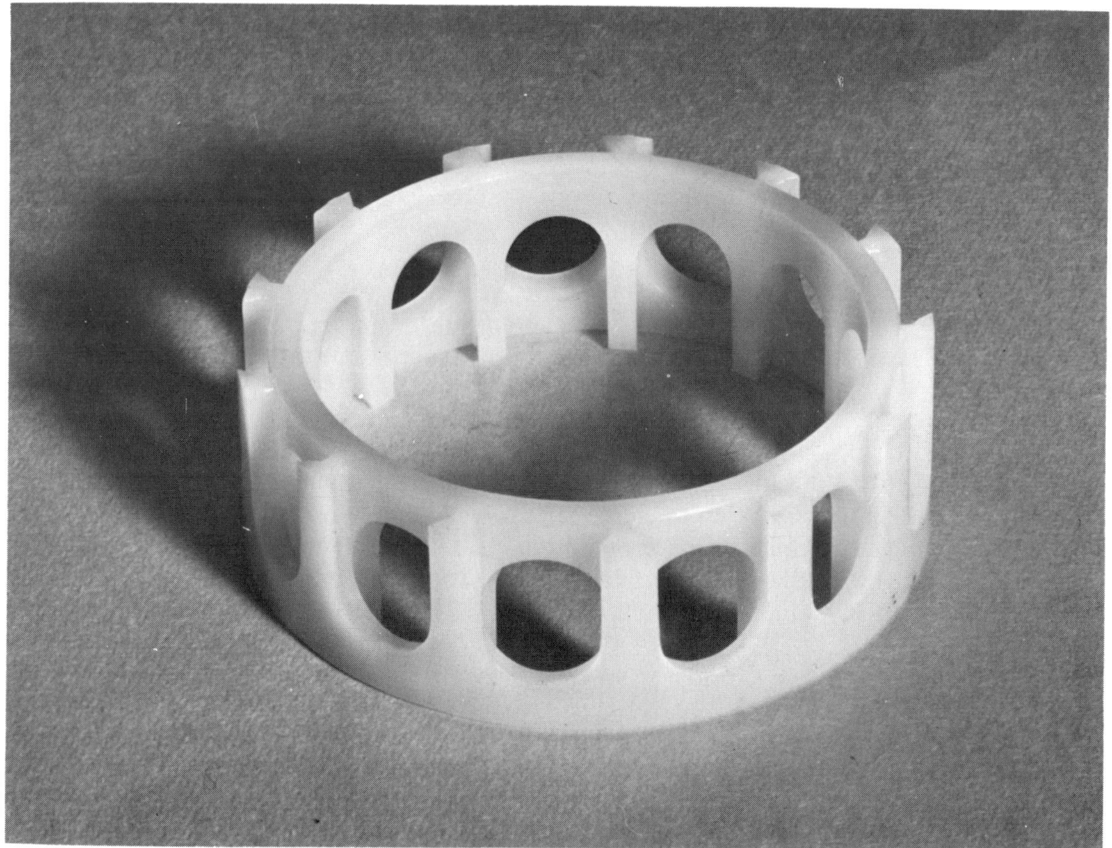

**Plate 10.** Retainer for an angular contact ball bearing moulded from acetal copolymer. Patent No. 28254/66. (Pollard Bearings Limited)

**Plate 11.** 'Tetron' bridge bearings with the stainless steel plates removed to show the PTFE bearing surface. (Made by P.S.C. Equipment Limited. PTFE Sheet made by Dalau Specialised Plastics Limited)

The ability of plastics to withstand and recover from fairly large strains is an advantage which can allow the use of designs which would be impossible in other materials. The spring-action hinge shown in Plate 14 could be made only from a plastics material because it relies for its successful functioning on the resilience and flex life of a relatively thin section.

**Plate 12.** PVC sheet used as cladding for the walls of the Clyde Tunnel. (Constructed for: The Corporation of the City of Glasgow. Consulting Engineers: Sir William Halcrow & Partners. Main Contractors: Charles Brand & Son Limited. Sub-contractors for the wall cladding: Mellowes & Co. Ltd.)

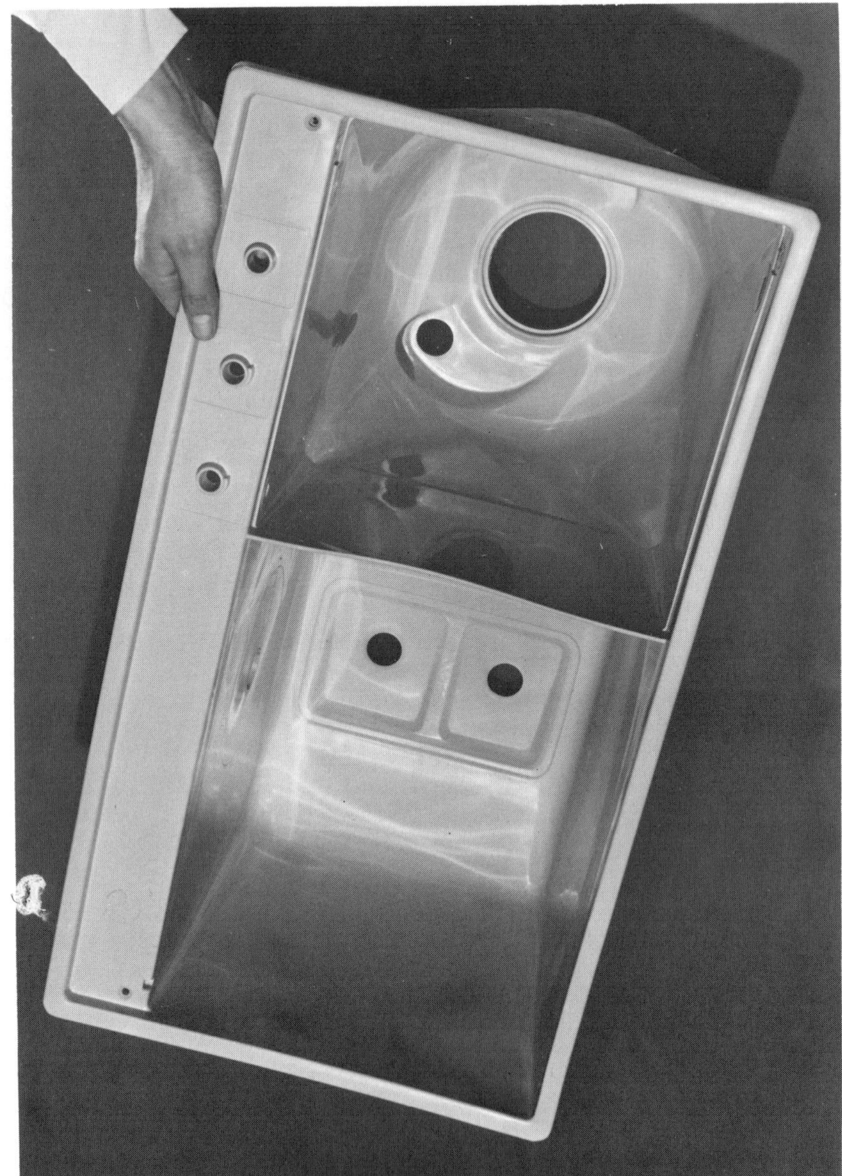

**Plate 13.** One-piece moulded polypropylene liner for a washing machine. (Moulded by Ets. Allibert, France for Soc. Ind. Hoover S.A.)

**Plate 14.** 'Springe' spring action hinge (Patent Number 1,013,106) moulded from acetal copolymer. (Designed by Schema Limited. Moulded by Tatra Plastics Limited)

## PRACTICAL DESIGN OF INJECTION MOULDED COMPONENTS

Although the design engineer can be involved with plastics in all their physical forms, it is perhaps with injection mouldings that he is most likely to be primarily concerned. It is, therefore, pertinent to discuss in detail some aspects of the practical design of injection mouldings. Before doing this, however, some general points must be made.

The cardinal rule when designing with plastics is to treat them as materials in their own right. In other words, they should not be substituted into designs originally conceived for, say, die-cast metals. The reasons for this are firstly, as this book has shown, that the properties of plastics are quite distinctive and very different from those of metals, and secondly, that it is seldom necessary for the plastic to match in properties the material it is replacing. Very often a metal article will not have been rationally designed at all and will be 'over-designed,' e.g. it will be stronger and more rigid than it needs to be. Also, the material, metal or otherwise, may not have been selected logically by a process of elimination, and may well possess properties not required at all, such as high temperature resistance, yet be deficient in some desirable features such as light weight, quietness in use and attractive finish—features that could be provided easily by a plastics material. What is important is that the properties of the material used should meet the essential requirements of the application.

An extension of this approach is that the properties of plastics must be considered as a whole against the conditions to be encountered during service. A good example is provided by the use of nylon for gear wheels. A preliminary inspection of the properties of nylon suggests that because of its relatively low rigidity and strength it would be unsuitable to take the place of materials such as zinc alloy or brass for gear wheels. However, it is the very fact that nylon has a low rigidity which makes it possible to use it in this application, in spite of its low strength. In a metal gear the rigidity of the metal imposes something approaching line contact between the meshing gear teeth. This small area of contact between the two wheels leads to a high pressure being developed and necessitates a material of sufficient hardness to resist this pressure, which would otherwise lead to permanent distortion of the gear. When using nylon gear wheels under the same total load conditions, the nylon wheel distorts somewhat, so that the tooth surface area increases and the pressure decreases. Furthermore, the bending of the teeth may result in more than one tooth being in mesh at any one time. This reduces the pressure loading on each tooth, and in suitable instances the tooth stresses can be brought well within the strength limitations of the nylon.

Another example again concerns the use of nylon. The drill housing shown in Plate 15 is moulded from glass-filled nylon, because this material provides the required mechanical strength and rigidity together with sufficient electrical insulation to afford complete safety to the drill user.

To derive full benefit from the injection moulding process, the designer must recognize its limitations as well as its advantages and appreciate the interplay between product design, tool design and the processing characteristics of the plastic selected. The greatest benefits are obtained when all these three aspects are considered together.

Discussion of impact strength in previous chapters has shown the need to avoid sharp changes of section and to use generously radiussed corners to obtain good shock resistance. There are, however, many other aspects of design which are not covered in the earlier sections, and which nevertheless will have a major influence upon the properties, appearance and cost of the final product. Some of these are discussed and illustrated in this section.

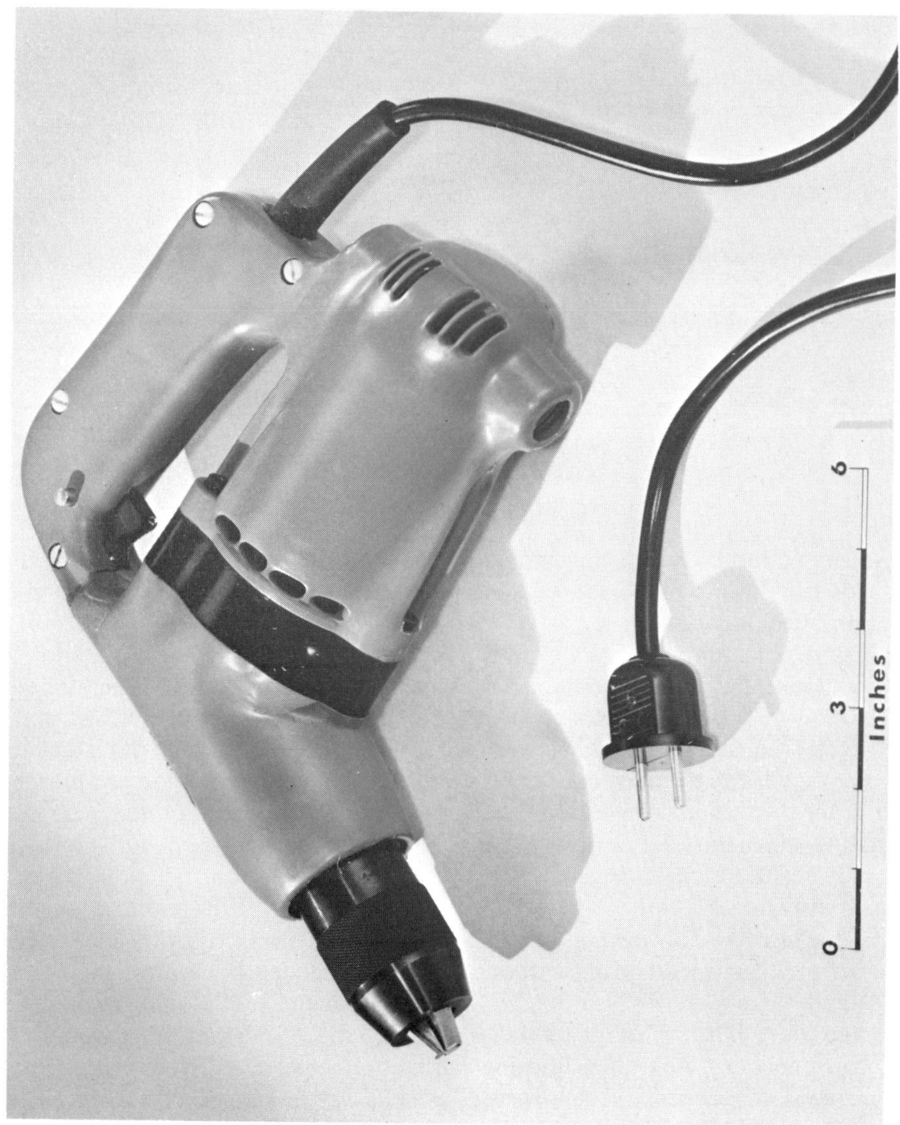


**Plate 15.** Housing for industrial drill moulded in glass-fibre-filled nylon 66. (Made by Star Utensili Elettrici S.p.A., Milan)

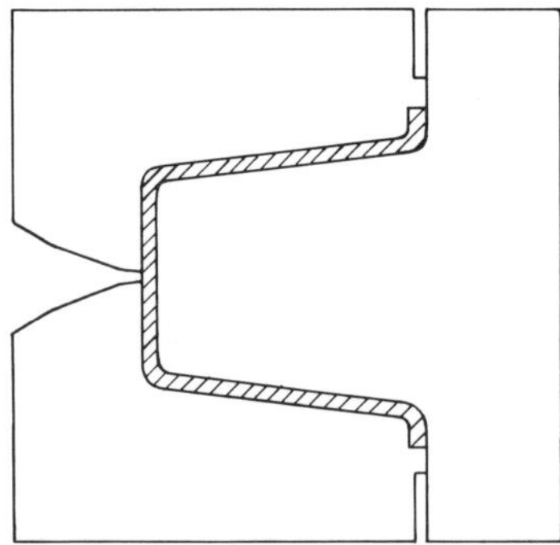

**Figure 15.1.** Simple injection moulding tool

## MOULD FEATURES

In its most simple form, an injection moulding tool consists of two parts which contain a cavity when closed and which are separated to permit removal of the moulding. This simple tool poses several problems for the designer.

The surface of the moulding will always carry a 'witness' of the closing line of the tool. If, for example, the component is a simple container, the closing line will be on the rim and the witness line will not be readily noticeable, provided that the tool has accurately mating surfaces and there is no 'flash', i.e. spewing out of material between the surfaces (Figure 15.1). On other articles, the tool closing line may leave a witness line which will mar the appearance of the product (Figure 15.2(a)) and hence there should be liaison between the component and tool designers in order that any such lines are in the least obvious position. Witness lines may be camouflaged by introducing a change in surface angle (Figure 15.2(b)), a styling groove or a styling bead along the affected line, as in Figure 15.2(c).

The tool, when closed, will always contain air which has to be displaced by the molten plastic. In the container tool described above, 'natural' venting of the air will occur at the rim as the mould is filled. If there were no venting, air entrapped in the tool would become heated adiabatically and cause 'burning' of the plastic. Figure 15.3 shows in B natural venting and in A a condition which may cause burning. Burning can be overcome in A if vents are let into the cavities at the apices of the cones; this is, however, an additional feature which will increase costs and which may leave slight discontinuities or flash marks on the surface. Wherever possible, both component and tool should be designed so that natural venting is achieved.

A simple 'open–shut' type of injection tool with a closing line normal to the direction in which the melt enters cannot produce components with undercuts. Because undercuts often appear to provide the most simple means of locating or assembling a component in service, this limitation is of major importance. The designer must ask himself whether undercuts are essential to the operation of the component. Frequently it will be found that the same results can be obtained by other means. However, if reentrant shapes are unavoidable, mechanisms such as collapsible cores and sliding side cheeks can be incorporated in the tool to permit withdrawal of undercuts

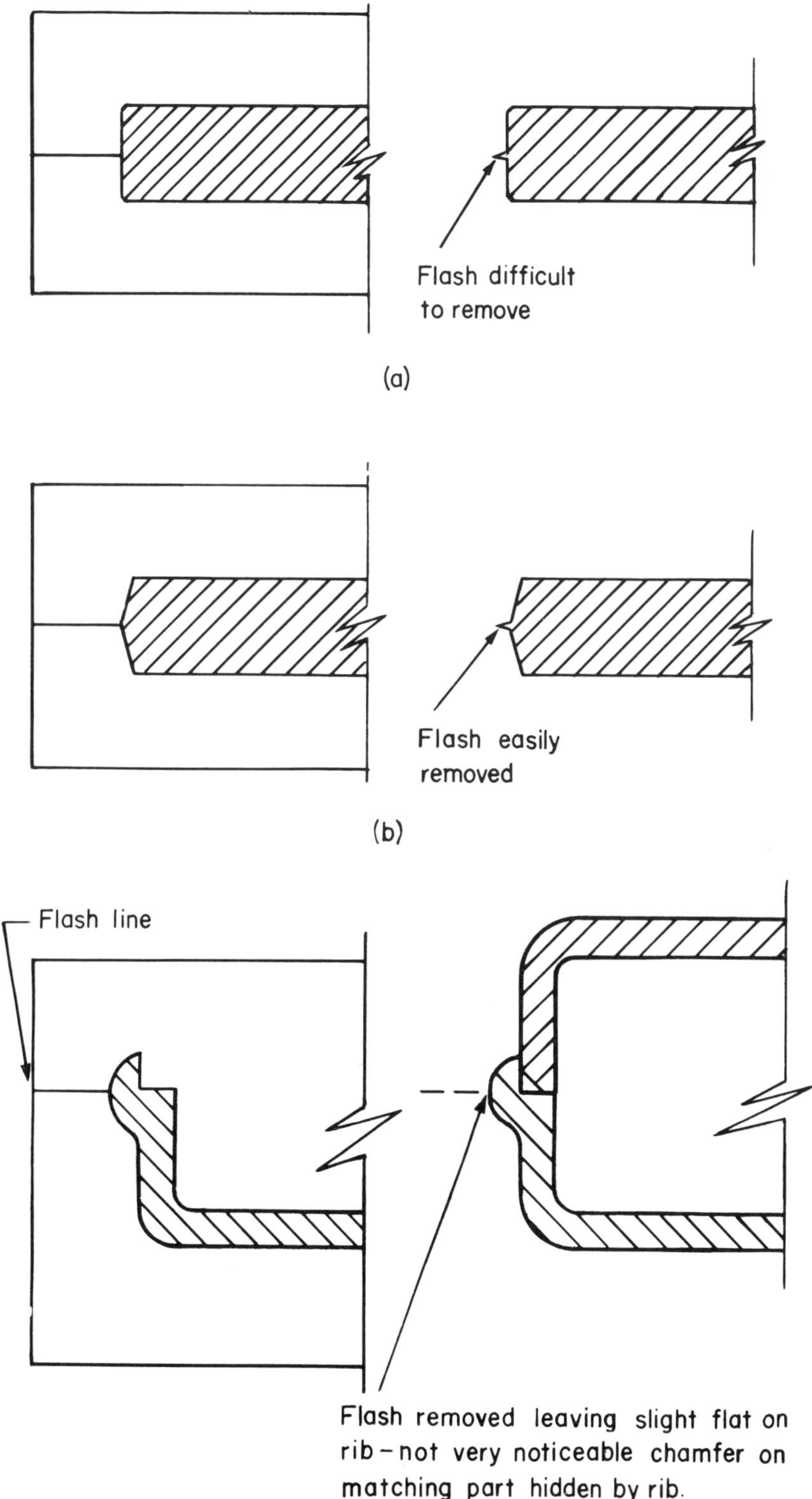


**Figure 15.2.** Witness lines

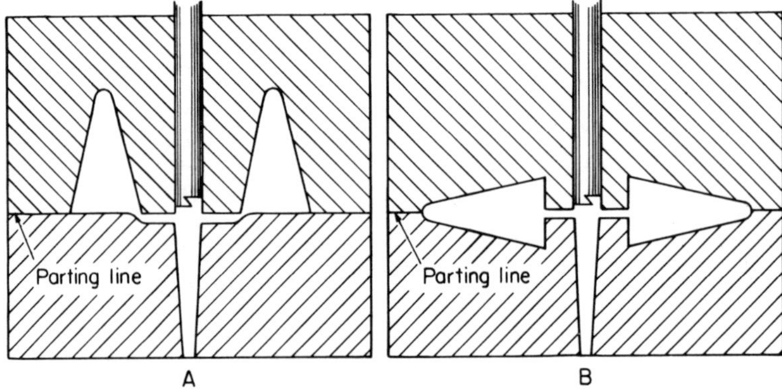


**Figure 15.3.** Arrangement of cavities in mould

(Figure 15.4(a)). Sometimes an undercut can be made extractable by having an inclined parting line between the two major parts of the tool (Figure 15.4(b)). These procedures, although technically well-proven, increase the cost of the tool and hence of the components made; wear of sliding mechanisms also may ultimately result in flash on the mouldings.

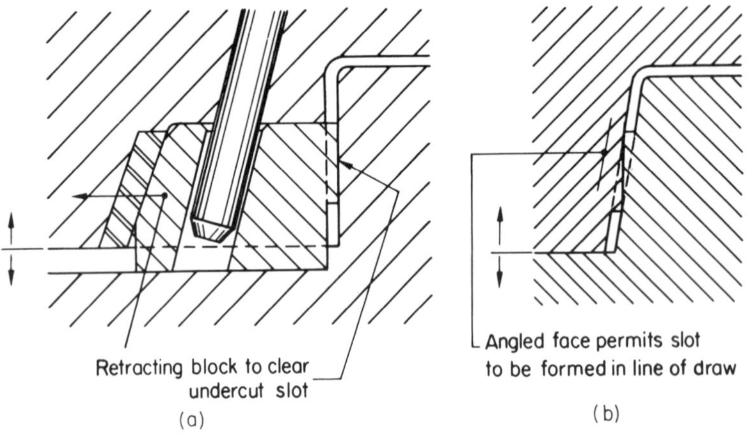


**Figure 15.4.** Moulding of undercuts

## ASSEMBLY OF COMPONENTS

There are five main methods of assembling thermoplastics components.

1. Cementing or solvent welding.
2. Heat welding.
3. Use of inserts.
4. Use of snap or interference fits.
5. Use of mechanical devices.

1. Some thermoplastics can be jointed to themselves by adhesives or by solvents. With acrylics and rigid PVC both are possible; with nylon and acetals special adhesives can be used. There are no adhesives or solvents which will give high strength bonds with polythene or polypropylene.

2. Many methods of heat welding are available but not all are applicable to all thermoplastics. Radio-frequency welding is commonly used for PVC but cannot be used with polythene or polypropylene. Ultrasonic welding is applicable to most of the more rigid thermoplastics but is less satisfactory with polythene and polypropylene. Jointing with gas-heated welding rods (Figure 15.5) is used with PVC, polythene and polypropylene and can be used with acetals but not with nylons or acrylics. Spin welding is applicable to most of the more rigid thermoplastics (Figure 15.6).

The designer must bear in mind these differences in behaviour between the different classes of thermoplastics. Full details of the methods available can be obtained from raw material suppliers.

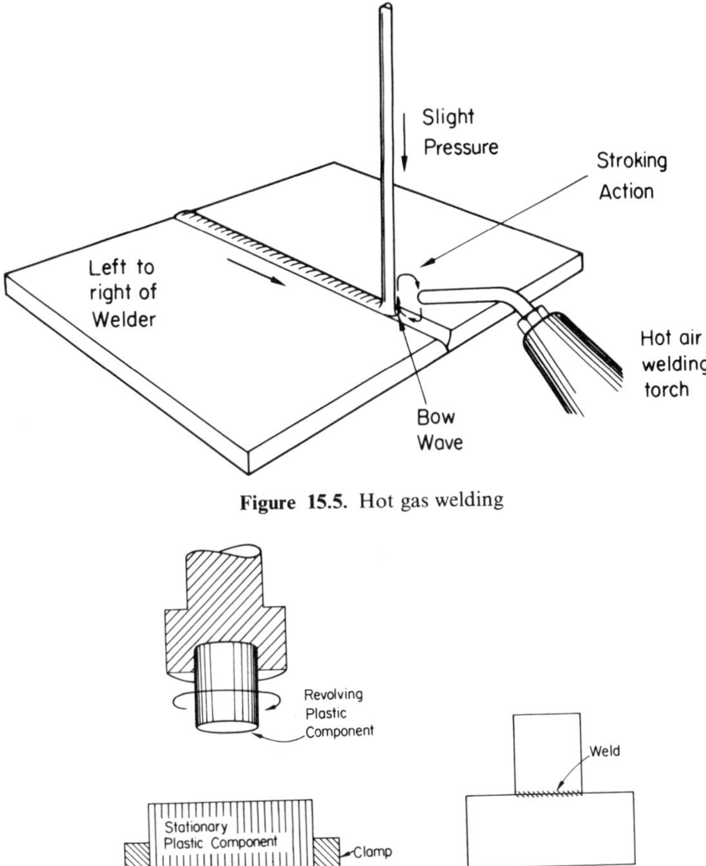


**Figure 15.5.** Hot gas welding

**Figure 15.6.** Spin welding

3. Many components are assembled by using metal inserts and on first inspection these may seem to provide the simplest answer to an assembly problem. However, the use of metal inserts is subject to two important restrictions.

(a) Not all thermoplastics will readily accept inserts. With the more flexible materials, such as low density polythene and plasticized PVC, it is often difficult to obtain sufficiently strong locking of the inserts. On the other hand, with some rigid thermoplastics, low extension at break may cause fracture in the areas immediately surrounding the inserts. Figure 15.7 shows typically good and poor examples of the use of inserts.

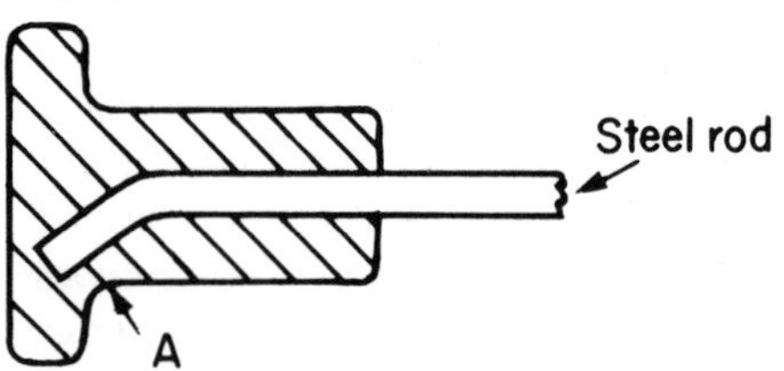


Bad design of insert. Leverage tends to cause breakage. Insert near to surface at A causes strain.

(a)

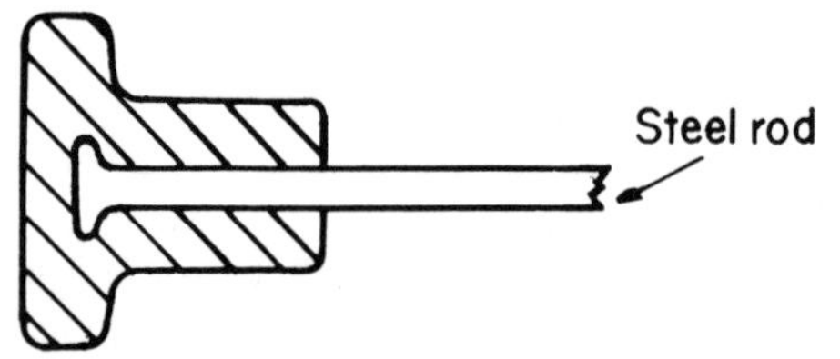


Good design. End of steel rod buried provides good anchorage and enables a smaller moulding to be used.

(b)

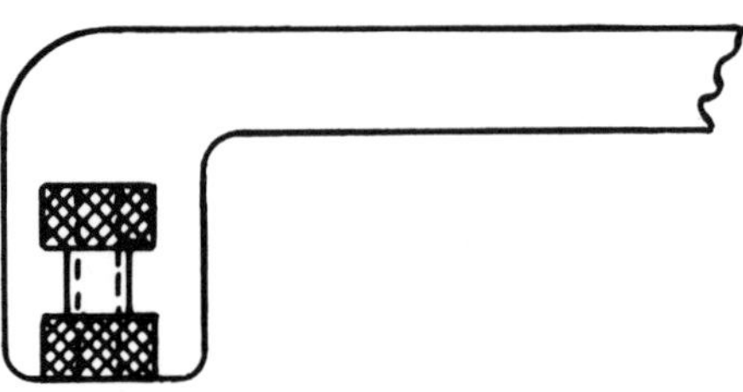

Grooves and knurling provide adequate anchorage with most plastics.

(c)

**Figure 15.7.** Moulded-in inserts

(b) Loading of inserts into a tool takes time and hence reduces output and increases part cost. As an alternative, it may be possible to have inserts of special design such as that shown in Figure 15.8 pressed into suitably sized holes after moulding rather than placed in the tool cavity.

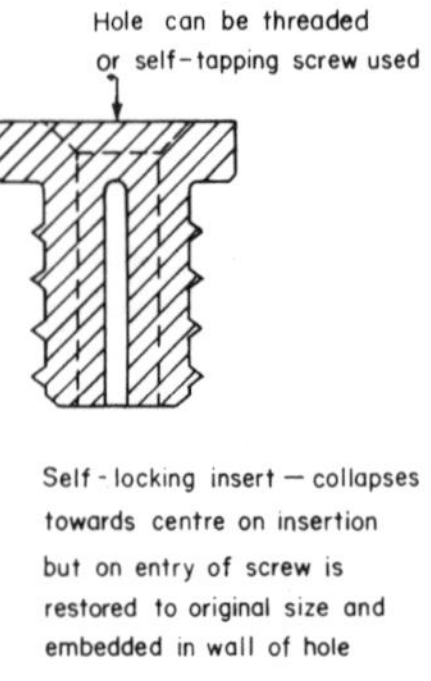


**Figure 15.8.** Self-locking inserts

4. With some plastics, polypropylene in particular, use may be made of snap or interference fits as alternatives to cementing, welding or using inserts. Examples of snap assembly methods are shown in Figure 15.9.

5. Various special fixing devices are available, their suitability depending on the type of plastic and the application; typical examples are shown in Figure 15.10. When plastics components are to be attached to metal parts and likely to be exposed to changing temperatures in service, allowance must be made for the difference in thermal expansion of the two materials, i.e. the

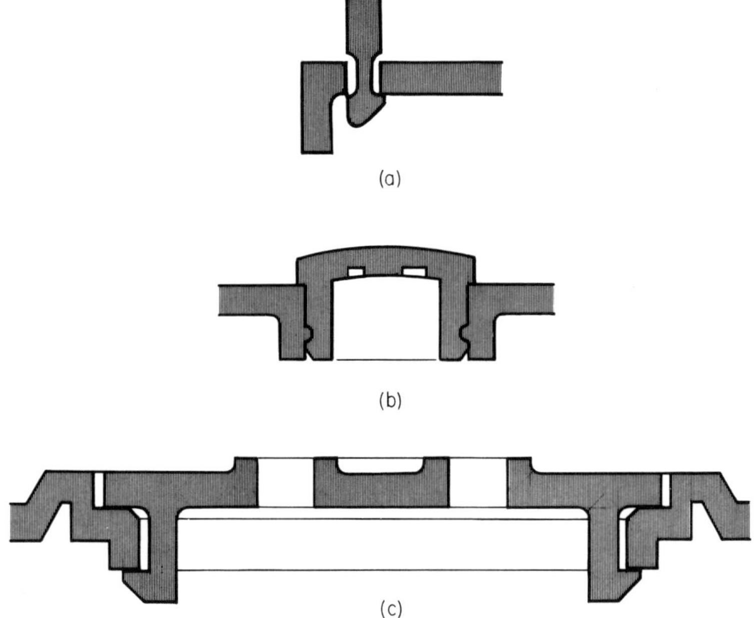


**Figure 15.9.** Snap assembly methods

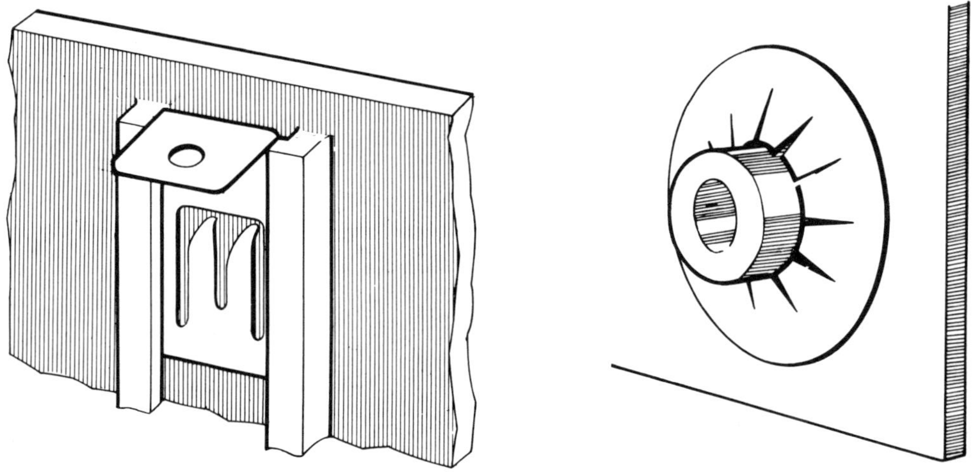

**Figure 15.10.** Fixing devices

fixing methods used must allow the plastic component to expand and contract without restraint, e.g. slots should be used instead of holes for locating screws. One method of fixing a resilient thermoplastic to metal is shown in Figure 15.11.

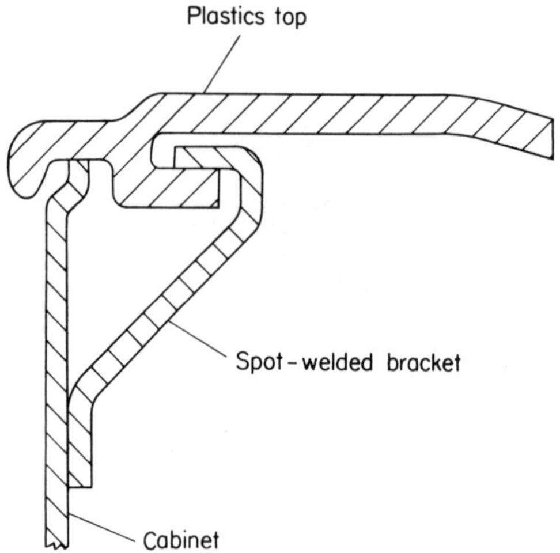


**Figure 15.11.** A method of fixing plastics components to metal assembly

## STIFFNESS

The simplest way of obtaining stiffness is by increasing section thickness. Thick sections are, however, costly because of high material usage and low production rates, the latter resulting from long heating and cooling times. Stiffness is therefore better obtained by other methods such as the use of ribs or of curvature. For example, the metal panel shown in Figure 15.12(a) could, as a moulding, be produced as shown in Figure 15.12(b).

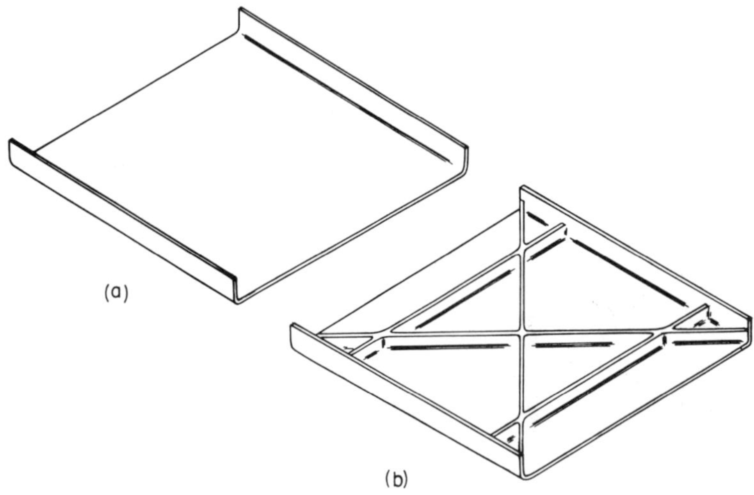


**Figure 15.12.** Obtaining stiffness in moulding by ribbing

It is not always convenient or desirable to use ribs—they can, for example, have a disruptive effect on flow—and curvature or stepped sections are often preferable alternatives. In Figure 15.13, the three sections have similar rigidity, but b and c would be easier to mould and eject. Plates 16 and 17 show further examples of the use of ribs and curvature.

Both devices necessitate higher tool costs, but in a long production run these may be more than offset by shorter moulding cycles and lower material usage.

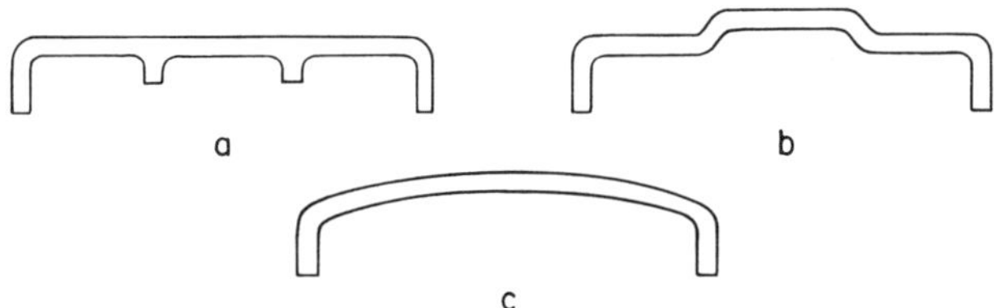


**Figure 15.13.** Obtaining stiffness in moulding by curvature

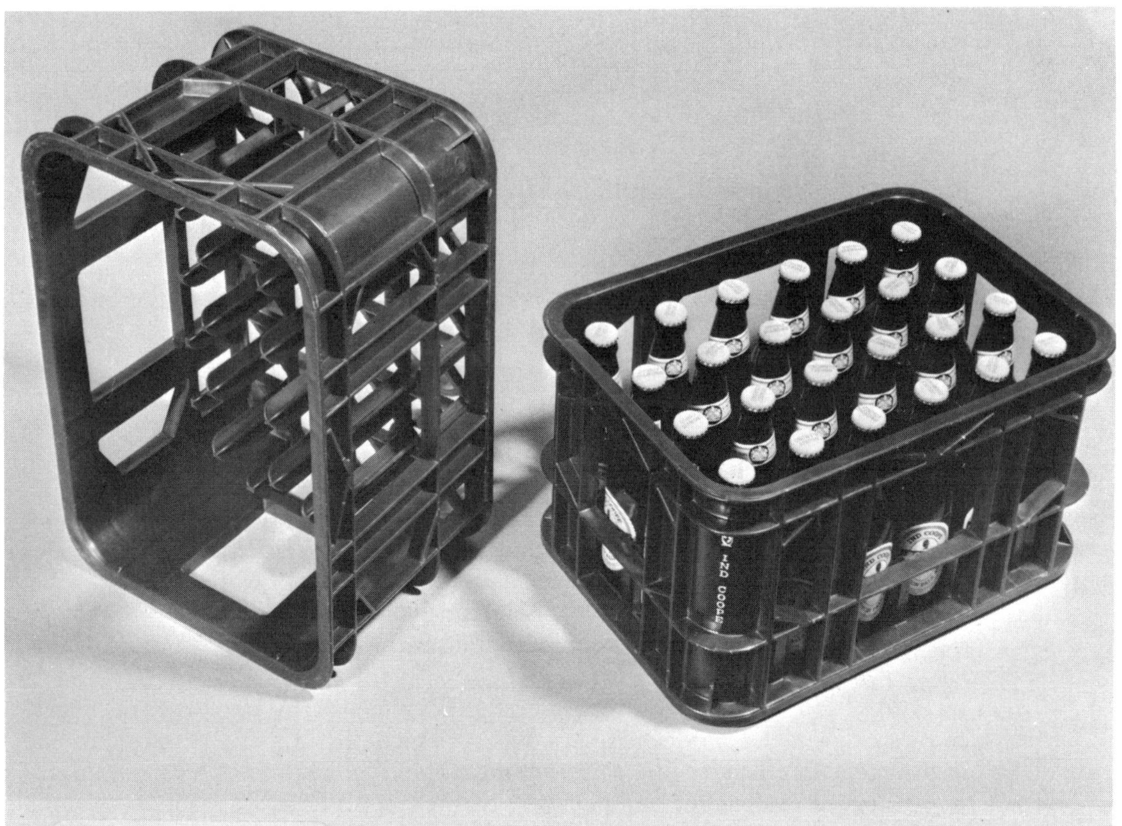


**Plate 16.** Beer crates moulded in polypropylene. (Moulded by: Marley Extrusions Limited. For Ind Coope Limited)

**Plate 17.** The shells of this suit case are moulded in polypropylene. (Moulded by Rolinx Limited. Suit case by Antler Limited)

## FLAT AREAS

The production of mouldings with large flat areas should not be attempted with certain thermoplastics, particularly polythene and polypropylene. With these partially crystalline polymers, there is a tendency for the molecules to line up in the direction of flow in the mould, thus giving rise to greater shrinkage in the direction of flow than in the transverse direction. Although this occurs with all plastics materials, the shrinkage difference is greater in polythene than in other plastics.

Figures 15.14 and 15.15 show how, with a single gate, differential shrinkage creates in effect a wedge-shaped excess of material which cannot be accommodated in a flat plane. The tool designer can overcome this distortion in some instances by using many gates to break up the flow pattern or a single gate along the length of one edge. Where flat areas are required, the tool designer must be consulted but equally, consideration should be given to 'doming' or other means of disguising and reducing the effects of differential shrinkage.

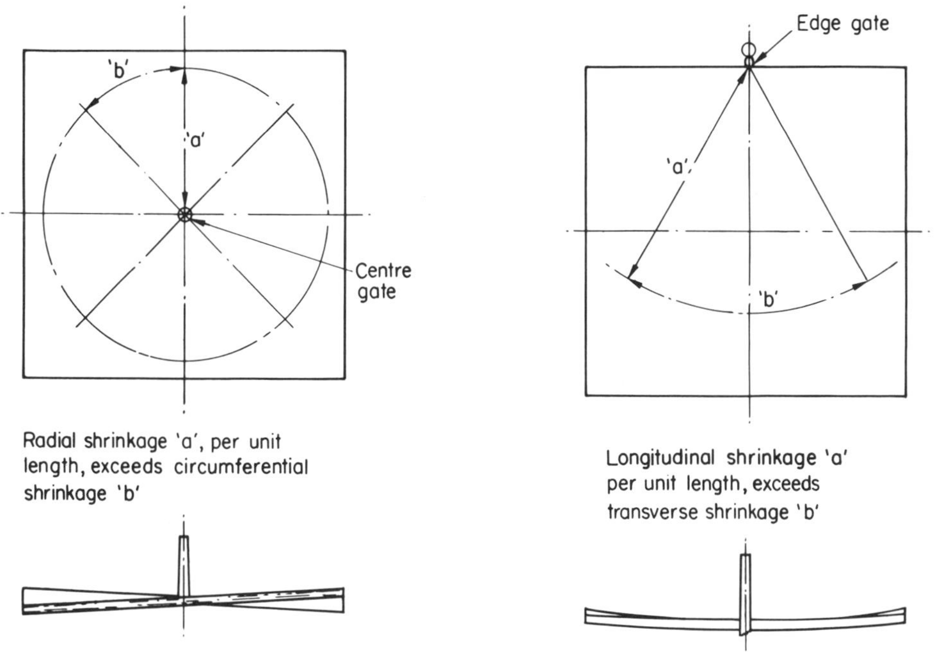


**Figure 15.14.** Warping of flat areas

**Figure 15.15.** Bowing of flat areas

## SECTION CHANGES

Localized thick sections do not usually cause distortion, but bowing may result when a thicker section extends over an appreciable length. Examples are shown in Figure 15.16. Figure 15.16(d) shows how edge bowing, as in 15.16(c), may be overcome by coring out the section to maintain a constant thickness.

Localized thick sections may display unsightly surface sinking or voiding in addition to bowing. Only the product designer can ensure freedom from this type of distortion by keeping section thicknesses as uniform as possible.

When faced with a design in which locally increased sections are unavoidable (a situation which should arise only very rarely) the tool maker can help to reduce distortion by providing maximum cooling near the thick section. Rapid local cooling will decrease the shrinkage and will thus lessen the variation in shrinkage which causes distortion. Such methods must be used with care, however, because they can cause local strain, which may be released in time or under exposure to high service temperatures.

## SINKS AND VOIDS

The contraction in volume which occurs during the solidification of plastics melts can lead to sinking and voiding unless component and mould design are carefully considered. The effects are most marked with crystalline polymers such as polypropylene. Sinks and voids both stem from the same cause although they differ in mechanism. Voids, or internal cavities, are created when the external 'skin' of the moulding is rapidly cooled and becomes sufficiently rigid to support the contraction of the underlying melt. Sinks, or surface depressions, occur usually above sections

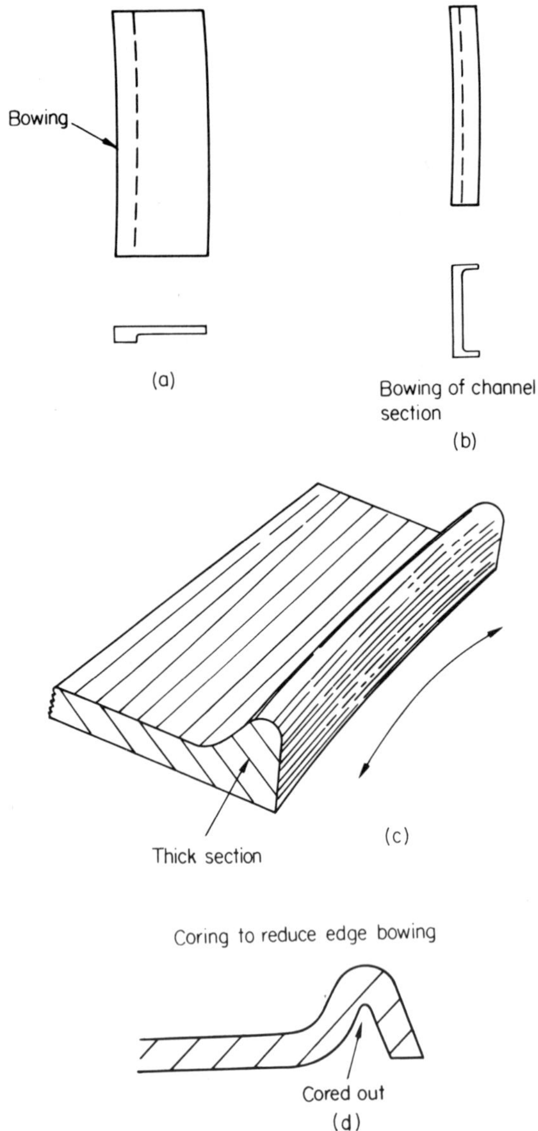


**Figure 15.16.** Bowing caused by extra shrinkage in thick sections

where there is local thickening and in which the internal mass contains sufficient heat to hold the surface in a plastic state and draw it inwards as cooling proceeds. As stated above, substantial changes in section thickness should be avoided wherever possible because the moulder can control these effects only to a very limited extent.

Ribs and bosses, which introduce locally increased section thicknesses, are main items for consideration. Figure 15.17(a) shows the type of section which results in sinking, and 15.17(b) and 15.17(c) show typical ways of minimizing or eliminating the problem.

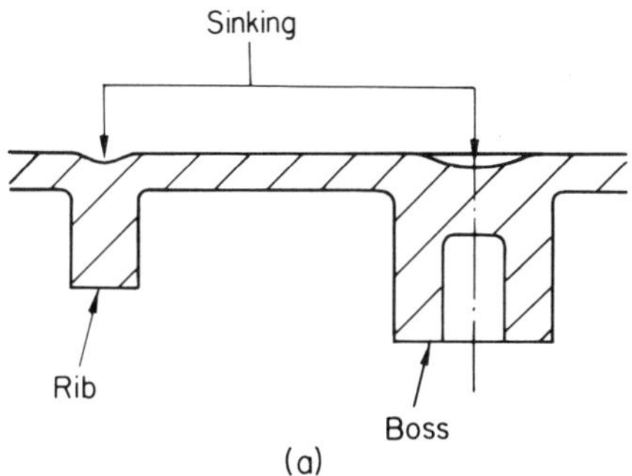


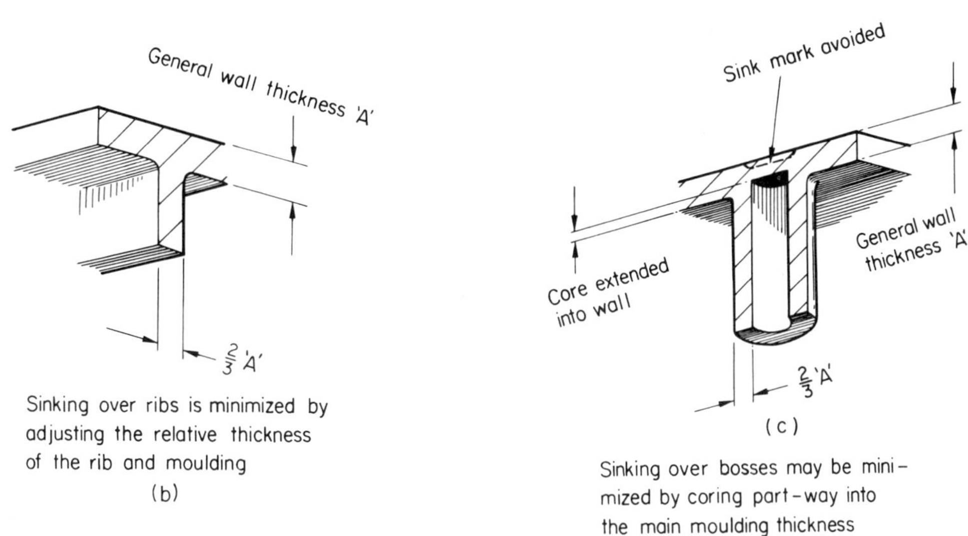


**Figure 15.17.** Sinking

## DIMENSIONAL TOLERANCES

The tolerances demanded of the moulder must take account of the nature of the material and the conditions in which the component is to be used.

The major factor affecting tolerance limits is shrinkage and this in turn is affected by many processing variables, among them being melt temperature, mould temperature and injection pressure. Some thermoplastics can be held to closer tolerances than others because the effects of small variations in moulding conditions on shrinkage are small. Also, some materials are less subject to further shrinkage ('after-mould' shrinkage) during the period immediately after removal of the component from the tool. The order of shrinkage to be expected for various plastics is given in Table 15.3.

It is not easy to produce mouldings to very close tolerances. Insistence on them will almost certainly put up tool costs, delay the achieving of satisfactory production, and necessitate very

TABLE 15.3 Mould Shrinkages for Plastics

| Plastic | Mould shrinkage (0·001 in/in) |
|---|---|
| Polythene | 10–20 |
| Polypropylene | 15–18 |
| 66 nylon | 13–23 |
| 610 nylon | 9–20 |
| 6 nylon | 6–18 |
| Glass-filled 66 nylon | 5–14 |
| Acetal copolymer | 20 |
| Acrylic | 4–8 |
| PVC (unplasticized) | 4 |

close control of moulding conditions. Moreover, it should always be remembered that close tolerances may either be unnecessary, e.g. because the resilience of the material allows snap fits to be used, or impractical because of the overriding effects of environment, e.g. swelling caused by absorbed water or expansion caused by a rise in temperature. Recommendations on tolerance allowances for thermoplastics are given in British Standard 4042.

Good design should clearly take into account such factors and assess their relative importance in the functioning of the finished article.

## CHOOSING THE RIGHT MATERIAL

In conclusion it may be helpful to outline a possible approach to the rational selection of materials.

The first step is to make an assessment of the essential properties of the component against which the various plastics materials can be measured. Suitable headings under which these properties could be grouped are strength, deformation, environment and special requirements, e.g. high clarity. Under strength one would need to assess the likelihood of brittle failure, the maximum stresses to be met in service and the significance of fatigue; under deformation, the nature of the loading, i.e. whether it is continuous or intermittent, the maximum permitted strain, the maximum service temperature and the desired life of the component; and under environment one must tabulate the oxidative, corrosive and abrasive elements present.

The second step, that of comparing the properties of the available materials, is one of elimination and is best carried out in two stages. The effects of environment should be considered first, partly because such effects are largely independent of the shape and dimensions of the part and are therefore easier to predict than mechanical performance, and partly because it is frequently possible to eliminate some plastics immediately because of their known behaviour. Any special requirements are best considered at this stage, because this will almost certainly reduce still further the number of materials worthy of further study.

Elimination of materials against the criteria of strength and deformation is much more difficult, although the availability of data such as those presented in the earlier chapters will do much to facilitate the task. Experience is invaluable in the initial selection of possible materials, but it should not be relied upon exclusively. It is of little use in assessing new materials, and of limited use in assessing the performance of established materials under different conditions.

The next stage is to form some estimate of finished component costs, and the comments in an earlier section of this chapter are applicable here. Running costs are more difficult to assess. Again, experience of similar applications is very useful, but it is probably of greater value to carry out some testing of prototypes.

Evaluation of prototypes is the vital final stage in material selection. Even if the functional requirements and material characteristics were known precisely enough to narrow the choice down to only one material, it may be advisable to compare the performance of several grades of this material. A small number of prototypes can be machined from rod or sheet stock, but this has the drawback that, if the article is subsequently to be made by injection moulding, the performance of such prototypes will be only a rough guide to the performance of the injection moulded article. This is because the fabrication conditions are different, and also because the materials are unlikely to be the same. For example, rod will probably be made by a slow-cooling extrusion process from a material of higher molecular weight.

It is much more preferable, therefore, to resort to a prototype injection moulding tool and this should always be done if the expenditure can be justified; moulded prototypes will be more authentic because their properties will be similar to those of the production part. The prototype tool can be either a simple one, designed to test vital parts only and to produce only a few mouldings, or a more complex tool to produce larger quantities of articles of the final shape. Such a tool can be used to make accurate measurements of shrinkage, a particular advantage with anisotropic materials like glass-filled nylon, and to determine the required size and shape of gating systems. Provided this tool is made from a hard-wearing material such as zinc alloy it can often be used for the first production run. In some cases, existing tools designed to be used for metals can be modified for use with plastics, and used as prototype tools.

The use of prototype tools has the great advantage also that likely moulding problems can be spotted and overcome at an early stage without the need for expensive tool modifications. The information obtained from prototype tool work is used to design the production tool, which again can be made from zinc alloy or, as is more usual, certainly for long runs, from steel.

Throughout the whole selection and evaluation process there will inevitably be interpenetration of the various stages and the choice of material will be influenced at all stages by the design of the component and by considerations of its method of manufacture. It must be emphasized too that early and complete cooperation among client, designer, toolmaker, moulder and raw material supplier is essential if plastics materials are to be used effectively and economically. In this way, the experience and knowledge of all can be used to evolve a well-designed article which can be made by the best possible means at the lowest possible price.

# BIBLIOGRAPHY—Suggested Further Reading


### *Introductory and General*

Van Vlack, L. H., *Elements of Materials Science*, Addison Wesley, 1959.
Wulff, J., (Ed.), *Structure and Properties of Materials* Vol. I—*Structure*, by Moffat, W. G., *et al.*, Wiley, 1964. Vol. III—*Mechanical Behaviour*, by Hayden, W. *et al.*, Wiley, 1965.
Winding, C. C. and Hiatt, G. D., *Polymeric Materials*, McGraw-Hill, 1961.
Miles, D. C. and Briston, J. H., *Polymer Technology*, Temple Press, 1965.
Brydson, J. A., *Plastics Materials*, Iliffe, 1966.
Moncrieff, R. W., *Man-made Fibres*, Heywood, 1963.
Hearle, J. W. S. and Peters, R. H., *Fibre Structure*, Butterworths, 1963.
*Landmarks of the Plastics Industry*, ICI Plastics Division, 1962.
Kaufman, M., *Giant Molecules*, Aldus Books, 1968.
Kaufman, M., *The First Century of Plastics*, Iliffe, 1962.
*Encyclopaedia of Polymer Science and Technology*, Interscience (*in preparation*).

### *Physical Properties and Testing*

Alfrey, T. and Gurnee, E. F., *Organic Polymers*, Prentice-Hall, 1967.
Gordon, M., *High Polymers: Structure and Physical Properties*, Iliffe, 1963.
Ritchie, P. D., (Ed.), *Physics of Plastics*, Iliffe, 1965.
Meares, P., *Polymers: Structure and Bulk Properties*, Van Nostrand, 1965.
Ferry, J. D., *Viscoelastic Properties of Polymers*, Wiley, 1961.
Nielsen, L. E., *Mechanical Properties of Polymers*, Reinhold, 1962.
Miller, M. L., *The Structure of Polymers*, Reinhold, 1966.
Beuche, F., *Physical Properties of Polymers*, Wiley, 1962.
Wetton, R. and Whorlow, R., (Eds.), *Polymer Systems: Deformation and Flow*, Macmillan, 1968.
Ogorkiewicz, R. M., (Ed.), *Thermoplastics: Effects of Processing*, Iliffe, 1969.
*Thermoplastics and Mechanical Engineering Design*, ICI Plastics Division, 1969.
Schmitz, J. V., (Ed.), *Testing of Polymers*, Vol. 1, Wiley, 1965.
Baer, E., (Ed.), *Engineering Design for Plastics*, Reinhold, 1964.
Faupel, J. H., *Engineering Design*, Wiley, 1964.
Rosen, B., (Ed.), *Fracture Processes in Polymeric Solids*, Wiley, 1964.
ASTM Standards of Plastics, Parts 26 and 27, *Am. Soc. Testing Mater.*, 1967.
BS 2782 Methods of Testing Plastics, *Brit. Std. Inst.*, 1965.
Haslam, J. and Willis, H. A., *Identification and Analysis of Plastics*, Iliffe, 1965.
Brown, W. E., (Ed.), *Testing of Polymers*, Vol. 4, Interscience, 1969.
*Plastics Materials Guide*, Heywood Temple Industrial Publications.
*Design Engineering Plastics Handbook*, Morgan Grampian.
Heap, R. D. and Norman, R. H., *Flexural Testing of Plastics*, Plastics Institute, 1969.

### *Production Processes*
#### *General*

Bernhardt, E. C., (Ed.), *Processing of Thermoplastic Materials*, Reinhold, 1959.
*Plastics Engineers Handbook*, Society of the Plastics Industry, USA, 1960.
Pearson, J. R. A., *Mechanical Principles of Polymer Melt Processing*, Pergamon Press, 1966.
*Working of Plastics*, OECD, 1964.

*Extrusion*

Fisher, E. G., *Extrusion of Plastics*, Iliffe, 1964.
Jacobi, H. R., *Screw Extrusion of Plastics*, Iliffe, 1963.
Schenkel, G. (transl. by Eastman, L. A. H.), *Plastics Extrusion Technology and Theory*, Iliffe, 1966.

*Blow Moulding*

Jones, D. A. and Mullen, T. W., *Blow Molding*, Reinhold, 1961.

*Moulding*

Butler, J., *Compression and Transfer Moulding*, Iliffe, 1959.
Bebb, R. H., *Plastics Mould Design:* Vol. I *Compression and Transfer Moulding*, Iliffe, 1962.
Walker, J. S. and Martin, E. R., *Injection Moulding of Plastics*, Iliffe, 1966.
Mink, W. (English Edn., Edited by Fisher, E. G.), *Practical Injection Moulding of Plastics*, Iliffe, 1964.
Glanville, A. B. and Denton, E. N., *Injection Mould Design Fundamentals*, Volumes I and II, The Machinery Publishing Co. Ltd., 1963.
Groves, W. R., *Plastics Moulding Plant*, Vol. I *Hydraulics, Compression and Transfer Equipment*, Iliffe, 1963.
Munns, M. G., *Plastics Moulding Plant*, Vol. II *Injection Moulding Equipment*, Iliffe, 1964.
Pye, R. G. W., *Injection Mould Design*, Iliffe, 1968.

*Sheet Shaping*

Estevez, J. M. J. and Powell, D. C., *Manipulation of Thermoplastic Sheet, Rod and Tube*, Iliffe, 1960.
Thiel, A. (transl. by Eastman, L. A. H.), *Principles of Vacuum Forming*, Iliffe, 1965.

*Machining and Welding*

Haim, G. and Zade, H. P., *Welding of Plastics*, Crosby and Lockwood, 1946.
Haim, G. and Neumann, J. A., *Manual for Plastic Welding*, Vol. II *Polyethylene*, Crosby and Lockwood, 1959.
Haim, G., *Manual for Plastic Welding:* Vol. III *Polyvinyl Chloride*, Crosby and Lockwood, 1959.
Neumann, J. A. and Bockhoff, F. J., *Welding of Plastics*, Reinhold, 1959.
Kobayashi, A., *Machining of Plastics*, McGraw-Hill, 1967.